Chemistry of the Environment

Thomas G. Spiro
Princeton University

William M. Stigliani
Center for Energy and Environmental Education
University of Northern Iowa

PRENTICE HALL, Upper Saddle River, New Jersey 07458

Library of Congress Cataloging-in-Publication Data

Spiro, Thomas G.
    Chemistry of the environment / Thomas G. Spiro, William M.
  Stigliani.
        p.   cm.
    Includes bibliographical references and index.
    ISBN 0-02-415261-7 (hardcover)
    1. Environmental chemistry.   I. Stigliani, William M.   II. Title.
  TD193.S7   1996
    540—dc20   95-51367
                                                              CIP

Senior editor: John Challice
Editorial/production supervision and interior design: ETP/Harrison
Copy editor: William N. Clark
Cover director: Jayne Conte
Cover photograph: NASA
Manufacturing manager: Trudy Pisciotti

© 1996 by Prentice-Hall, Inc.
Simon & Schuster/A Viacom Company
Upper Saddle River, NJ 07458

Printed in the United States of America

10   9   8   7   6   5   4   3

ISBN 0-02-415261-7

Prentice-Hall International (UK) Limited, *London*
Prentice-Hall of Australia Pty. Limited, *Sydney*
Prentice-Hall Canada Inc., *Toronto*
Prentice-Hall Hispanoamericana, S.A., *Mexico*
Prentice-Hall of India Private Limited, *New Delhi*
Prentice-Hall of Japan, Inc., *Tokyo*
Simon & Schuster Asia Pte. Ltd., *Singapore*
Editora Prentice-Hall do Brasil, Ltda., *Rio de Janeiro*

*Dedicated to
our wives Helen and Marie,
and to our children and their generation*

Contents

Part II Atmosphere   105

Preface

This book is about environmental issues and the chemistry behind them. It is not a methods book, nor is it a catalog of pollutants and remediation options. It aims to deepen knowledge of chemistry and of the environment and to show the power of chemistry as a tool to help us comprehend the changing world around us. The book can be used in a one- or two-term environmental chemistry course. The instructor in a one-term course will want to pick a limited set of the book's topics for special emphasis; in a two-term course there would be time to address other topics and to explore the underlying chemical principles in more detail. The book assumes familiarity with basic chemistry, but none of the chemical themes are so advanced as to preclude the book being used to augment a general chemistry course. In fact, we believe that integrating the environmental story would greatly strengthen the teaching of introductory chemistry.

Chemistry of the Environment is an outgrowth of our earlier *Environmental Issues in Chemical Perspective* (SUNY Press, 1980; reprinted by Kendall/Hunt Publishing Co., 1990), which in turn grew out of an environmental chemistry course that we developed at Princeton University in the 1970s. In the aftermath of the first Earth Day in 1970, interest in environmental issues was high. Our students were happy to help us learn about the chemistry behind the headlines, whether it involved threats to the ozone layer from the supersonic transport plane (then the major concern) or from the newly considered chlorofluorocarbons (Molina and Rowland published their now-famous warning in 1974), or the promise and peril of the plutonium-breeder reactor, or the possibility of environmental poisoning by dioxins or heavy metals. In revisiting our earlier book, we were struck by how many of the issues identified in the 1970s are still concerns today. Interest may wax and wane, but the environment and its vulnerabilities are here to stay.

Although our new book shares many issues with the old one, it has been completely rewritten. Facts and figures have been updated, much new material has been added, and we have striven for better writing and a higher level of integration of the material. In this we have received invaluable help from our special editor, Dr. Judith Swan, an expert in scientific communication. Keeping the reader firmly in mind, she has artfully pushed and pulled our drafts into shape, sharpening our arguments and making the exposition more compelling.

We are indebted as well to the following colleagues for reviewing parts of the manuscript and offering many helpful suggestions: Drs. Ann Kinzig, Hiram Levy, Robert Socolow, and Valerie Thomas. We are also grateful to Dr. Trace Jordan, who, while a graduate student at Princeton, identified many important references and participated enthusiastically in the project.

Thomas G. Spiro
William M. Stigliani

Introduction

"Here's a short chemistry lesson," wrote Bill McKibben in a recent *New York Times Magazine* feature story.* "Grasp it and you will grasp the reason the environmental era has barely begun. . . ." McKibben's lesson is about the difference between two molecules, carbon monoxide (CO) and carbon dioxide (CO_2). Today's automobiles release half a pound of carbon as CO per gallon of gasoline burned, around half as much as they did a generation ago, and the rate is going down with continuing improvements in technology. As a result, the air is now cleaner than it used to be in Los Angeles and in many other cities. But the same gallon of gasoline releases almost five and a half pounds of carbon as CO_2, and this rate cannot be decreased. The atmospheric concentration of CO_2 is increasing worldwide, and bringing global warming with it, according to a consensus of international scientific opinion. The two molecules capture two sides of the environmental coin, local versus global effects of human activity. Environmental quality has improved in many localities, thanks to environmental controls and new technologies, but the global problems are just beginning to be addressed, and they are much more difficult to solve. CO is a side-product of combustion, and is subject to emission controls, but CO_2 is the end-product of combustion and is the inevitable accompaniment of our reliance on fossil fuels. "CO versus CO_2," says McKibben, "one damn oxygen atom, and all the difference in the world."

To us this is a wonderful illustration of the power of chemistry to illuminate environmental issues. Chemistry is all around us, and it really does make a difference. The chemical cycles of the planet are increasingly disturbed by human activities, and these disturbances can degrade the quality of life, as when auto emissions overwhelm the atmosphere's capacity to clean the air over our cities. We are capable of ameliorating these perturbations, as the experience of Los Angeles demonstrates. But first we must understand the chemistry. In the case of Los Angeles, initial attempts to alleviate smog back in the 1960s actually made things worse. Standards were imposed on CO and hydrocarbon levels in auto emissions, and auto makers met those standards by increasing the air/fuel ratio to burn the

*Bill McKibben, "Not So Fast," *New York Times Magazine*, July 23, 1995, pp. 24–25.

fuel more completely. But smog levels *increased* because higher air/fuel ratios made combustion hotter, thereby increasing the nitrogen oxide emissions. Only then was it discovered that nitrogen oxides and hydrocarbons are *both* key actors in smog formation and that both have to be controlled. This kind of surprise is not unusual in environmental affairs. The world is a marvelously complex place, chemically, as in other ways. We are just beginning to understand how it works.

This book tells the environmental story in chemical language. It is grounded in the flows of chemicals and energy through nature on the one hand, and through our industrial civilization on the other. The units of the book, Energy, Atmosphere, Hydrosphere, and Biosphere, reflect this holistic perspective. Environmental issues frequently cut across these divisions, and the resulting interconnections add richness to the story. For example, leaded gasoline is linked to the issue of auto emission controls, a subject that arises in the Atmosphere section, but it is also a major health hazard, as discussed in the Biosphere section.

Interconnections are even more numerous at the level of the underlying chemistry. For example, the reactivity of dioxygen, O_2, is a leitmotif for all parts of the book. Thus energy flow through industrial civilization (as well as through the biosphere itself) depends on the oxygen-oxygen bond being relatively weak, so that energy is released when O_2 combines with organic molecules. Yet, because of its unusual electronic structure, O_2 is quite unreactive until it encounters a free radical or a transition metal ion. These O_2 activators determine most aspects of atmospheric chemistry, including how smog is formed. They are also vitally important for the biosphere, since O_2 metabolism gone awry is a threat to the integrity of biological molecules and has been implicated as a factor in cancer and aging.

We hope these interconnections fascinate the reader as much as they fascinate us, and we hope the tapestry we weave will provide a satisfying context for understanding the chemical world we live in and the environmental issues we face.

Part I

Energy

A. INTRODUCTION

The question of energy use underlies virtually all environmental issues. The harnessing of energy for the manifold needs of industrial civilization has driven economic development, and access to affordable energy has been the key to a better life for people around the world. At the same time the environmental costs of human energy consumption are becoming ever more apparent: oil spills, the scarring of land by mining, air and water pollution, and the threat of global warming from the accumulation of carbon dioxide and other greenhouse gases. Increasingly, maintaining an expanding supply of cheap energy seems to clash with concern for the environmental costs of such expansion. In this part of the book, we explore the background of energy production and energy consumption, and examine the prospects for meeting the energy needs of society while protecting the environment.

1. Natural Energy Flows

It is instructive to view human energy use against a backdrop of the continual and massive flow of energy that occurs at the surface of Earth. This flow is diagrammed in Figure 1.1; the magnitudes of the energy fluxes are given in units of 10^{20} kilojoules (kJ) per year. (See Appendix A for a comprehensive listing of the common forms and units of energy.) A tiny part of Earth's energy budget derives

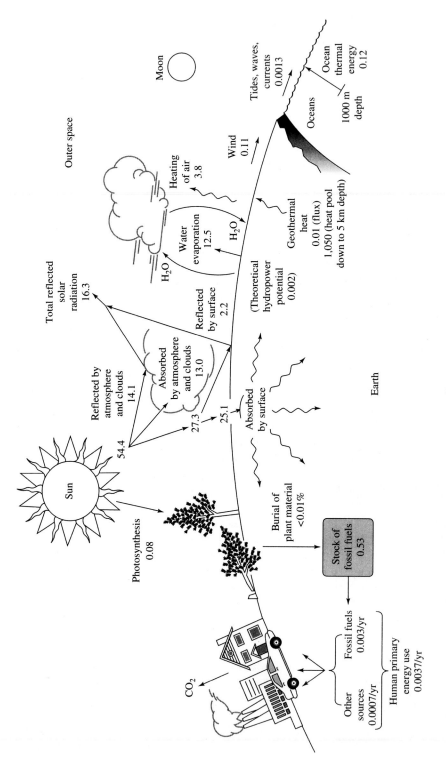

Figure 1.1 Annual energy fluxes on Earth (in 10^{20} kilojoules) [not included are fluxes of outbound in-frared radiation from Earth's surface (see discussion in Part II, and Figure 2.2)].

from nonsolar sources: tidal energy, which arises from the gravitational attraction between the moon and Earth, and geothermal heat, which emanates from Earth's molten core. (Human-generated nuclear energy is nonsolar, but it is also minimal compared to total energy flows.) The rest of the energy on Earth comes from the sun, either directly or indirectly.

The sun radiates a nearly unimaginable number of kilojoules every year—1.17×10^{31}. A very small fraction of the total, 54.4×10^{20} kJ per year, is intercepted by Earth, which is 150 million kilometers (93 million miles) from the sun. Of this quantity, about 30% is reflected or scattered back into space, either from Earth's surface, or its atmosphere. This fraction is called the *albedo*, and it contributes significantly to the overall energy balance of Earth.

The rest of the light is absorbed by Earth or its atmosphere and is converted to heat before being reradiated back into space. This heat flow drives Earth's weather system via wind, rain, and snow. About half of the absorbed energy flows through the hydrological cycle, the massive evaporation and precipitation of water upon which we depend for our freshwater supplies. While it takes 4.2 joules (1 calorie) to heat a gram of water by $1°C$, much more energy is needed to vaporize the same gram of water (2.5 kJ, the *latent heat* of water at $15°C$, which is the average annual global temperature). The latent heat is released when a gram of water vapor condenses into rain. This is the reason that rainfall is associated with storms; even a modest rainfall releases a huge amount of energy. We extract a very small fraction of the energy in the hydrological cycle through dams and hydroelectric power generation, which constitute an indirect tapping of the solar energy flow.

About 0.15% of the incident sunlight is used by green plants and algae in photosynthesis. We depend on this fraction of the solar flux for our food supply and for a habitable earth. Some of the energy we use is provided by burning wood and other biomass forms (garbage, cow dung), and most of the rest is obtained by mining the store of photosynthetic products buried in ages past, in the form of fossil fuels.

2. Human Energy Consumption

Compared to the enormous energy flux provided by the sun, human energy utilization is puny. In 1990 the total primary energy consumed by all humans amounted to 3.7×10^{17} kJ, equivalent to only 0.01% of the solar heat absorbed by Earth's surface. Figure 1.2 shows, however, that world energy consumption is growing rapidly and nearly tripled between 1960 and 1990.

The shaded areas in the figure correspond to projections of future energy consumption in the world and the United States, based on a range of different assumptions. The higher world projection assumes that current trends continue into the first decade of the 21st century; energy consumption, particularly in the developing world, increases without major improvements in energy efficiency or further adoption of renewable sources of energy. In contrast, the low world projection assumes the opposite prevails: revolutionary improvements in energy efficiency around the globe and the wide-scale application of renewable energy sources.

With respect to the United States, the high projection corresponds to a scenario, developed during the Bush Administration, in which there are modest gains in energy efficiency

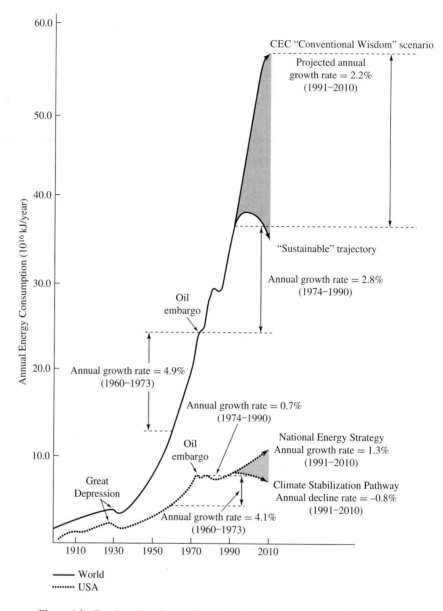

Figure 1.2 Trends and projections in U.S. and world energy consumption. *Sources: National Energy Strategy: Powerful Ideas for America* (1st ed. 1991/1992), (Washington, DC: U.S. Government Printing Office); Union of Concerned Scientists et al. (1991). *America's Energy Choices: Investing in a Strong Economy and a Clean Environment* (Cambridge, Massachusetts: Union of Concerned Scientists); Commission of the European Communities (CEC) (1989). *Energy in Europe: Major Themes in Energy* (Brussels: Commission of the European Communities); J. Goldemberg et al. (1987). *Energy for a Sustainable World* (Washington, DC: World Resources Institute).

and adoption of renewable energy, but with continued primary reliance on fossil fuels; economic growth offsets to some extent the gains made in energy efficiency. The lower projection assumes the same level of economic growth, but with incentives that provide for large improvements in energy efficiency and use of renewable energy (see discussion on Energy and Well-Being on pp. 96–99).

Table 1.1 shows some of the quantities that we have been discussing. It also shows that in 1990, the total energy consumed by the burning of fossil fuels amounted to 3.0×10^{17} kJ. This is about 8% of the annual energy conversion by green plants (net primary productivity—see Fossil Fuels, p. 12). The energy content of the food consumed by human beings in the same year was much less, only 1.9×10^{16} kJ. On a per capita basis, this was 9,200 kJ or 2,200 kilocalories per person per day, a quantity which is close to the minimum required for the daily energy needs of humans. Some people consume much more than this, while a large fraction of the world's population has an inadequate diet. (It should be noted that measurements of nutritionists are usually in Calories with a capital C, which are 1,000 ordinary calories. The average human diet is usually given as 2,200 Calories. This notation can be a source of confusion.)

In 1990, the United States, with only 5% of the world's population, consumed 23% of the world's external energy (Table 1.1), but for the past 15 years, U.S. energy consumption has been growing at a much slower rate (0.7% per year) than has worldwide energy consumption (2.8% per year—see Figure 1.2). Between 1960 and 1973, the annual growth rates were much higher, 4.1% for the United States and 4.9% for the world. But the shock of the 1973 oil embargo set energy savings measures in motion, especially in the United States and in other industrialized countries (see discussion on pp. 91–92). In the developing world, energy consumption is still increasing rapidly, partly because of economic development itself, and partly because of high rates of population growth.

Anything that increases at a constant rate exhibits exponential growth (see Appendix B for a discussion of exponential growth and decay), a type of growth that cannot be maintained indefinitely. Something that increases at a rate of 4% a year takes only 18 years to

TABLE 1.1   GLOBAL ENERGY FLUXES

Sources	Rates (10^{20} kJ/year)
Energy radiated by the sun into space	1.17×10^{11}
Solar energy incident on Earth	54.4
Solar energy affecting Earth's climate and biosphere	38.1
Energy used to evaporate water	12.5
Energy in wind	0.109
Solar energy used in photosynthesis	0.0836
Energy used in net primary productivity	0.0372
Energy conducted from Earth's interior to its surface	0.0100
Total primary energy consumed by humans, 1990	0.00368
Energy content of fossil fuels consumed, 1990	0.00297
Energy in tides and waves	0.00126
Total energy consumed in the United States, 1990	0.000837
Energy content of food consumed by humans, 1990	0.000188

double, and then another 18 years to double again. Nothing in the natural world can grow through too many doubling periods before it runs into some constraint. The present era of rapidly increasing energy use represents a transition to a new level of energy consumption. There remains considerable debate, however, about what that level will be and at what rate it is to be attained. What the future holds is a matter of considerable uncertainty, as the divergent projections in Figure 1.2 indicate.

What are the sources of our external energy? We are overwhelmingly dependent on the fossil fuels—oil, gas and coal. Figure 1.3 shows the distribution of energy sources for the United States. A century and a half ago we relied almost exclusively on wood for fuel, but wood was supplanted by coal to power the industrial revolution. Then over the past fifty years, gas and oil took over as the dominant energy sources, which they remain today. (This, too, is a transitory phase of history, because the fossil fuel deposits will one day run out.) After fossil fuels the largest source of energy is nuclear power, which has made substantial inroads in the past twenty years. Only a small fraction of our energy consumption derives from hydropower, although it is quite important in the regions having large dams. In a few places, geothermal plants and windmills are starting to contribute to the electricity supply, and solar heating and electricity-producing installations are in operation here and there, but the energy totals from these sources are too small to be seen on the graph.

If we compare industrialized and developing countries with respect to energy sources (Figure 1.4), we see that a) the developing countries use much less energy in aggregate, although their rate of energy growth is high, and b) they rely much more on biomass—wood and agricultural wastes—than do the industrialized countries. Nevertheless, the developing world is increasingly dependent on oil and coal, as is the industrialized world. Fossil fuel deposits are not evenly distributed around the globe. As seen in the resource maps of Figure 1.5, the great majority of the available oil is in the fields of the Middle East. Because the industrialized world imports most of its oil from the Middle East, it is an area of great geopolitical importance, as the Gulf War of 1990 reminded us forcefully. The distribution of gas, and especially of coal, is somewhat more even around the globe.

What about the future? There are many energy source trade-offs before us. The sun pours enough energy on us to fulfill our needs, if we can learn to extract it efficiently and economically. There are many forms of this "renewable" energy: wind and hydropower, biomass, solar electricity, and direct solar heating. Some of these are already being harnessed on a limited scale, and new developments are occurring rapidly. The geothermal energy flowing from Earth's molten core also has a limited role to play. Nuclear fission can extend our fuel supply substantially by using the energy locked in the uranium nucleus, if we can overcome the serious problems of nuclear safety, proliferation, and waste disposal. And nuclear fusion waits in the wings, promising nearly inexhaustible energy from the fusion of hydrogen atoms, the same energy source that powers the sun, if we can solve the formidable problem of maintaining the operating temperature at millions of degrees.

Table 1.2 summarizes the current estimates of the amount of energy available from these different energy sources within the United States. These estimates are necessarily uncertain because what is "available" depends on how much money and energy are required to extract the resource under consideration. For example, as rich pools of oil near the land surface become depleted, it becomes more expensive to obtain the remaining oil in deeper or thinner deposits, or under the oceans. A limit would be reached, of course, when the energy

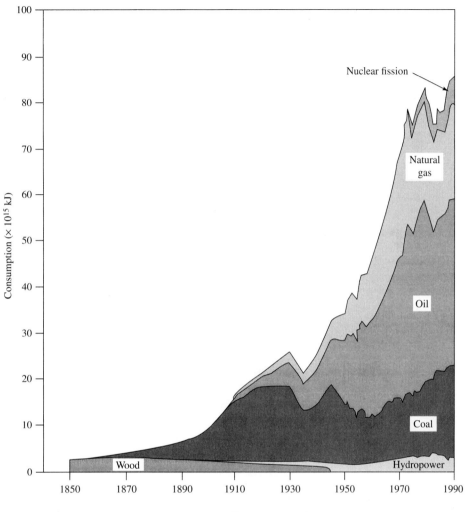

Figure 1.3 Historical trends in U.S. energy consumption by fuel type. *Sources:* U.S. Bureau of the Census (1975). *Historical Statistics of the United States, Colonial Times to 1970* (Washington, DC: U.S. Department of Interior); U.S. Geological Survey (Yearbooks, 1880–1932). *Mineral Resources of the United States* (Washington, DC: U.S. Department of Interior); U.S. Bureau of Mines (Yearbooks, 1932–1950). *Mineral Yearbook* (Washington, DC: U.S. Department of Interior); Energy Information Administration (1991). *Annual Energy Review 1990* (Washington, DC: Department of Energy).

required to extract and refine the oil exceeds the energy content of the oil itself, but long before this point, the process becomes uneconomical compared to other energy options. This limit of practicality itself changes with advancing technology, as exploration methods improve, and extraction techniques become more efficient. As a result, estimates of fossil fuel

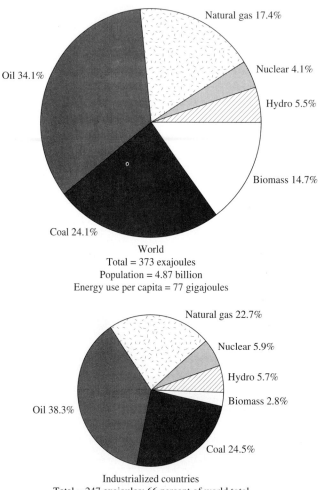

World
Total = 373 exajoules
Population = 4.87 billion
Energy use per capita = 77 gigajoules

Industrialized countries
Total = 247 exajoules; 66 percent of world total
Population = 1.22 billion; 25 percent of world total
Energy use per capita = 202 gigajoules

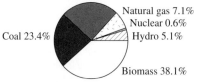

Developing countries
Total = 126 exajoules; 34 percent of world total
Population = 3.65 billion; 75 percent of world total
Energy use per capita = 35 gigajoules

Figure 1.4 Primary energy use for the world (top), industrialized countries (middle), and developing countries (bottom); (1 exajoule = 10^{18} joules, 1 gigajoule = 10^9 joules). *Source:* D. O. Hall et al. (1993). Biomass for energy: supply prospects. In *Renewable Energy, Sources for Fuels and Electricity,* T. B. Johansson et al., eds. (Washington, DC: Island Press). Copyright © 1993, Island Press. Reprinted with permission.

TABLE 1.2   ENERGY RESOURCES OF THE UNITED STATES IN 1990 (IN BILLIONS OF BARRELS, OIL EQUIVALENTS)

	Nonrenewable resources	
	Proven (supply years)[a, b]	Proven + additional (supply years)[b, c]
Petroleum	32.0 (2.2)[d, e]	52.0–97.0 (3.5–6.5)[f]
Natural Gas	31.6 (2.1)[d, g]	100–230 (6.8–15.5)[h]
Coal	1100 (74.3)[d, i]	1900 (128)[j]
Fission	11.0–61.0 (0.7–4.1)[d, k]	65.8–260 (4.4–18.0)[d, k]

	Renewable resources	
	Feasible (% total)[l, m]	Upper bound (% total)[m, n]
Wind	0.6 (4.2)	
(class 5+)		1.1 (7.1)
(class 3–4)		18.5 (124.9)
Biomass	3.0 (20.4)	3.6 (24.5)
Municipal Solid Waste	0.1 (0.9)	0.1 (0.9)
Hydro	0.7 (4.4)	0.7 (4.7)
Geothermal	0.7 (4.4)	91.0 (614.6)
Solar	0.8 (5.7)	17.2 (116.2)
Total	5.9 (39.9)	132.2 (892.7)

Notes: [a] *Proven* resources are those that have been identified as extractable at current market prices and with conventional technology.

[b] *Supply years* refer to the number of years the nonrenewable resource would last if all energy needs were drawn from that resource at a rate of U.S. energy use equal to energy consumption in 1990, which totaled 14.8 billion barrels of oil equivalents (boe).

[c] *Additional* resources are 1) those that are currently undiscovered, but likely to be in place, or 2) those extractable at higher costs, or 3) those extractable with advanced technology.

[d] *Source:* Energy Information Administration (1991). *Annual Energy Review, 1990* (Washington, DC: U.S. Department of Energy).

[e] Includes natural gas liquids, the U.S. reserve of which is estimated to be about 5.5 billion boe.

[f] *Source: National Energy Strategy: Powerful Ideas for America* (1st ed. 1991/1992) (Washington, DC: U.S. Government Printing Office). This higher projection reflects the projected impact of the implementation of advanced oil recovery technology made possible by new investments in federal and private-sector R&D, development of the coastal plain of the Arctic National Wildlife Refuge, and implementation of the Outer Continental Shelf leasing program. The report projected that, assuming the R&D program were successful, the economically recoverable reserve would increase by between 20 billion barrels (at $20 per barrel) and 65 billion (at $50 per barrel). Estimates do not include the substantial U.S. reserves of oil shale (now recoverable at production costs more than double those of crude oil production), and tar sands (production costs about 30% higher than for crude oil).

[g] Based on conversion factor of 1,000 cubic feet natural gas = 0.18 boe.

[h] Larger estimates take into account additional reserves that could become available at higher wellhead gas prices, or with the implementation of advanced technologies. These estimates do not take into account gas resources contained in tight gas sands, which could increase the resource base substantially. See: W. M. Burnett and S. D. Ban (1989). Changing prospects for natural gas in the United States. *Science* 244: 305–310; J. P. Longwell (1991). U.S. production of liquid transportation fuels: costs, issues, and research and development directions. In J. W. Tester et al., eds., *Energy and Environment in the 21st Century* (Cambridge, Massachusetts: The MIT Press). S. I. Freedman (1991). The role of natural gas in electric power generation. In J. W. Tester et al. *ibid.*

[i] Assumes that about one-half of total coal reserve is economically recoverable; based on a conversion factor of 1 metric ton coal = 5.05 boe.

[j] Assumes that nearly 90% of the coal reserve is recoverable, as discussed in Longwell, *op. cit.* These additional reserves, however, would be available only at a much higher price.

[k] Lower value corresponds to uranium resources at up to $30 per pound; upper value is for up to $100 per pound; based on conversion factor of 1 metric ton U_3O_8 = 87,250 boe. Estimates do not include implementation of the breeder reactor, which could extend the resource base by a factor of around 50.

[l] The values given in this column as "feasible" are in billion boe/yr, obtained from the "climate stabilization scenario" discussed in Union of Concerned Scientists et al. (1991). *America's Energy Choices: Investing in a Strong Economy and a Clean Environment* (Cambridge, Massachusetts: Union of Concerned Scientists). They correspond to reasonable estimates of renewable energy supplies that could be available in the United States by the year 2030. Combined with a greater degree of energy conservation than was available in 1990, the percentage of the supply that could be met by the renewables would be even greater. The estimates take into account adoption of greater end-use energy efficiency, efficient new power supplies, infrastructure changes, and renewable energy investments. An added benefit would be reductions of CO_2 emissions in the United States, relative to 1988, of 25% in 2005, and over 70% in 2030.

[m] % *total* refers to the percentage of total energy needs in 1990 that could have been supplied by renewable resources.

[n] Values for "upper bound" are in billion boe/yr. The source of data is the same as that cited in note (l). These figures represent plausible upper bounds on the amount of renewable energy available in 2030, at a cost competitive with projected fossil-fuel prices; where possible, environmental and siting constraints were taken into account.

resources have tended to increase with the passage of time, and may yet increase further. The numbers in Table 1.2 indicate that the amount of oil that can still be recovered economically in the United States is equivalent to around five times the total U.S. energy consumption in

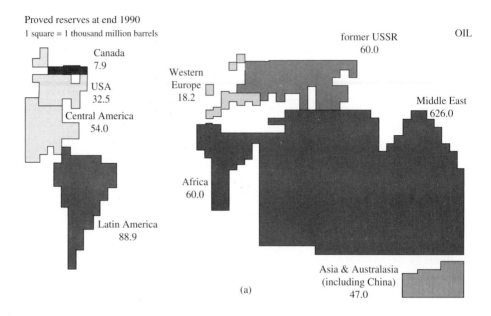

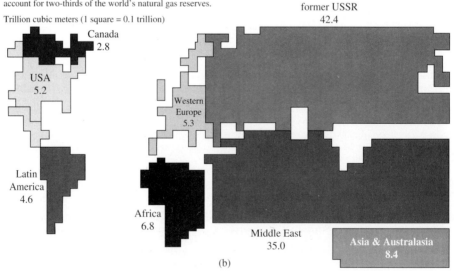

Figure 1.5   Global reserves of fossil fuels, (a) oil (b) natural gas (c) coal. *Sources:* British Petroleum Statistical Review, 1987; Energy Information Administration (1991). *Annual Energy Review 1990* (Washington, DC: U.S. Department of Energy).

Unlike oil and gas, the world's reserves of coal are rather more evenly distributed around the globe.   COAL
Nevertheless, the USA, the USSR, and China are endowed with very much larger reserves than any other area.

Thousand million tonnes (bracketed figures are share of total represented
by bituminous coal and anthracite)

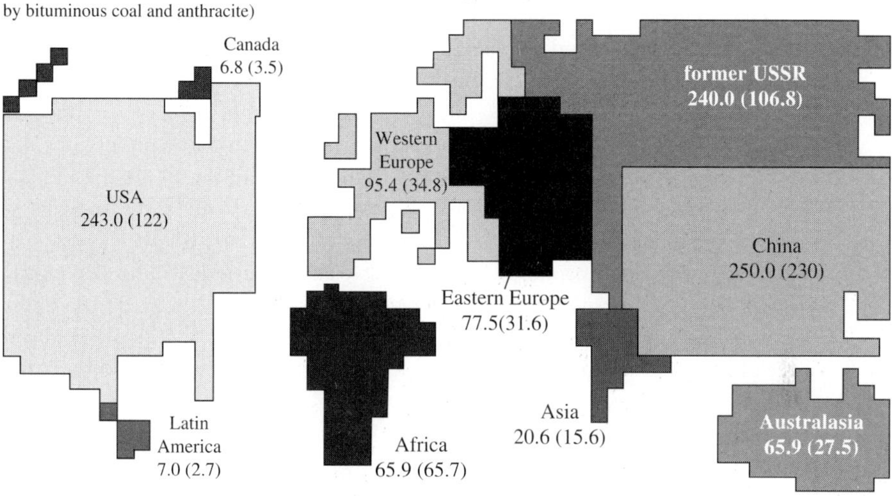

Canada
6.8 (3.5)

former USSR
240.0 (106.8)

Western
Europe
95.4 (34.8)

USA
243.0 (122)

China
250.0 (230)

Eastern Europe
77.5(31.6)

Latin
America
7.0 (2.7)

Africa
65.9 (65.7)

Asia
20.6 (15.6)

Australasia
65.9 (27.5)

(c)

Figure 1.5   (*continued*)

1990 (and about 25 times the amount of domestic crude oil consumed in 1990.) Accessible natural gas deposits in the United States contain about twice as much energy as petroleum. The rest of the world has much more of these fuels (Figure 1.5), but it is clear that as world oil and gas consumption continues to rise, the age of petroleum will be measured in decades. Coal is a good deal more abundant, and is likely to last us for centuries. But coal is a much less versatile fuel and does not easily substitute for petroleum. What do we do when the oil is no longer available at a reasonable cost?

As momentous as this question may seem, it is actually being overshadowed by more immediate concerns about the environmental consequences of using fossil fuels. Burning the reduced carbon of ages past is driving up the CO_2 content of the atmosphere and contributing to greenhouse warming of Earth. And the combustion of fossil fuels produces acid rain, as well as urban and regional air pollution. Increasing determination to clear these pollutants from the air of U.S. urban areas, particularly Los Angeles, is driving the design of cars and the formulation of gasoline. The California Air Resources Board has mandated that 2% of new cars and light trucks should have zero emissions by 1998, and the percentage increases to 5% by 2001 and 10% by 2003. It is not possible to reduce emissions to zero for any combustion process, and so this requirement means that the vehicles are to be powered by electricity. How is this electricity to be delivered, and what does the answer mean for the ultimate fuel that produces the electricity? These are the sorts of questions that will increasingly come to the fore as we grapple with the intertwined questions of energy technology, environmental impacts, and economics. Renewable sources, which have fewer environmental problems associated with their use, may play a significant role if costs diminish. The figures in Table 1.2 suggest that within the next several decades it would be feasible to supply about 40% of the U.S. energy needs (based on 1990 consumption) with renewable sources, and their potential contribution is even greater.

The issues of energy supply and resource substitution are profoundly influenced by the rate at which energy is used. Reducing energy consumption lowers environmental impacts, stretches out the resource base, and buys time for the development of new, more benign technologies. Different ways of projecting the energy consumption curve produce large differences in the estimate of future energy demand. The high and low projections in Figure 1.2 are both attempts at realistic forecasting, based on quite different assumptions about technology and behavior. The validity of these assumptions is hotly contested, and energy policy will be the focus of intense debate for a long time to come. But the experience of the last twenty years, in which economic activity, at least in the industrialized countries, has increased much faster than energy consumption (see pp. 96–97), leaves little doubt that new technology and conservation measures can increase substantially the efficiency with which we use energy. We will return to this subject at the end of Part I, after looking more closely at the various energy sources.

B. FOSSIL FUELS

1. Carbon Cycle

Only about 0.3% of the energy in the sunlight reaching Earth's surface is converted by photosynthesis to chemical energy in the form of carbohydrates. Carbohydrates are so named because they contain two atoms of hydrogen and one of oxygen for every atom of carbon. Their chemical formula is $(CH_2O)_n$ where n is a definite integer, often a very large one, as in starches. The carbohydrates are formed by the reduction of carbon dioxide and the simultaneous oxidation of water to form molecular oxygen. The overall reaction of photosynthesis is

$$CO_2 + H_2O = CH_2O + O_2$$

The products of this reaction are less stable than are the reactants, by an amount of energy corresponding to about 460 kJ per mole of carbon. This is the energy extracted from the sunlight. It can be released by reversing the reaction, either by combustion, or, in biology, by respiration. Respiration provides aerobic organisms with the energy needed for all vital functions. Green plants use up about half of their carbohydrates for their own energy needs. The remainder is converted to other biological molecules or it accumulates in the growing plant tissue; the energy it represents is the *net primary productivity*.

The processes of photosynthesis and respiration are balanced closely, and the cycling of carbon between the carbon dioxide of the atmosphere and the reduced organic compounds of biological organisms is essentially a closed loop. However, a very small fraction of plant and animal matter, estimated at less than 1 part in 10,000, is buried in Earth and removed from contact with atmospheric oxygen (Figure 1.6). Over the millennia this small fraction added up to a large amount of reduced carbon compounds. Some of the buried carbon compounds accumulated in deposits, and were subjected to high temperatures and pressures in

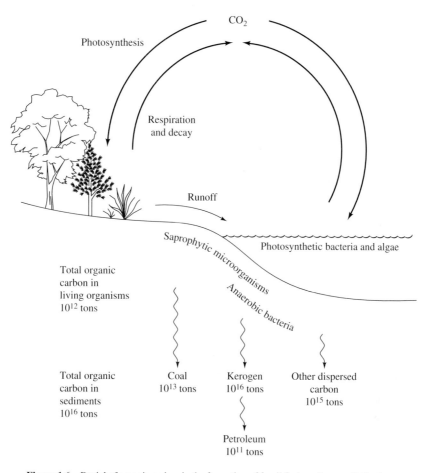

Figure 1.6   Burial of organic carbon in the formation of fossil fuels.   *Source:* G. Ourisson et al. (1984). The microbial origin of fossil fuels, *Scientific American* 251(2): 44–51. Copyright © 1984 by Scientific American, Inc. All rights reserved.

Earth's crust. They became coal, oil, and gas, which we now use to fuel our industrial civilization. We are therefore living off the store of solar energy of past ages. The total energy estimated to be available in easily recoverable fossil fuels is about 5×10^{19} kJ, roughly 14 times the annual net primary productivity.

Additionally, this process of burying reduced carbon compounds is believed to have produced essentially all the oxygen in the atmosphere. The original atmosphere of Earth was devoid of oxygen. With the evolution of green plants, some 600 million years ago, oxygen was released during photosynthesis; the burial of organic matter allowed oxygen to accumulate. We might wonder whether we are in danger of using up our oxygen supply if these buried carbon compounds are now mined and burned as fuel. But oxygen is very abundant, accounting for 21% of the atmosphere, and only a small fraction of it would be consumed (see problem 4, Part I to calculate the fraction). Most of the buried carbon is not in recoverable deposits, but is widely dispersed in Earth's crust. However, burning fossil fuel makes a big difference in the amount of carbon dioxide, whose concentration in the atmosphere is very much lower than that of oxygen. This has serious implications for warming up Earth's surface through the greenhouse effect, as discussed in Part II of this book.

2. Origins of Fossil Fuels

Petroleum and natural gas deposits are of marine origin. In the oceans, photosynthesis is estimated to produce 25 to 50 billion tons of reduced carbon annually. Most of this is recycled to the atmosphere as carbon dioxide, but a minute fraction settles to the bottom, where oxidation is negligible. This biological debris is covered by clay and sand particles and forms a compacted organic layer in a matrix of porous clay or sandstone. Anaerobic bacteria digest the biological matter, releasing most of the oxygen and nitrogen. The molecules most resistant to digestion are the hydrocarbon-based *lipids* (see pp. 202–203), and the saturated hydrocarbons found in oil have structural and carbon number distributions similar to those found in the lipids of living organisms. All petroleum deposits contain derivatives of the hydrocarbon hopane ($C_{30}H_{52}$), attesting to the importance of bacterial processing, since bacteria contain hopane derivatives in their membranes (Figure 1.7).

As the sediment becomes more deeply buried, the temperature and pressure rise. Bacterial action decreases, and organic disproportionation reactions are thought to occur. These reactions release large quantities of methane and light hydrocarbons as gases, and the gases accumulate in pockets, under impermeable rock. Oil is produced from the remaining heavy organic compounds, which migrate as a water emulsion, from which the water is squeezed out as the sediment is compacted. The oil is trapped in porous layers of rock. As shown in Figure 1.8, the process of gas and petroleum formation has spanned hundreds of

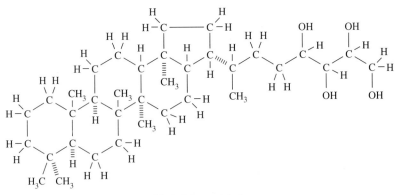

Bacteriohopanetetrol

Figure 1.7   Structure of a geohopanoid, derived from the hydrocarbon hopane.   *Source:* G. Ourisson et al. (1984). The microbial origin of fossil fuels. *Scientific American* 251: 44–51. Copyright © 1984 by Scientific American, Inc. All rights reserved.

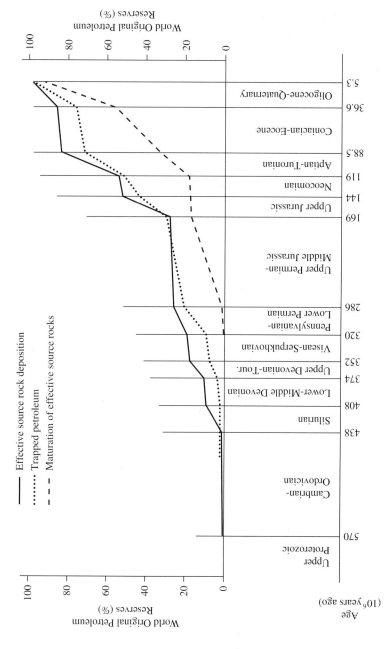

Figure 1.8 Cumulative chart of effective source rock deposition, source rock maturation, and petroleum and gas trapped in the stratigraphic succession, given as a percentage of the world's original reserves. The time of formation of the fuel is the interval between the time in which it is trapped (the middle line in the figure), and the time of maturation (indicated by the dashed lower line). *Source:* H. D. Klemme and G. F. Ulmishek (1991). Effective petroleum source rocks of the world: stratigraphic distribution and controlling depositional factors (1991). *The American Association of Petroleum Geologists Bulletin 75*: 1809–1851. Reprinted with permission from American Association of Petroleum Geologists.

millions of years. However, the time span of our use and depletion of this resource is likely to be on the order of a century and a half (approximately 1900 to mid-twenty-first century.)

In contrast to oil, coal is of terrestrial origin. Coal deposits are the remains of plant matter from the huge, thickly wooded swamps that flourished 250 million years ago during a period of mild and moist climate. Woody plants are made up mainly of lignin and cellulose. While aerobic bacteria rapidly oxidize cellulose (a carbohydrate) to carbon dioxide and water when the plant dies, lignin is much more resistant to bacterial action. It is a complex, three-dimensional polymer based on benzene rings (Figure 1.9). The building units are coniferyl and sinapyl alcohol for lignins from coniferous and deciduous plants, respectively.

In swamps, the lignin accumulates under water, compacting into peat. Over the geological ages, the peat layers of the primeval swamps metamorphosed into coal. Depression and thrusting of Earth's crust buried the deposits and subjected them to high pressures and temperatures for long periods of time. Under those conditions, the lignin gradually lost its oxygen atoms via the expulsion of water and carbon dioxide gas, and additional bonds formed among the aromatic groups, producing a hard, black, carbon-rich material, which we mine as coal. If this metamorphosis goes even further, the eventual product is graphite, a form of pure carbon in which the carbon atoms are arranged in layers of fused benzene rings.

3. Fuel Energy

In this section we consider the chemical energy stored in fuels. Where does this energy reside? It resides in the bonds that hold the atoms together. But there is a seeming paradox in this statement, since breaking bonds does not release energy. Quite the opposite, an input of energy is required to break a bond. A balloon filled with H_2 is quite stable as long as no air is admitted. The H-H bond is strong; it takes 432 kJ to dissociate a mole of H_2 into H atoms. But an explosion occurs if H_2 is mixed with O_2, and a spark is lit. The explosion results from the reaction

$$2H_2 + O_2 = 2H_2O \tag{1.1}$$

Thus, the chemical energy of H_2 resides in its propensity to react with O_2. Two molecules of H_2O are formed in the reaction, each with a pair of O-H bonds, and the energy gained by forming these bonds more than offsets the energy lost by breaking the bonds of two H_2 molecules and one O_2 molecule. Table 1.3 lists several values for bond energies, while Table 1.4 shows the energy account for the reaction. The net energy released is 482 kJ, quite enough to cause an explosion.

Similar bond energy accounting can be used to estimate energy releases for the fossil fuels, as illustrated in Table 1.4. For example, natural gas is mostly methane, whose reaction with O_2 is

$$CH_4 + 2O_2 = CO_2 + 2H_2O \tag{1.2}$$

If we add up all the bond energies of the products and subtract those of the reactants, we estimate the energy release as 810 kJ. (The number may not be exactly right because the bond energies represent average values for the bond between two atom types, which may vary somewhat from one molecule to another. But this variation is not enough to invalidate comparisons of the type being made here.)

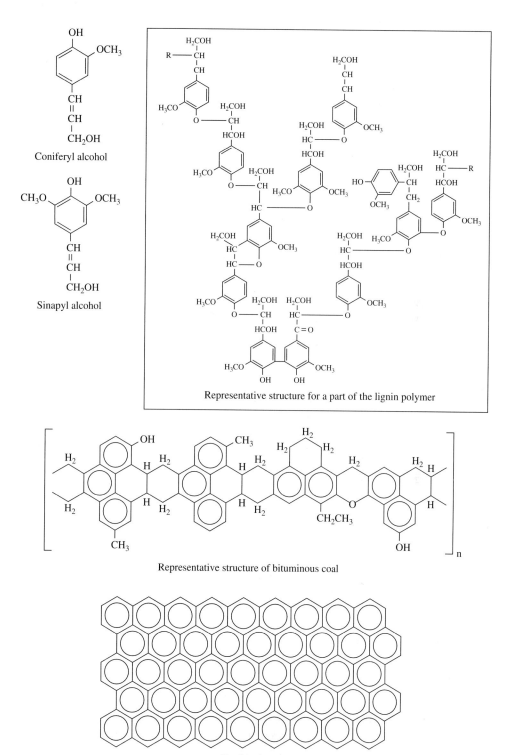

Coniferyl alcohol

Sinapyl alcohol

Representative structure for a part of the lignin polymer

Representative structure of bituminous coal

Graphite layer

Figure 1.9 Structural units of coal formation.

17

TABLE 1.3   SOME AVERAGE BOND DISSOCIATION ENERGIES

Bond	Enthalpy (kJ/mol)
H–H	432
O=O	494
O–H	460
C–H	410
C–O	360
C=O	799
C≡O	1071
C–C	347
C⋯C*	519
N=O	623
N≡N	941

*Aromatic, 1.5 bond order

We see that the energy release for reaction (1.2) is much larger than for reaction (1.1). Does this mean that methane explodes more violently than hydrogen? Not at all. The number of oxygen molecules is different in the two reactions, as written. If we compare the energy released per mole of oxygen reacted, the methane value comes down to 405 kJ and is smaller than the hydrogen value. So the reaction of an oxygen molecule with methane is slightly less violent than the reaction with hydrogen. On the other hand, a mole of methane has a much higher energy content than a mole of hydrogen (see the "per mole of fuel" column in Table 1.4), since one mole of oxygen reacts with two moles of hydrogen but 0.5 moles of methane. Because a mole of any gas occupies about the same volume (at a given temperature and pressure), a cubic meter of methane has over three times the energy content of a cubic meter of hydrogen. But if it is weight that counts, then hydrogen wins (see the "per gram of fuel" column in Table 1.4). The energy content per gram of fuel is over twice as high for hydrogen as for methane, since its molecular weight is eight times smaller. This is why rockets are fueled with liquid hydrogen. The weight of the fuel is a large fraction of the weight of the rocket, and the lighter the fuel, per unit of energy, the better.

The energy content of other fossil fuels can be similarly estimated. Table 1.4 shows schematic reactions for petroleum and coal. Since neither is a pure substance, we choose representative composition and bonding arrangements. Petroleum is largely made up of saturated hydrocarbons, and we therefore consider the combustion reaction of a representative CH_2 group in a hydrocarbon chain.

$$2(-CH_2-) + 3O_2 = 2CO_2 + 2H_2O \tag{1.3}$$

Although there are two C–C bonds connecting this group to its neighbors, we include the C–C bond energy only once in the energy account, because each bond joins two neighbors. As written, the reaction is estimated to release 1220 kJ. Per mole of oxygen, however, the energy released is 407 kJ, nearly the same value as for methane. Per gram of fuel, the energy is 43.6 kJ, somewhat less than methane. (The H/C ratio of saturated hydrocarbons is greater than 2/1, especially for shorter molecules, because of the methyl groups at the ends of the

TABLE 1.4   COMBUSTION ENERGETICS, ESTIMATED* FROM BOND ENERGIES

					Energy content (kJ)				Moles CO_2
					Reaction enthalpy (kJ)	per mol O_2	per mol fuel	per gram fuel	per (1000 kJ)
Hydrogen	$2H_2 + O_2$			$= 2H_2O$					
	$2(H–H)$	$=$	864	$2(2O–H)$	$= 1840$				
	$O=O$	$=$	494						
			1358		1840	482	482	241	120	0
Gas	$CH_4 + 2O_2$			$= CO_2 + 2H_2O$					
	$4C–H$	$=$	1640	$2C=O$	$= 1598$				
	$2(O=O)$	$=$	988	$2(2O–H)$	$= 1840$				
			2628		3438	810	405	810	51.6	1.2
Petroleum	$2(–CH_2–) + 3O_2$			$= 2CO_2 + 2H_2O$					
	$2(2C–H)$	$=$	1640	$2(2C=O)$	$= 3196$				
	$2(C–C)$	$=$	694	$2(2O–H)$	$= 1840$				
	$3(O=O)$	$=$	1482						
			3816		5036	1220	407	610	43.6	1.6
Coal	$4(–CH–) + 5O_2$			$= 4CO_2 + 2H_2O$					
	$4(C–H)$	$=$	1640	$4(2C=O)$	$= 6392$				
	$4(C\!\cdots\!C)$	$=$	2076	$2(2O–H)$	$= 1840$				
	$5(O=O)$	$=$	2470						
			6186		8232	2046	409	512	39.3	2.0
Ethanol	$C_2H_5OH + 3O_2$			$= 2CO_2 + 3H_2O$					
	$5(C–H)$	$=$	2050	$2(2C=O)$	$= 3196$				
	$C–C$	$=$	347	$3(2O–H)$	$= 2766$				
	$C–O$	$=$	360						
	$O–H$	$=$	460						
	$3(O=O)$	$=$	1482						
			4699		5956	1257	419	1257	27.3	1.6
Cellulose	$(–CHOH–) + O_2$			$= CO_2 + H_2O$					
	$C–H$	$=$	410	$2C=O$	$= 1598$				
	$C–O$	$=$	360	$2O–H$	$= 920$				
	$C–C$	$=$	347						
	$O–H$	$=$	460						
	$O=O$	$=$	494						
			2071		2518	447	447	447	14.9	2.2

*For each reaction, the bond-energy accounting is shown in columns under the reactants and the products.

chains. On the other hand, petroleum has a significant fraction of aromatic molecules with H/C ratios less than 2/1. For crude oil, the average heating value is 45.2 kJ per gram of oil, very close to the value calculated for reaction (1.3), while for gasoline, the value is slightly higher, 48.1 kJ per gram of gas, reflecting a higher H/C ratio.)

The hydrocarbons in coal are largely aromatic in nature, and the H/C ratio is 1/1 or slightly less. We consider a representative C–H group in an aromatic ring

$$4(-CH-) + 5O_2 = 4CO_2 + 2H_2O \tag{1.4}$$

The C atom is connected to neighboring C atoms by bonds with 1.5 bond order. Again the C–C bond energy is entered only once in the energy account in order to avoid duplicating bonds between neighbors. The energy released is 2046 kJ for the equation as written, 409 kJ per mole of O_2, and 39.3 kJ per gram of fuel, slightly less than petroleum. Actual heating values for coal are less than this, mainly because they contain significant amounts of water and minerals. Typical values for a hard coal (bituminous, or anthracite) are 29–33 kJ/gram, while soft coals (sub-bituminous, lignite) have heating values near 17–21 kJ/gram.

Bond-energy accounting can be applied equally to biomass-derived fuels (see pp. 69–72), such as ethanol:

$$C_2H_5OH + 3O_2 = 2CO_2 + 3H_2O \tag{1.5}$$

As shown in Table 1.4, the energy released is 419 kJ per mole of O_2, slightly higher than the fossil fuels, but the energy per gram of fuel, 27.3 kJ, is significantly lower than that for fossil fuels. The reason is that the O atom in ethanol is already in a chemically reduced state and does not contribute to the combustion energy, although it does add to the weight. A car will get fewer miles on a tank of ethanol than gasoline, since although the density of ethanol (0.79 g/cc) is about 12% higher than gasoline (~0.70 g/cc), the energy density is nearly 40% lower.

Finally, we can estimate the combustion or respiration energy for carbohydrates, the reverse of the photosynthesis reaction:

$$(-CHOH-) + O_2 = CO_2 + H_2O \tag{1.6}$$

In the energy account for this reaction (Table 1.4), we assume that each fuel unit has one C–H, one C–O, and one O–H bond, and is connected to its neighbors by C–C bonds. This leads to a slight underestimate of the energy, because in carbohydrates up to half of the O atoms are actually connected to two C atoms, and there are fewer C–C but more C–H bonds. For example, the combustion energy for glucose, which has the formula $(CHOH)_6$, is 2803 kJ per mole of fuel, whereas six times the 447 kJ obtained for reaction (1.6) is 2682 kJ. The energy per gram for carbohydrates is only about a third of that for hydrocarbons, since there are so many O atoms in the carbohydrate molecules. This is a familiar fact of nutrition and dieting; fats, which are mainly hydrocarbon in composition, have many more Calories per gram than do carbohydrates (see Part IV, pp. 263–264).

4. Petroleum

a. Composition and refining.   Oil is a complex mixture of hydrocarbons, molecules that contain mostly carbon and hydrogen. There are also small quantities of sulfur (up to 10%), oxygen (up to 5%) and nitrogen (up to 1%), bound in complex organic molecules. Several metallic elements—V, Ni, Fe, Al, Na, Ca, Cu, and U—are present in traces. Most of the hydrocarbon molecules are saturated (no multiple bonds), but an appreciable fraction, around 10%, are aromatic (contain benzene rings). The molecules range widely in size and are separated in refineries on the basis of their boiling points. Figure 1.10 is a diagram of the distillation process for dividing petroleum into its various fractions, and indicates the uses to which these fractions are put.

In addition to distillation towers, oil refineries are equipped with reactors that transform the molecules chemically in order to match the quantities of the various fractions to the needs of the marketplace. Since gasoline is the most valuable fraction, several reactions increase the percentage of the gasoline fraction. A particularly important chemical transformation is "cracking," whereby a larger hydrocarbon, in the kerosene or gas-oil range, is broken down to two smaller hydrocarbons in the gasoline range. This is accomplished at high temperature (400–600°C) with the aid of a catalyst, an aluminosilicate material impregnated with potassium.

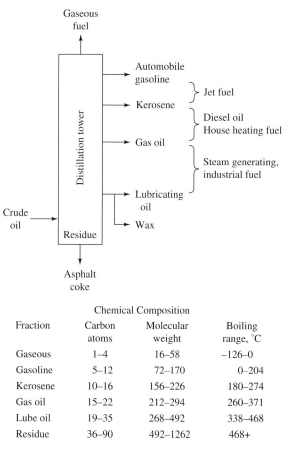

Fraction	Chemical Composition		
	Carbon atoms	Molecular weight	Boiling range, °C
Gaseous	1–4	16–58	−126–0
Gasoline	5–12	72–170	0–204
Kerosene	10–16	156–226	180–274
Gas oil	15–22	212–294	260–371
Lube oil	19–35	268–492	338–468
Residue	36–90	492–1262	468+

Figure 1.10   Crude oil refining.

$$C_{(m+n)}H_{2(m+n)+2} = C_mH_{2m} + C_nH_{2n+2} \tag{1.7}$$

alkane   alkene   alkane
(kerosene or gas-oil size)   (gasoline size)
[catalytic cracking reaction]

Another way to enhance the gasoline fraction is to build up a mid-size molecule from two smaller ones, in the process of "alkylation."

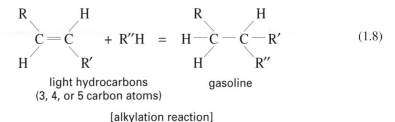

$$\tag{1.8}$$

light hydrocarbons   gasoline
(3, 4, or 5 carbon atoms)

[alkylation reaction]

This process is catalyzed by strong acids. The cracking and alkylation reactions increase the gasoline fraction of crude oil, typically 20% by volume, to 40–45%.

The conditions of the alkylation process are arranged to produce hydrocarbons that have a high degree of branching; these have higher octane ratings (less tendency to pre-ignite, or "knock," during piston compression—see p. 179) than straight-chain hydrocarbons. Even higher octane ratings are obtainable with aromatic hydrocarbons (benzene derivatives), and an additional process carried out in the refinery is catalytic reforming, whereby straight-chain alkanes are converted to aromatics:

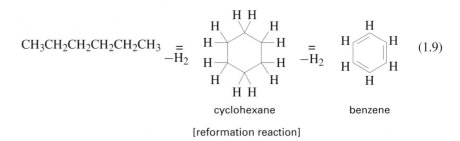

$$\tag{1.9}$$

cyclohexane   benzene

[reformation reaction]

These reactions are run at high pressure (15–20 atm) and temperature (500–600°C) with a Re-Pt-Al$_2$O$_3$ catalyst.

A final reaction is the controlled oxidation of the hydrocarbons, which can produce oxygen-containing fuel molecules called "oxygenates." These are now required to be added to gasoline in order to reduce carbon monoxide emissions (pp. 181–183). The most widely used oxygenate is MTBE, methyl, tertiary-butyl ether.

$$\tag{1.10}$$

MTBE is made by adding methanol to isobutene, using an acidic catalyst. The reaction is run with excess methanol to suppress isobutene dimerization and at low reaction temperature (<100°C) to suppress the formation of dimethyl ether. MTBE is produced in refineries where isobutene is generated in the catalytic cracking units. To implement the U.S. Clean Air Act, however, MTBE production will have to be greatly increased, requiring additional sources of isobutene. The principal alternative source is from the butane fraction of natural gas. At present, this is extracted along with propane as liquified petroleum gas (LPG). The world production of LPG is very large and some LPG could be directed as a feedstock for isobutene manufacture.

b. Advantages.   The overwhelming advantage of petroleum is that it is a liquid, and therefore easily transported. The liquid fuels of the petroleum age have made possible the development of efficient means of transport, from the airplane to the automobile to diesel locomotives. Petroleum fuels are relatively clean, since the refinery produces hydrocarbon fractions, leaving sulfur- and metal-containing compounds in the residue. The entire system of oil extraction, transport, refining and delivery has been developed to a high level of integration and efficiency, creating a formidable challenge for fuel alternatives.

c. Disadvantages

1) Oil Spills.   The extraction of petroleum from the ground inevitably produces contamination from spills. Coastal waterways are particularly vulnerable because of the fragility of their ecosystems combined with the importance of shipping in petroleum transport. Figure 1.11 illustrates the geography of oil transport. Tankers ply the shipping lanes continually, loading and unloading their cargo at the ports of the world. Tanker accidents are the most notorious instances of petroleum pollution. Oil washes up on beaches, smothers birds and sea animals, and contaminates fish. Yet these obvious spills are a relatively small part of the total output, accounting for only about an eighth of the petroleum finding its way into the sea. As detailed in Table 1.5, there are many sources of petroleum pollution, adding up to an annual total estimated at 3.2 million tons. The largest single category is normal tanker operations, during which the discharge of ballast and tank-washing water carries an estimated 0.7 million tons of petroleum into the oceans annually, while the discharge of bilge and fuel oils adds another 0.3 million tons. Runoff and wastes from municipal and industrial sources adds another 1.2 million tons.

Interestingly, a significant quantity of petroleum, estimated at 0.2 million tons a year, enters the sea without any help from humans, through natural seeps at continental margins (see Figure 1.12). Thus, oil is a natural constituent of the marine environment. This oil does not accumulate, because it is metabolized by microbes, which have evolved to exploit seepage oil as their food source. In fact, hydrocarbon-metabolizing bacteria are ubiquitous in nature, because hydrocarbons are continuously released by plants and algae. The total hydrocarbon input to the sea from marine biota is estimated to be 180 million tons annually, dwarfing the petroleum inputs from all sources. These same microbes eventually break down the oil molecules spilled by human activity, but the process can take a long time, during which there can be considerable ecosystem damage. It is possible to speed up the process

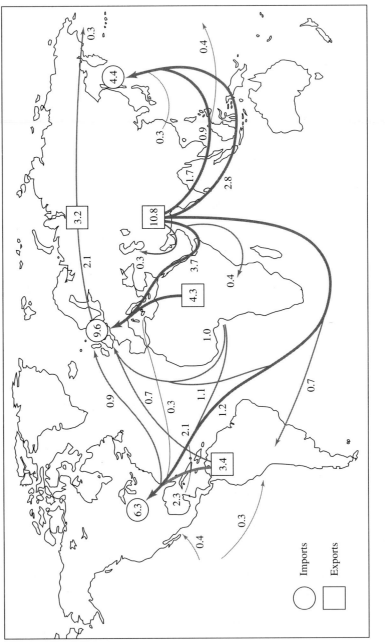

Figure 1.11 Movement of oil by sea in 1987 (in millions of barrels per day). *Source:* M. K. Tolba et al., eds. (1992). *The World Environment, 1972–1992, Two Decades of Challenge* (London: Chapman & Hall on behalf of The United Nations Environment Programme).

Imports ◯

Exports ▢

TABLE 1.5   ESTIMATED INPUTS OF PETROLEUM HYDROCARBONS INTO THE MARINE ENVIRONMENT (10^6 TONS PER YEAR)

Source	Probable range	Best estimate
Natural sources		
Marine seeps	0.02–2.0	0.2
Sediment erosion	0.005–0.5	0.05
(Total natural sources)	(0.025)–(2.5)	(0.25)
Offshore production	0.04–0.06	0.05
Transportation		
Tanker operations	0.4–1.5	0.7
Dry-docking	0.02–0.05	0.03
Marine terminals	0.01–0.03	0.02
Bilge and fuel oils	0.2–0.6	0.3
Tanker accidents	0.3–0.4	0.4
Nontanker accidents	0.02–0.04	0.02
(Total transportation)	(0.95)–(2.62)	(1.47)
Atmosphere	0.05–0.5	0.3
Municipal and industrial wastes and runoff		
Municipal wastes	0.4–1.5	0.7
Refineries	0.06–0.6	0.1
Nonrefining industrial wastes	0.1–0.3	0.2
Urban runoff	0.01–0.2	0.12
River runoff	0.01–0.5	0.04
Ocean dumping	0.005–0.02	0.02
(Total wastes and runoff)	(0.585)–(3.12)	(1.18)
TOTAL	1.7–8.8	3.2

Source: National Research Council (1985). *Oil in the Sea* (Washington, DC: National Academy Press).

under favorable conditions. The fate of the oil spilled in the March 1989 *Exxon Valdez* disaster at Prince William Sound, Alaska, attests to the cleansing power of nature. The spill resulted in the release of 35,500 metric tons of North Slope crude oil into the sound. As shown in Figure 1.13, by the autumn of 1992 all of the original floating oil had disappeared. About 50% was biodegraded either in situ on the beaches, or in the water column; 20% evaporated and underwent photolysis in the atmosphere; 14% was recovered or dispersed; 13% remained in subtidal sediments; 2% remained beached on intertidal shorelines; and less than 1% remained dispersed in the water column. One of the more effective remediation steps taken in the wake of the accident was to fertilize oil-soaked beaches with a preparation designed to stick to the sand and provide the natural bacteria with nitrogen and phosphorus to supplement their hydrocarbon diet. The oil disappeared from these beaches much faster than from untreated beaches.

    The lighter hydrocarbon fractions are the most toxic to sea creatures, because they have the highest solubility in water. These lighter molecules are also the most volatile and

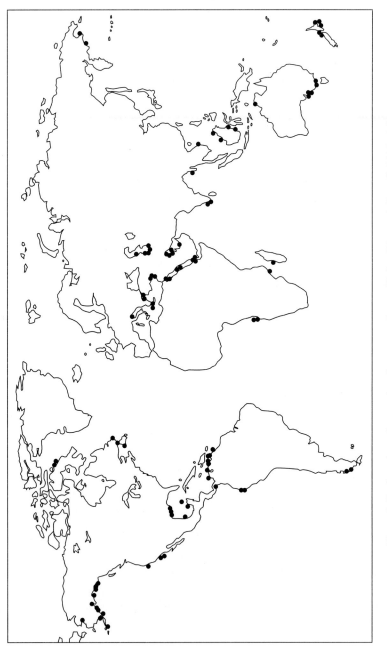

Figure 1.12 Locations of reported marine seeps. *Source:* National Research Council (1985). *Oil in the Sea* (Washington, DC: National Academy Press).

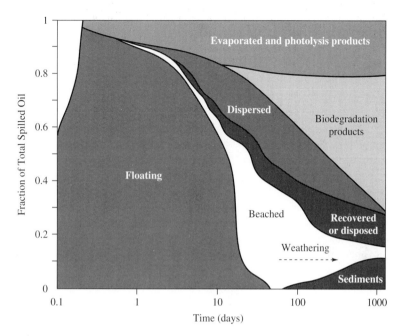

Figure 1.13 Overall fate of oil spilled in *Exxon Valdez* accident over time period from March 1989 through Autumn 1992. *Source:* D. A. Wolf et al. (1994). The fate of the oil spilled from the *Exxon Valdez. Environmental Science and Technology* 28: 561A–568A. Copyright © 1994 by American Chemical Society. Reprinted with permission.

evaporate rapidly. Between one-third and two-thirds of this fraction can evaporate in a few days. The remaining molecules are heavier and tend to form an emulsion with seawater, sometimes called a "mousse." The emulsion eventually forms tar balls, which can persist for a long time. All of the evaporated molecules, and a substantial fraction of those left behind, are oxidized through the action of sunlight and oxygen (photo-oxidation), processes that we will encounter in discussing air pollution and photochemical smog (pp. 171–175).

  2) Emissions.   The burning of all fossil fuels produces CO_2 and adds to the greenhouse warming of Earth (pp. 118–127). Because the C/H ratio declines from about 1/1 for coal to about 1/2 for petroleum and to 1/4 for methane, the amount of CO_2 produced per kJ of energy decreases in the same order. The last column of Table 1.4 lists the moles of CO_2 released per thousand kJ of energy produced by the combustion of each of the fuels; this number is obtained by dividing the number of molecules of CO_2 in each of the equations by the reaction enthalpy and multiplying by 1000. For the same energy production, coal and petroleum release significantly more CO_2 than methane, by about 60% and 33%, respectively. For completeness, CO_2 values are also listed in Table 1.4 for ethanol and carbohydrates, but biomass-derived fuels do not alter the atmosphere's CO_2 content, since the released CO_2 exactly balances the CO_2 that was taken from the atmosphere in producing the biomass.

  All combustion processes produce the pollutant nitric oxide, NO. Although the N_2 and O_2 of the atmosphere do not react at ordinary temperatures, they do react at the high temperatures of a furnace or an automotive engine. The reaction

$$N_2 + O_2 = 2NO$$

requires an input of energy, which is provided by the combustion process. The N≡N and O=O bond energies (Table 1.3) sum to more (1,435 kJ/mol) than the pair of N=O bonds (1,247 kJ/mol), by 188 kJ/mol. We say the reaction is *endothermic*, by this much energy. The energy deficit can be made up with a heat source, which drives the reaction to the right. Once the exhaust gases leave the combustion zone and cool off, the reaction should shift back to the left; the decomposition of NO into N_2 and O_2 is *exothermic*. However, the reverse reaction is very slow at low temperatures, because the NO bonds first have to break before the atoms can rearrange to N_2 and O_2. Consequently, NO does not decompose to any significant extent. Rather, it is slowly oxidized further to nitrogen dioxide, NO_2, and then to nitric acid, HNO_3, which is absorbed by raindrops and washed out of the atmosphere. Combustion sources, both stationary and mobile, are major contributors to acid rain, through the production of nitric acid (pp. 207–208).

In addition, nitrogen dioxide is a key ingredient in smog production because of its photochemical activity (pp. 171–175). The products of incomplete combustion, carbon monoxide (CO) and unburned hydrocarbons, are the other key ingredients in smog production. In both cases the main culprit is the automobile, and enormous effort has gone into curtailing automotive emissions. Because of the ever-increasing volume of traffic, however, these efforts are approaching diminishing returns. As mentioned in the introduction, the requirement that "zero-emission" vehicles be introduced by 1998 in Southern California is a harbinger of changes to come.

Thus in summary, petroleum stands as a major contributor to the greenhouse effect, acid rain, and urban air pollution.

5. Gas

Although natural gas represents as large an energy resource as petroleum, it has historically been considered a by-product of oil exploration and production. Indeed, the magnitude of the potential gas supply has only recently become appreciated, thanks to better information about gas-bearing formations, and to improved recovery techniques. Natural gas provides a substantial fraction of the U.S. energy budget (nearly 25%), but its use has mostly been limited to heating and cooking. New technologies may expand its roles, however, especially in view of the superior environmental attributes of gas. A particularly promising technology is the gas turbine electric generator. Stemming from the development of advanced jet engines for aircraft, the gas-fired turbine is a heat engine that can run at very high temperatures, and is therefore capable of high efficiency (see pp. 79–80). An attractive arrangement is the "combined-cycle" system, in which the waste heat of the turbine is used to run a second, steam-powered turbine, thereby boosting the overall efficiency.

a. Advantages. Natural gas is a clean fuel, requiring very little processing. It is readily transported overland through pipelines. Its CO_2 emission rate, per unit of energy, is lower than other fossil fuels, especially coal, as already noted. In addition, gas contributes less to smog formation than does gasoline, because unburned CH_4 molecules are considerably less reactive, with respect to the free radical chemistry responsible for smog, than are hydrocarbon molecules having more than one carbon atom. There is considerable interest in substituting gas for gasoline in automotive transport.

b. Disadvantages.   Natural gas is much harder to carry around than are liquid hydrocarbons. To pack enough of it into a reasonable space for mobile power sources requires high pressures or low temperatures or both. Compressors and/or refrigerators are required, and the storage tank must be thick-walled and/or insulated. In addition, it requires a distribution system capable of transferring the gas under pressure. These requirements are formidable obstacles to the replacement of petroleum by natural gas in automotive transport. The problems are more manageable for fleets of trucks or buses, which can carry large tanks and which are supplied at a central depot. A number of fleets are currently running on natural gas. Countries with abundant gas reserves, notably New Zealand, Canada, and Russia, are actively promoting the use of natural gas in vehicles.

Although natural gas produces less CO_2 than other fossil fuels, methane is itself a potent greenhouse gas. Its infrared absorption bands fall in the window of the CO_2 and H_2O spectra (see pp. 118–122), and, because it is less reactive than other hydrocarbons, it has a long atmospheric lifetime. A methane molecule contributes about twenty times as much to the greenhouse effect as a CO_2 molecule. Consequently, leaks of methane are a serious environmental concern. These can occur at the gas wells, during transfers, and from power sources. Escape of methane during idling of gas-powered vehicles, for example, could nullify their CO_2-related greenhouse advantage.

6. Coal

Coal deposits vary significantly in the extent to which the original woody plant tissue has been metamorphosed. Hard coals have undergone greater transformation than soft coals. Table 1.6 gives percentages of various coal constituents for deposits in different regions of the United States. Lignite is the softest coal; its name recognizes the close similarity to the

TABLE 1.6 COMPOSITION AND HEAT CONTENT OF COMMON COALS FOUND IN THE UNITED STATES

Rank	Location by state	Chemical analysis				Heating value (kJ/g)
		Moisture	Volatile matter	Fixed carbon	Ash	
Anthracite	Pa.	4.4%	4.8%	81.8%	9.0%	30.5
Bituminous						
Low volatile	Md.	2.3	19.6	65.8	12.3	30.7
High volatile	Ky.	3.2	36.8	56.4	3.6	32.7
Sub-bituminous	Wyo.	22.2	32.2	40.3	4.3	22.3
Lignite	N. Dak.	36.8	27.8	30.2	5.2	16.2

Source: U.S. Bureau of Mines (1954). *Information Circular, No. 769* (Washington, DC: U.S. Department of Interior).

parent wood component, lignin. Over a third of the lignite mass is moisture, while the remaining carbonaceous material is almost evenly divided between "volatile matter," hydrocarbons that are released upon heating, and "fixed carbon," the nonvolatile carbon fraction. Sub-bituminous coal is harder than lignite, containing about 20% moisture and 40% fixed carbon, but softer than bituminous coal, which contains very little moisture. The hardest coal is anthracite, which is about 80% fixed carbon. The heating value of the coal varies with the fraction of reduced carbon and hydrogen, and is much lower for soft than hard coals because of their high moisture content.

The different coals have variable amounts of ash, the mineral residue left after complete combustion, reflecting different amounts of minerals incorporated during the metamorphic processes. Some of this mineral is pyrite, FeS_2. In addition, some sulfur is bound in the complex organic molecules of coal. When coal is burned, both inorganic and organically bound sulfur is oxidized to SO_2, a significant air pollutant.

a. Advantages.   The coal resource base is very large, and coal is relatively cheap to mine and transport by rail. This is coal's great advantage.

b. Disadvantages.   Coal is, of course, much less convenient to carry around and handle than petroleum. Its use in transportation disappeared when diesel replaced the steam locomotive. In technologically advanced countries coal is no longer used directly for space heating, but in the rest of the world, the burning of coal in stoves and furnaces fouls the air with soot and SO_2, contributing significantly to respiratory distress. The main use of coal is in large electricity generating plants, where it is burned efficiently and relatively completely. Tall smokestacks disperse emissions widely, lessening the local air pollution. However, the SO_2 and NO emitted from the burning of coal in power plants are the main sources of acid rain (pp. 207–208), a problem of large regional dimensions. As with the other fossil fuels, the emitted CO_2 enters the global greenhouse budget; because of its lower C/H ratio, coal emits more CO_2 per unit of energy produced than either gas or oil (see Table 1.4).

NO emission can be significantly reduced by carrying out the combustion in two stages. In the first stage, the fuel is burned at a high temperature, but the amount of air is limited ("sub-stoichiometric"), so that combustion is incomplete, and there is no excess O_2 with which the N_2 can react. In the second stage, combustion is completed in excess air, but the temperature is kept relatively low to limit the N_2 reactivity. In this way the emission of NO has been reduced by as much as 90% in some power plants.

Extraction of coal adds significant costs to the environment and to human health. Traditional coal mines, which follow coal seams deep into the earth, are hazardous places to work, and the coal dust produces black-lung disease in miners. These problems have been alleviated to some extent through improved safety measures, better ventilation, and by spraying water to reduce dust during drilling operations. The drainage from mines, which is highly acidic, once contaminated local streams, but federal regulations in the United States now require collection of drainage in settling and treatment ponds. Strip-mining has permitted surface extraction of shallow seams of coal, at the expense of great gashes in the earth, often on steep and erodible hillsides. United States regulations now require strip-mine operators to restore the original contour of the land, replace topsoil, and replant grasses, legumes, and trees.

   c. Coal-derived fuels.   Technologies are available for converting coal to clean fuels via chemical reactions, to produce gaseous or liquid hydrocarbons. The basic requirement is to increase the H/C ratio of the coal. For example, direct reaction of coal with H_2 can yield methane

$$C + 2H_2 = CH_4 + 74.9 \text{ kJ} \tag{1.11}$$

   But this "hydrogasification" reaction requires a high operating temperature, 800°C, for adequate reaction rates. Moreover the reaction is inefficient because, being exothermic, it is thermodynamically unfavored at the required high temperature. A more efficient route is provided by "methanation" of CO:

$$CO + 3H_2 = CH_4 + H_2O + 206.3 \text{ kJ} \tag{1.12}$$

   This reaction is even more exothermic, but it can operate at a lower temperature, 400°C in the presence of a nickel catalyst. Reaction conditions can be altered to produce liquid hydrocarbons via what is called Fischer-Tropsch chemistry

$$nCO + (2n+1)H_2 = C_nH_{2n+2} + nH_2O \tag{1.13}$$

again using metal catalysts. Still another possibility is to make methanol, via

$$CO + 2H_2 = CH_3OH \tag{1.14}$$

using other catalysts and reaction conditions. Methanol is an alternative liquid fuel (p. 71).
   But all these conversion reactions require CO and H_2. What is their source? They can be produced by treating coal with water at a very high temperature, 900°C:

$$C + H_2O = CO + H_2 - 131.4 \text{ kJ} \tag{1.15}$$

   This reaction, called the "steam reforming" reaction, produces only as much H_2 as CO, but extra H_2 can be produced by the "water-gas shift" reaction

$$CO + H_2O = CO_2 + H_2 + 41.4 \text{ kJ} \tag{1.16}$$

   If we multiply equation (1.15) by two and add equations (1.16) and (1.12), the result is

$$2C + 2H_2O = CH_4 + CO_2 - 15.1 \text{ kJ} \tag{1.17}$$

   In theory, all the heating value of coal can be transferred to methane [or alternatively to gasoline, via reaction (1.13), or methanol, via reaction (1.14)], with an energy expenditure of only 15.1 kJ. In reality, the energy costs are much higher, because the individual stages of the process are not well matched thermodynamically. The heat released in reaction (1.12) cannot be recovered to drive reaction (1.15), because the required temperature is much higher for the latter than for the former. Instead the energy input to reaction (1.15) must be provided by burning extra coal. The required energy for one mole of methane is twice the enthalpy of reaction (1.15), 262.8 kJ, which is about 32% of the energy content of the methane (see Table 1.4). Thus the energy-conversion efficiency of the process can be no better than 68%, and is actually lower than this because of other losses. The cost of coal-derived fuels is therefore high. Moreover, they contribute disproportionately to the greenhouse effect, because excess CO_2 is released by burning the extra coal required for the energy input to the conversion process. More CO_2 is released in the conversion and burning of coal-derived fuels than would be released in the production of equivalent energy from the coal itself.

The fossil fuel era will certainly extend well into the twenty-first century. But what fuels will replace them? The following sections discuss the relative merits of alternative fuels that could reduce our current dependence on fossil fuels.

C. NUCLEAR ENERGY

1. Fission

Nuclear power is presently the most highly developed alternative to energy supplied by coal. Apart from geothermal energy, the one significant form of energy on earth that is not related to the sun either directly or indirectly is energy that resides in the nuclei of atoms. Nuclei are made up of collections of protons and neutrons called *nucleons*, which are held together by strong nuclear forces. As indicated in Table 1.7, protons are positively charged, while neutrons are neutral. The number of protons determines the number of negatively charged electrons that surround the nucleus, which determines the chemical properties of the element. The mass of the atoms is determined by the total number of nucleons. Isotopes of the elements, such as He-3 and He-4, have the same number of protons but different number of neutrons. The nuclear forces increase as the number of nucleons increases, but so does the electrostatic repulsion among the positively charged protons. Figure 1.14 is a graph of the

TABLE 1.7   SIMPLE NUCLEAR PARTICLES

Type	Schematic representation	Charge	Mass*	Chemical symbol[†]
Neutron	●	0	1.0087	$_0^1 n$
Proton	⊕	+1	1.0078	$_1^1 p$
Helium-4 (alpha particle)		+2	4.0026	$_2^4 He$
Helium-3		+2	3.0160	$_2^3 He$

*In atomic mass units (amu), where 1 amu $= 1.6606 \times 10^{-24}$ g.

[†]The superscript is the *mass number*, equivalent to the number of protons and neutrons in the nucleus; the subscript is the *atomic number*, equivalent to the number of protons. $_2^4 He$ and $_2^3 He$ are isotopes.

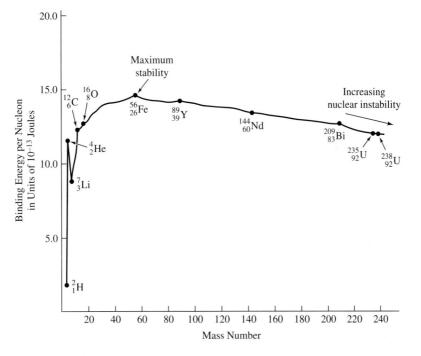

Figure 1.14   Nuclear binding energy curve.

total binding energy per nucleon with increasing mass of the nuclei. The highest nuclear stability is associated with atoms of intermediate atomic weight, in the vicinity of the element iron. For heavier elements, the repulsive forces gradually become more important, and elements heavier than bismuth, with 83 protons, are unstable. As illustrated in Figure 1.15, these heavier elements split off alpha (α) particles, nuclei of the element helium; alpha particles contain two protons and two neutrons. A few of the heaviest isotopes can also undergo spontaneous fission, splitting into two *daughter* atoms of intermediate atomic weights—fission releases a great deal of energy.

Isotopes of the elements can also be unstable with respect to the number of neutrons they contain. The ratio of neutrons to protons needed for stability increases slowly with increasing atomic number, as shown in Figure 1.16. Isotopes with too many neutrons can convert a neutron to a proton with the emission of a beta (β) ray, a high-energy electron. This corresponds to moving diagonally across the graph of Figure 1.16 toward the stability curve. Both alpha particle emission and nuclear fission produce isotopes with too many neutrons, and the products are therefore beta emitters. Unstable isotopes can also release some of their energy in the form of high-energy electromagnetic radiation, called gamma (γ) rays. The rate of nuclear decay, whether by alpha, beta, or gamma emission, varies from one isotope to another. It is generally expressed as the half-life, the time required for half of the nuclei to decay, as illustrated in Figure 1.17 for plutonium. Plutonium no longer exists naturally in Earth's crust but is produced in nuclear reactors. Half the plutonium that was initially formed on Earth

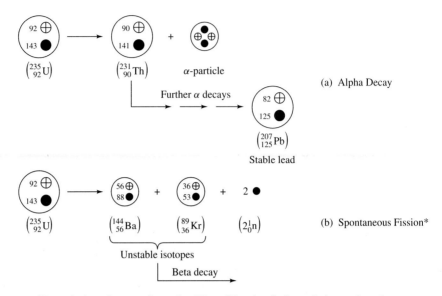

(a) Alpha Decay

*Example shows just one of more than 30 possible pairs of primary fission products that can occur when U-235 splits up.

Figure 1.15   Reactions of heavy nuclei.

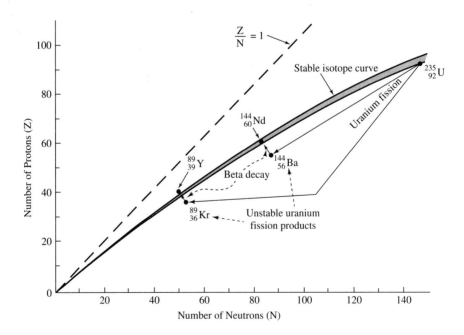

Figure 1.16   Proton-neutron ratio for stable isotopes: beta emission [beta particles (β^-) are high-energy electrons emitted by the reaction: $(^1_0n) \rightarrow$ proton ($^1_1p^+$) + β^-].

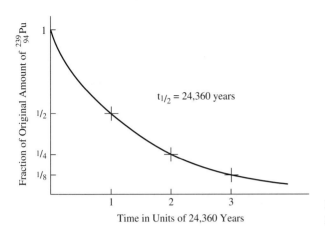

Figure 1.17   Half-life ($t_{1/2}$) decay curve of plutonium-239.

decayed after 24,360 years. After 100,000 years, that is, approximately four half-lives, the quantity of plutonium was reduced to one-sixteenth of its initial value. Thus, plutonium and other unstable elements have long since disappeared over the course of Earth's history. However, a few of them decay so slowly that they are still present in significant abundance. Among these, uranium is the only element that can undergo spontaneous fission.

Even among uranium atoms, spontaneous fission is an extremely rare event. However, the isotope of uranium, the one containing 235 nucleons and called uranium-235 or U-235, can be induced to undergo fission with high probability if it encounters a neutron. In addition to the two fission fragments, two or three neutrons are produced in this process, as illustrated in Figure 1.18. If these neutrons, in turn, strike other atoms of uranium-235, they can produce further fission. Since the ratio of neutrons released to neutrons absorbed in each fission event is greater than one, a chain reaction can be built up in which many atoms in a

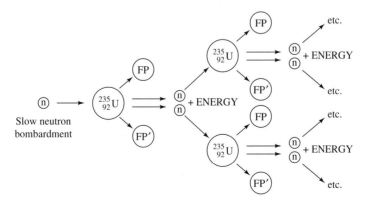

Figure 1.18   Chain reaction induced with thermal (slow) neutrons (FP and FP′ represent various fission products).

piece of uranium-235 undergo fission in a short period of time with a large release of energy. This is the basis of the first atom bomb, which was tested in New Mexico and exploded over Hiroshima in 1945.

Whether a chain reaction will occur depends mainly on the amount of uranium present, since some neutrons escape from the system before they are absorbed by the uranium nuclei. This tendency decreases as the surface-to-volume ratio of the total sample decreases, that is, the mass of the uranium increases. A *critical mass* is that amount of uranium for which the probability, k, that a neutron produced in a fission reaction induces another fission reaction is equal to unity. This is the break-even point for a self-sustaining chain reaction. A fission explosion is set off by assembling a *supercritical mass* ($k > 1$) of uranium-235 (or plutonium-239, a man-made fissionable isotope—see pp. 44–46) very rapidly, using a chemical explosive for the trigger. For pure uranium-235, the critical mass is 15 kg, while for pure plutonium-239, it is 4.4 kg. (Smaller masses can be critical if surrounded by material that reflects escaping neutrons, making them available for additional fission events. Reflectors are incorporated into the design of nuclear weapons.)

2. Naturally Occurring Radioisotopes

The most important radioisotopes found in Earth's crust, along with their half-lives, are:

	$t_{1/2}$ (years)		$t_{1/2}$ (years)
^{232}Th*	1.4×10^{10}	^{235}U	7×10^{8}
^{238}U	4.5×10^{9}	^{40}K	1.27×10^{9}

*The left superscript gives the mass number of the isotope, the number of neutrons plus protons. Sometimes the atomic number (number of protons) is also given as a left subscript, for example, $^{235}_{92}$U. Since the element with atomic number 92 is uranium, the atomic number is implicit in the symbol U.

The first three are α emitters, while ^{40}K is a β emitter. ^{232}Th and ^{238}U are both quite abundant in Earth's crust. ^{235}U has become scarce, since 6.4 of its half-lives have elapsed since the formation of Earth. ^{40}K is a form of potassium that occurs in low abundance (0.001%), but because potassium is an important constituent of biological tissues, ^{40}K provides a significant fraction of the background radiation to which we are normally subject.

a. Decay chains: the radon problem.   Alpha emission often leaves a product isotope that is itself unstable with respect to β or further α emission. Consequently, the decay of the heavy elements generally proceeds in a sequential cascade, as shown for ^{238}U in Figure 1.19. The long-lived ^{238}U produces a much shorter-lived (24.1 day half-life) isotope of thorium, ^{234}Th. Two successive β decays then produce another uranium isotope, ^{234}U, which is less stable than ^{238}U, but still long-lived ($t_{1/2} = 245,000$ years). It decays to a succession of more-or-less short-lived isotopes, ending with the stable lead isotope ^{206}Pb. All of these daughter isotopes accumulate in a sample of matter that contains ^{238}U. Their concentrations build up to a steady state, in which their production rate equals their decay rate. For the rapidly decaying isotopes, this concentration is very low, but for relatively long-lived iso-

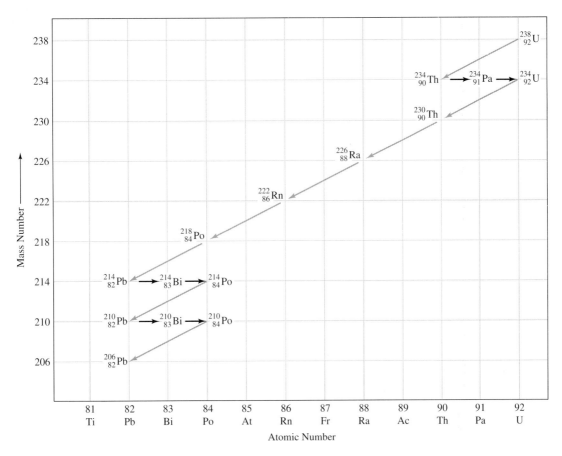

Figure 1.19   Uranium-238 decay chain through lead-206.   *Source:* P. Buell and J. Gerard (1994). *Chemistry in Environmental Perspective* (Upper Saddle River, New Jersey: Prentice Hall). Copyright © 1994 by Prentice Hall. Reprinted with permission.

topes, the concentrations can be significant. For example, radium, ^{226}Ra, with a 1600 year half-life, accumulates sufficiently in uranium deposits to support extraction for commercial applications. The other two very long-lived α emitters, ^{235}U and ^{232}Th, have their own sequence of daughter isotopes, all of which can be found in Earth's crust.

One isotope in the ^{238}U decay chain is of special environmental significance, namely ^{222}Rn. Radon is a noble gas, falling just below xenon in the periodic table. Consequently, radon isotopes do not form chemical bonds, and are free to escape from the site where they are formed. Born in uranium-containing rock, ^{222}Rn can travel an appreciable distance before it decays, with a 3.82 day half-life. It can enter nearby buildings by seeping through foundations, and accumulate to significant levels in the air. It can also infiltrate wells and enter the water supply.

In 1984, a nuclear plant worker in Pennsylvania set off a radiation-monitoring alarm on his way *into* work. His clothes were found to be radioactive and the source of the contamination was traced to his home, in which a high radon level was found. This incident led to the recognition that naturally accumulating radon can be a significant radiation hazard. Since uranium is a relatively common element, and is distributed widely in Earth's crust, the problem is widespread. Radon can build up to high levels in houses built on rock with high uranium content, and where the ground is sandy and therefore permeable to gases. The radon level increases when the building foundation is porous, and when the air pressure in the building is less than the pressure of the gases in the soil. This pressure difference, which depends on indoor and outdoor temperatures, on wind velocity, and on a host of other factors, is a key variable, and probably accounts for large variations in radon levels even among nearby houses. Fortunately, inexpensive tests for radon are readily available, and remediation is generally straightforward. The usual method is to lay vent pipes next to or under the building and to pump soil radon directly out of doors with an exhaust fan. There is continuing debate, however, about just how serious the problem is and how worried people should be.

b. Radiocarbon dating.   Unstable isotopes can also be created by the interaction of stable nuclei with neutrons or with high-energy particles. Cosmic rays are charged particles from outer space that enter Earth's atmosphere at high velocity. Their collisions with nitrogen nuclei in N_2, the major constituent of the atmosphere, produce ^{14}C ($t_{1/2} = 5,730$ years), which reacts with O_2 to form $^{14}CO_2$. This isotopic CO_2 enters the biological carbon cycle. All living things therefore contain the same fraction of ^{14}C as that produced in the atmosphere by the steady rain of cosmic rays. When life stops, so does the exchange of carbon in the atmosphere, and the ^{14}C content of preserved organic matter gradually decreases. This is the basis of the radiocarbon dating method which can be applied to cloth, wood or bones, and which has been of great help to archaeology. The ^{14}C disintegration rate is determined for a sample of the material in question, and is compared with the rate for a modern material with the same carbon content. From the ratio of the rates and the known half-life, the age of the sample is readily calculated (see problem 8, Part I).

3. Radioactivity: Biological Effects of Ionizing Radiation

When unstable nuclei decay, α, β, or γ rays are released with very high energy, in the range of millions of electron volts (Mev). On encountering molecules in their path, they knock electrons out of the atomic shells. Enough energy is deposited in the ionized molecules to break chemical bonds and induce reactions. In biological tissues the result is generalized damage and the production of reactive chemical fragments (for example, free radicals). Table 1.8 summarizes the penetration depth for the different rays and their relative effectiveness in producing damage.

Some biological molecules are more susceptible than others to the effects of ionizing radiation. The nucleic acids, which make up each cell's genetic apparatus, are particularly vulnerable. A single ionization in the cell's nucleus can produce an error in the genetic instructions for assembling the protein constituents of the cell. Certain kinds or combinations of such errors are believed to transform normal cells into cancerous ones, and there are well-established correlations between radiation exposure and the incidence of cancer. If the nuclei of reproductive cells are damaged, the result may be genetic mutations and, thus, the transmission of hereditary disorders to succeeding generations. At high levels, radiation damages

TABLE 1.8   PATHS OF ENERGETIC PARTICLES IN BIOLOGICAL TISSUE

Type of radiation	Range in biological tissue	Relative biological effectiveness*
alpha	0.005 cm	10–20
beta	3 cm	1
gamma	~20 cm	1

Some hazardous radioactive isotopes			
Element	Type of radiation	Half-life	Site of concentration
$^{239}_{94}$Pu	alpha	24,360 years	Bone, lung
$^{90}_{38}$Sr	beta	28.8 years	Bone, teeth
$^{131}_{53}$I	beta, gamma	8 days	Thyroid
$^{137}_{55}$Cs	beta, gamma	30 years	Whole body

*Accounts for the fact that cell damage increases as the density of the damage sites increases.

all cells, particularly those which divide rapidly—white blood cells, platelets, and the cells of the intestinal linings—leading to a variety of symptoms, collectively called radiation sickness.

The potential for damage is crucially dependent on the location of the isotope emitting the radiation, relative to target molecules. If the isotope is external to the body, then the issue is what kind of shielding is needed to absorb the rays. If the isotope is ingested, however, then the issue becomes one of transport and elimination, as well as the decay rate. The most worrisome isotopes are those that lodge in particular tissues and are long-lived enough to do significant damage. Table 1.8 lists four isotopes of particular concern. ^{239}Pu is part of the nuclear fuel cycle, while ^{90}Sr, ^{131}I, and ^{137}Cs are fission products. They all live long enough to do significant damage, and they all concentrate in particular tissues. Plutonium and strontium lodge in bone, primarily, because their chemistry mimics that of calcium. Cesium spreads through all tissues, along with potassium, which it mimics. Iodine is a natural constituent of the thyroid hormone thyroxine, and ^{131}I concentrates in the thyroid gland, which it can damage; indeed, controlled doses of ^{131}I are used therapeutically to counter hyperthyroidism.

a. Alpha rays. An α ray, being a doubly charged helium nucleus, produces intense damage over a short distance. It ionizes half the atoms in its path, losing about 30 ev (electron volts) per collision. A 6-Mev α ray would therefore produce about 200,000 ionizations among the first 400,000 atoms it encounters, before its energy is dissipated. In a substance with the density of water or biological tissue, the distance traveled, or range, is about 0.05 mm. This is less than the thickness of the skin's protective outer layer of dead cells. If, however, an α emitter is ingested into the body, it produces a high density of localized damage, with an appreciable potential for cancer induction. The likeliest route of ingestion is inhalation of dust particles that carry radioisotopes.

Uranium miners are at high risk of developing lung cancer. The uranium itself is relatively harmless because of its very slow decay. Some of the daughter isotopes, which accumulate in the mines, are much more radioactive and therefore more hazardous. Most of the cancer effect is attributed to radon and its daughters. Radon itself, being an unreactive gas, is expelled from the lungs as fast as it is inhaled (and with a 3.8 day half-life, little of it decays in the lungs). The daughters are isotopes of reactive elements, however (see Figure 1.19), and these are incorporated into dust particles that the miners breathe. The dust particles stick to the lining of the lung, and the radioactive daughter isotopes therefore have time to do their damage. The cancer rate has been correlated with the levels of radon (and its daughters) to which the miners are exposed.

The same mechanism of biological damage applies to radon and its daughters in houses. At the levels found in some houses, the radon exposure is calculated to be close to the low end found in uranium mines (taking into account the longer time people spend in houses than in mines). This is the basis of concern over the cancer-causing potential of radon in houses. Indeed, the extrapolation from concentrations known to cause cancer (uranium mines) to those commonly encountered by the population at large (houses), is less than that for any other known or suspected carcinogen. Nevertheless, there is continuing skepticism as to the extent of the danger, because an association between radon levels in houses and cancer incidence has not yet been established in the epidemiological data. It is often suggested that the uranium miner cancer incidence is not a reliable guide, because mines are much dustier than homes. This debate is representative of those encountered for all environmental carcinogens (see pp. 296–299). The evidence is often equivocal, and extrapolations to usual exposures are uncertain, allowing individuals to evaluate the risks very differently.

b. Beta and gamma rays, and neutrons.   A β ray, being an energetic electron, is much lighter than an α ray and is only singly charged. It also loses about 30 ev per collision, but it ionizes only about 1 in 1,200 atoms in its path. The damage density is therefore lower, but the range of a β ray is larger; a 6-Mev β ray travels 3 cm in water or biological tissue. External β radiation is therefore hazardous, and β-emitting isotopes must be shielded.

A γ ray is a high-energy photon and has a different mode of interaction with matter than do charged particles. The probability of a γ ray hitting an atom in its path is quite low, but when it collides, it transfers a large amount of energy, and the ionized electron carries away enough energy to ionize many other electrons (secondary ionizations). Because of this, γ rays do not have a well-defined range but rather a distribution of path lengths. For 6-Mev γ rays, the median path length at which half the γ rays have stopped is 20 cm in water or biological tissue. Energetic γ emitters require heavy shielding.

Finally, neutrons interact with matter by penetrating the electron shells and reacting directly with the nuclei of matter, displacing them and causing ionization or producing radioisotopes, which in turn release ionizing radiation. Neutrons decay spontaneously ($t_{1/2}$ = 12 minutes) into protons and electrons, and are therefore of concern only in the immediate vicinity of nuclear reactions.

c. Radiation exposure.   Aside from the location of the radiation source and the type of radiation, exposure depends on the concentration of a given radioisotope and its half-life. The shorter the half-life, the greater the disintegration rate, and the more intense the exposure. On the other hand, if the half-life is very short, then the exposure is also brief.

Radioactive disintegrations are measured in *curies* (Ci); one Ci is 3.7×10^{10} disintegrations per second. In the case of radon in houses, the U.S. Environmental Protection Agency has set an action level at 4 p Ci/l (4 picocuries per liter), corresponding to 0.15 disintegrations per second for each liter of house air. Exposure to radiation, on the other hand, is measured in several different units. A *roentgen* (R) is the amount of radiation that, on passing through 1 cm³ of air (at 0°C and 1 atm pressure), would create one electrostatic unit (2.08 $\times 10^9$ times the charge on an electron) each of positive and negative charges. A *rad* is the amount of radiation that deposits 100 ergs of energy in a gram of material; for biological tissues, 1 R is about equivalent to 1 rad. Finally a *rem* (roentgen-equivalent-man) is the amount of radiation that produces the same biological effect in a person as 1 R of x-rays. For β and γ rays, 1 rad is equivalent to 1 rem, but for α rays, because of their greater ionizing ability, 1 rad is equivalent to 10–20 rem, depending on their energy. (There are new international units, the *gray*, equivalent to the rad, and the *sievert*, equivalent to 100 rem.)

To how much radiation are we exposed? Table 1.9 lists the average annual doses for people in the United States. The total is about 360 mrem (millirem), of which the great

TABLE 1.9   AVERAGE ANNUAL EXPOSURE (1990) TO RADIATION FOR PEOPLE IN THE UNITED STATES

Source of radiation	Dose (mrem)	Percent of total dose
Natural		
Radon gas	200	55
Cosmic rays	27	8
Terrestrial (radiation from rocks and soil other than radon)	28	8
Inside the body (naturally occurring radioisotopes in food and water)	39	11
Total natural	294	82
Artificial		
Medical		
X-rays	39	11
Nuclear medicine	14	4
Consumer products (building materials, water)	10	3
Other		
Occupational (underground miners, x-ray technicians, nuclear plant workers)	<1	<0.03
Nuclear fuel cycle	<1	<0.03
Fallout from nuclear weapons testing	<1	<0.03
Miscellaneous	<1	<0.03
Total artificial	64	18
Total natural plus artificial	358	100

Source: National Council on Radiation Protection and Measurement (NCRP87b), (1990). (Washington, DC: National Academy Press).

majority is from natural sources. Over half the total comes from radon, with significant additional increments from cosmic rays, radiation from rocks and soil (other than radon) and radioisotopes that occur naturally in the body (principally ^{40}K, see p. 36). The remaining 18% is from artificial sources, mainly x-rays and radioisotopes used in medicine. Consumer products, principally building materials, account for 10 mrem, or 3% of the total. Fallout from nuclear weapons testing (a serious concern during the period of heavy testing) is now estimated to contribute less than 1 mrem, as does the entire nuclear fuel cycle. A similar small dose is produced by occupational exposures. But this entry in the table is a little misleading, since it is an average dose for the entire U.S. population. In the one million or so workers engaged in the affected occupations, the average annual dose is 230 mrem, about the same as the average radon dose, while for those in particularly hazardous occupations, such as uranium miners, the dose can be much higher.

There are other variables that can increase an individual's exposure from the average. The large variability in radon levels in houses has already been mentioned. Each dental or chest x-ray adds 10 mrem to an individual's total, while a gastrointestinal tract x-ray adds 200 mrem. The cosmic ray entry in Table 1.9 is for people living at sea level, but the cosmic ray intensity increases at high altitudes. (It is twice as high at 2000 m and four times as high at 3000 m.) For this reason individuals living at high altitudes, or who are frequent fliers (3 mrem exposure for a five-hour flight at 9000 m) have a higher exposure to cosmic rays. None of these factors are terribly worrisome, however, when viewed in the context of the 360 mrem/yr background exposure. This background also puts radiation hazard from the nuclear industry in context. For the population at large, this hazard is seen to be negligible, provided that all the radioisotopes stay where they are supposed to stay. This is, of course, a major proviso.

4. Fission Reactors

a. Pressurized light-water reactor.   Since the grim dawn of the nuclear age, scientists have dreamed of harnessing the power of the nucleus for peaceful purposes, and from this impetus the nuclear power reactor program arose. Figure 1.20 shows the operating principles of the most common nuclear reactor design—the pressurized light-water reactor. The fuel rods of this reactor contain pellets of uranium or uranium oxide. Naturally occurring uranium is made up mostly of ^{238}U, which does not fission. ^{235}U constitutes only 1 out of 140 uranium atoms. In order to build up a chain reaction in the light-water reactor, the uranium is concentrated in the ^{235}U isotope to a level of 2–3%. Control rods containing cadmium or boron, which absorb neutrons effectively, are lowered automatically among the fuel rods to a level that adjusts the neutron flux so that the chain reaction is maintained but does not run out of control.

Surrounding the fuel and control rods is a bath of water, which acts both as a coolant to carry away the energy generated in the fission reaction, and as a moderator, that is, a substance that slows down the neutrons to increase the fission probability. The probability of fission occurring when a neutron collides with a ^{235}U atom is maximal when the neutron energy is close to the *thermal energy*, the average energy of the surrounding molecules. The fission neutrons are released with high energy (velocity), and they must be slowed down to propagate the chain

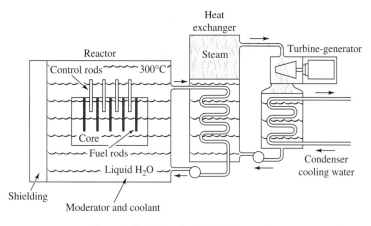

Figure 1.20   Pressurized light-water reactor.

reaction effectively. The lighter the atoms of the moderator, the greater the energy removed per collision. With its two hydrogen atoms per molecule, water is an effective moderator.

An alternative is to use heavy water, D_2O, instead of H_2O, as the coolant and moderator. D_2O is a less effective moderator because of the larger deuterium mass, but deuterium absorbs neutrons much less than hydrogen. As a result, it is possible to maintain a chain reaction even with unenriched uranium (0.7% ^{235}U) if D_2O is used as the moderator. This is the operating mode of the CANDU reactor designed in Canada. The relative merits of light and heavy water reactors depend in large measure on the relative costs of separating D_2O from H_2O versus separating ^{235}U from ^{238}U. The costs of the latter have been somewhat hidden by the pre-existence of large ^{235}U separation plants constructed for nuclear weapons production, which have been used to supply the fuel for commercial reactor programs.

The water in the reactor circulates through a heat exchanger that generates steam from a secondary water coolant, and this steam is then used to drive a turbine to generate electricity. Basically, the reactor is a conventional steam generator in which the heat source is fissioning uranium instead of burning coal. The amount of energy concentrated in uranium is vastly greater than that in coal. Whereas the highest heating value found for coal is about 33 kJ/g, a gram of ^{235}U can release 7.2×10^7 kJ. One gram of ^{235}U is equivalent to about 2.5 metric tons of high-grade coal.

b. Isotope separation.   Because isotopes of a given element have the same chemistry, their separation must be based on the slight differences in their properties resulting from their differences in mass. The greater the mass ratio of the isotopes, the easier it is to exploit these differences. The rates of chemical reactions involving hydrogen, for example, are appreciably faster than those involving deuterium, and electrolysis of water to H_2 and O_2 leaves a liquid that is increasingly rich in D_2O.

The mass ratio of ^{235}U and ^{238}U is too close to unity to utilize reaction rate differences, but physical separations that are sensitive to this ratio can be operated in many successive stages to produce gradual enrichment. The main method in use is gaseous diffusion. The gaseous compound UF_6 is passed through a succession of porous diffusion barriers. At each of these, the lighter $^{235}UF_6$ is enriched by a factor equal to the square root of the $^{238}UF_6/^{235}UF_6$ mass ratio, that is, $\sqrt{352/349} = 1.004$. The enrichment after n diffusion barriers is $(1.004)^n$. The factor of three enrichment from the natural abundance of ^{235}U, 0.7%, to the 2.1% minimally needed for light-water reactors, takes 263 diffusion stages (that is, $3 = (1.004)^{263}$). The plants required for this process are very large and expensive, and high inputs of energy are required to force UF_6 through so many barriers.

The application of centrifugal force offers a somewhat more efficient route to isotope separation because, for a given centrifugal velocity and radius, the force is directly proportional to the mass, rather than to its square root. Gas centrifuge and gas nozzle techniques are both based on this principle.

The most promising alternative approach, however, is laser isotope enrichment. Lasers are devices that produce light of very well-defined (monochromatic) energy. When a photon of appropriate energy is absorbed by a molecule, it undergoes a transition to an excited state, in which it may be considerably more reactive than it is in its ground state. The excited-state energy levels depend slightly on the isotopic composition, and with a sufficiently well-tuned laser it is sometimes possible to excite molecules that contain one isotope, while exciting only a small fraction of the molecules that contain other isotopes. If the excited-state reactivity can be properly exploited, then large enrichments are possible with a single pass. Laser enrichment has the potential of substantially lowering the complexity and cost of ^{235}U separation.

Low-cost isotope separation would improve nuclear power's economics, although the effect is limited to the fuel cost, which is a minor component of the cost of delivering power. It would also place nuclear technology within the means of many currently non-nuclear nations and complicate the problems of nuclear weapons proliferation.

c. Breeder reactor.   Although ^{235}U represents an extremely concentrated form of energy, there is not a great deal of it present in the world. However, it is possible to extend the nuclear fuel supply by converting the dominant uranium isotope, ^{238}U, to another fissionable isotope, ^{239}Pu. This is accomplished simply by irradiating uranium with neutrons via the nuclear reactions diagrammed in Figure 1.21. Absorption of a neutron by ^{238}U produces a very unstable isotope, ^{239}U, which produces ^{239}Np by beta emission. ^{239}Np undergoes a similar transformation to ^{239}Pu. Being fissionable, this isotope can be used in a power reactor just as ^{235}U can. Indeed, conventional reactors actually gain some of their energy from ^{239}Pu, which is inevitably produced when the reactor neutrons encounter ^{238}U, the main uranium isotope present even in enriched reactor fuel.

Not much ^{239}Pu is produced in this way, because the probability of the ^{238}U absorbing a neutron is maximal for fast neutrons, not the thermal neutrons for which a pressurized light-water fission reactor is designed. Efficient production of ^{239}Pu requires a "breeder" reactor, operating with fast neutrons. Figure 1.21 shows a schematic diagram for a fast breeder reactor. The main difference with respect to an ordinary fission reactor is that the water coolant is replaced by liquid sodium. Being much heavier than hydrogen, the sodium atoms slow the neutrons to a much smaller extent, while liquid sodium efficiently carries away heat from the reactor. The primary sodium coolant transfers its heat to a secondary liquid sodium coolant, which in turn transfers heat to a steam generator that runs the turbine for producing electric-

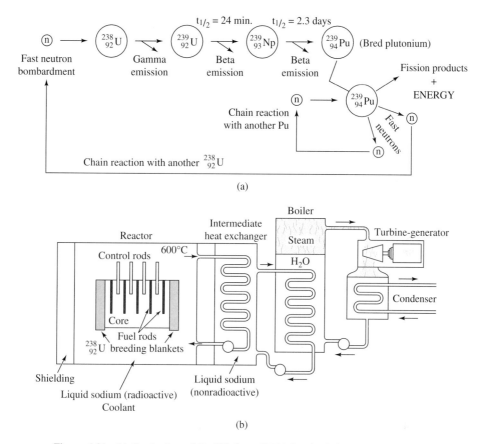

Figure 1.21   (a) Production of Pu-239 from U-238 bombarded with fast neutrons; (b) liquid sodium breeder reactor.

ity. The purpose of the secondary sodium coolant is to prevent any accidental contact of the primary sodium coolant with water, since some of the sodium atoms do absorb neutrons and become radioactive.

Surrounding the reactor is a blanket of ordinary uranium, in which the ^{239}Pu is bred by the fast neutrons. At the same time, the fast neutrons make the induced fission reaction less efficient, so the reactor fuel must be enriched to the extent of 15–20% with a fissionable isotope, either ^{235}U or ^{239}Pu.

There are inevitable losses in the system, but use of the breeder reactor would still stretch the supply of uranium fuel by at least a factor of 50 and would transform nuclear fission into an energy resource much larger than coal. The technology of the breeder reactor is far more complex than that of the ordinary fission reactor, however, and it is still being developed. The most formidable problem seems to be associated with the plumbing of the heat exchanger, since leaks of liquid sodium can be disastrous. Just such a leak resulted in a 1994 explosion at a French breeder reactor, which killed a worker. The French breeder program is the most advanced in the world, but technical difficulties and poor economics have stalled its development as of this writing.

d. Reprocessing.   The ^{235}U in a fuel rod of a pressurized light-water reactor cannot be completely used up, because of the buildup of fission products, which themselves absorb the neutrons and eventually slow the chain reaction down. After about a year, the fuel rods must be replaced with new ones.

The spent fuel can be reprocessed by chemical extraction of the fission products, separation of the accumulated plutonium, and reconcentration of the uranium. The uranium can be refabricated into new fuel rods. What to do with the plutonium is a difficult issue (see following section on weapons proliferation), lacking a functioning breeder program. The current choice is to blend plutonium and uranium into MOX (mixed uranium-plutonium oxide) fuel rods. These can be burned in conventional reactors, although the differing nuclear properties of the two elements restrict the amount of MOX fuel to one-third of the reactor core. However, MOX fuel is currently much more costly to fabricate than conventional ^{235}U fuel.

At the fuel reprocessing plant, the fuel rods are chopped up and dissolved in acid, and the resulting solution is subjected to successive solvent extraction and ion exchange steps in order to separate the elements. The chemistry is straightforward, but the technology is complicated by the need for remote handling of intensely radioactive material. Because of technical and safety problems, as well as policy considerations, no reprocessing plant has been in operation in the United States for over two decades. Currently, reprocessing plants are in operation only in France and England, but they reprocess fuel from other countries as well. A large facility is planned for Japan, early in the next century.

5. Hazards of Nuclear Power

a. Reactor safety.   Under normal operating conditions, the radiation released by nuclear reactors is very low, but the potential for accidental releases is a serious concern because the average reactor contains as much radioactive material as that released by the Hiroshima atomic bomb. The danger is not that a nuclear explosion could be set off; the fissionable material in reactor fuel is too dilute to become explosive itself. But a great deal of heat is generated by an operating fuel rod, even after control rods have been lowered to stop the chain reaction (which happens automatically in case of an accident). This heat is carried away by the water circulating through the reactor. However, if the water leaks out or boils away and is not replenished, then the temperature can rise to disastrous levels, and allow highly radioactive materials to be released. There are emergency backup systems to replenish the water, but these systems can fail through equipment malfunction or human error.

Both malfunction and human error were responsible for the two major accidents that have marked the era of nuclear power generation. In March 1979, the fuel rods in the nuclear plant at Three Mile Island, near Harrisburg, Pennsylvania, melted down when cooling water was lost. A pump in the primary cooling system failed, the auxiliary pumps were not operational, and the emergency cooling system was momentarily turned off, due to operator error. By the time it was turned back on, a few minutes after the pump failure, a large gas bubble had formed, as a result of thermal decomposition of water into hydrogen and oxygen. The bubble prevented the cooling water from reaching the fuel rods, which partially melted. There were fears that the hydrogen and oxygen would explode and breach the containment wall. But fortunately, this did not happen, and the core eventually cooled after having vented a small amount of radioactive gas.

Much more serious was the accident on April 26, 1986 in the Ukrainian town of Chernobyl, a name now synonymous with nuclear disaster. This time a reactor literally caught fire and blew apart, releasing a great cloud of radioactive debris that rained out radio-isotopes over much of Europe and parts of Asia. The reactor continued to burn for about 10 days, eventually releasing 10 million curies of radioactivity into the environment. Reactors of the Chernobyl class, which are still in use throughout the former Soviet Union, are of a different design than the light-water reactor described above. They employ graphite, instead of water, as the moderator. Graphite contains only carbon atoms, and it slows neutrons quite effectively; it can sustain a chain reaction when only 1.8% of the fuel is the fissionable ^{235}U. The heat is carried away by flowing water around the individual fuel rods. Because the rods are cooled individually, they can be replaced one at a time, without shutting down the reactor. Consequently, this type of reactor has one of the highest rates of productive time online in the nuclear industry.

Unfortunately, a graphite reactor is also inherently hazardous. The plumbing is very complicated, and the graphite, being carbon, is flammable. That is why the Chernobyl reactor burned so vigorously after the accident. Most important, graphite reactors have a dangerous property known as a "positive reactivity coefficient." When power levels dip very low, the reactor becomes unstable, and can race out of control. This is what happened on the day of the accident during a test of the reactor's response to a simulated power failure. The loss of power, in combination with control rod malfunctions and operator misjudgments, allowed the reactor to surge out of control, vaporizing the water, blowing off the roof of the reactor and igniting the graphite. Two people died in the explosion and a thousand others were injured. Over the next several months, 29 individuals died of radiation effects. Long-term effects of the radiation can be expected but are hard to gauge because of large variations in exposure. Some areas of Europe downwind from the reactor experienced heavy rain as the radioactive cloud was passing overhead, and registered fallout 100 to 1,000 times greater than during the peak years of nuclear weapons testing. The milk from cows eating the fallout-laden grass was contaminated and had to be condemned. Even several years after the accident, the radioactivity in reindeer in Scandinavia and sheep in some areas of northern England exceeded levels permissible for meat.

Taken as a whole, the safety record of the nuclear industry is actually quite impressive. Only one terrible disaster has occurred among the hundreds of nuclear reactors in operation over the last four decades, and the human toll was not worse than in a number of non-nuclear large-scale industrial accidents. Nevertheless, the possibility of another Chernobyl casts a pall over the industry, and the burden of taking heavy precautions to avoid conceivable accidents adds substantially to the costs of nuclear plant construction and maintenance.

There are new reactor designs currently under development that promise to improve safety margins by a factor of ten or more. These reactors would be "passively stable" because they are designed to shut themselves down automatically in the event of an accident. For example, in a "passively stable light-water cooled reactor" (Figure 1.22), emergency cooling

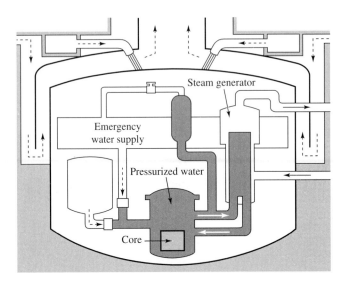

Figure 1.22   Passively cooled light-water nuclear reactor.

water would be driven by gravity and by nitrogen pressure rather than by electric pumps, and the containment shell would be cooled in an emergency by evaporating water that is fed by gravity from large tanks located above the containment vessel. Moreover, this new reactor has a simplified design that requires far fewer valves, pumps, ducts, and cables than do current reactors. A more radical departure from current practice is the "advanced modular high-temperature gas-cooled reactor," in which the problem of overheated fuel is obviated by coating fuel pellets with a protective layer of very hard material, silicon carbide; these pellets can withstand temperatures exceeding 1600°C, a temperature higher than the fuel could generate. The reactor would run at 600°C, making it thermodynamically efficient (see pp. 77–79), and would utilize helium gas as the heat transfer agent. It would be buried in the ground and would be small enough to ensure that any excess heat would be conducted away by the surrounding soil.

    b. Weapons proliferation.   Since nuclear energy made its first appearance as the atom bomb, an overriding issue for civilian nuclear power is how to avoid diversion of nuclear fuel into weapons manufacture. ^{235}U is not the major problem in this regard, since a bomb requires highly enriched ^{235}U > 93%, whereas conventional fuel rods are only slightly enriched, 2–3%. Weapons-grade fuel requires isotope enrichment, a technically demanding process that involves a major commitment of resources. Since highly enriched uranium is used only in weapons, safeguards against ^{235}U diversion are relatively straightforward. Evidence that Iraq was planning the clandestine production of a ^{235}U weapon was one of the factors contributing to the 1990 Gulf War.

    ^{239}Pu is another matter entirely. Plutonium and uranium, being different elements, are readily separated by chemical means. Since ^{239}Pu is the major plutonium isotope created in a uranium reactor, no isotope enrichment is needed to produce weapons-grade fuel. The plutonium extracted in a reprocessing plant can be used in nuclear explosives. India produced and tested a nuclear bomb using plutonium recovered from reprocessed reactor fuel. It has been claimed that as little as 5 kg can be fashioned into a crude but usable weapon by groups or even individuals working from published government documents. (Some of the ^{239}Pu produced in reactor fuel is converted to ^{240}Pu and ^{241}Pu as the fuel rod continues to be irradiated with neutrons. These isotopes emit neutrons spontaneously and can set off a chain reaction prematurely, reducing the yield of a Pu bomb. But a simple bomb made of reactor-grade Pu could still have a yield of a few kilotons, about a third as powerful as the Hiroshima bomb.)

As nuclear power grows, and fuel reprocessing expands, the amount of plutonium in circulation will increase. Accurately accounting for the plutonium will become more difficult, and opportunities for diversion will multiply. Figure 1.23 is a schematic diagram of the estimated flows of fissionable material, as of 1991, from uranium mining, through the world's reactors and on to reprocessing plants. This is a snapshot in time, since most of the uranium now in reactors was mined and processed a long time ago, and reprocessing of the fuel discharged in 1991 is still in the future. Of the ~74 tons of Pu in this discharged fuel, about 10 tons will be recovered in reprocessing, and about 4 tons will be recycled as MOX fuel, with the rest held in storage. By the year 2000 it is estimated that about 190 tons of Pu will have been reprocessed.

Ironically, the issue of Pu diversion has been magnified by the fruits of disarmament. During the decades of the Cold War, the United States and the Soviet Union stocked their arsenals with tens of thousands of nuclear weapons. The end of the Cold War has meant that most of these weapons can be dismantled, to the world's vast relief. Yet the dismantled weapons contain large stockpiles of nuclear fuel. The amounts are shown in Figure 1.24. About 260 tons of Pu are contained in military stockpiles, an amount equalling about one half of the total Pu in spent reactor fuels. An even larger amount of fissionable material exists as highly enriched uranium (HEU). The HEU can be eliminated as weapons material simply by diluting it with natural uranium. Indeed, once political and economic arrangements are in place, the diluted HEU can be added to the supply of conventional reactor fuel. But, like reprocessed Pu, the Pu from weapons cannot be diluted isotopically and must therefore be safeguarded. This is a serious problem, in view of the difficulty of ensuring against theft.

There is no current consensus on what to do with the accumulating Pu. Nuclear power advocates want to burn it in reactors, thereby extending the nuclear fuel supply. The governments of France, Japan, and Russia, in particular, are committed to this course as a method of assuring energy independence. At present, however, Pu is costlier than uranium as a nuclear fuel, and this situation is unlikely to change in the near future. For this reason, it is argued (in many quarters) that the Pu should be disposed of as a waste, along with the rest of the nuclear wastes from reactors and reprocessing (see next section). Indeed, it is further argued that uranium fuel should not be reprocessed at all, but used only on a once- through basis, in order to avoid the Pu safeguard problem and because reprocessing is currently uneconomical. This course of action would, of course, forego a significant source of energy. Whether this energy will be needed, in light of future alternative sources, is a matter of conjecture. An intermediate course of action would be to disperse the Pu in glass rods that would be placed in recoverable storage. The Pu would then not be readily available for weapons manufacture, but could be re-extracted for nuclear fuel when it is needed. This storage method is currently undergoing technical evaluation.

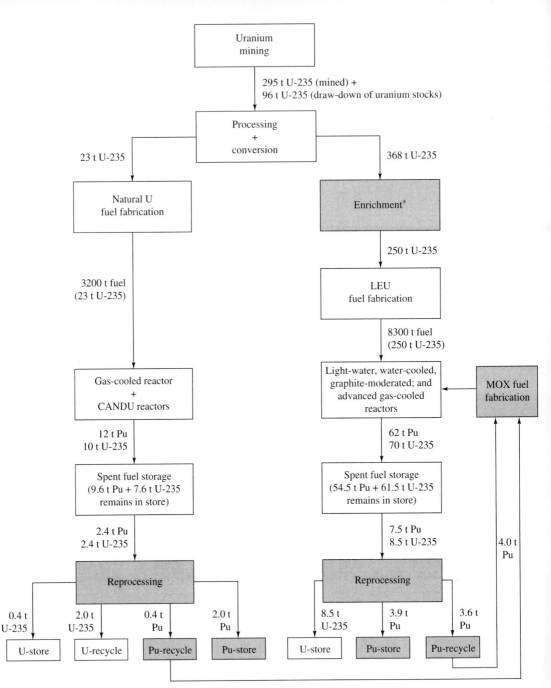

Figure 1.23 Flows of fissile material through the world civil fuel cycles in 1991 (LEU is *low-enriched uranium*). *Source:* F. Berkhout and H. Feiveson (1993). Securing nuclear materials in a changing world. *Annual Review of Energy* 18: 631–665. Copyright © 1993 by Annual Reviews Inc. Reproduced with permission.

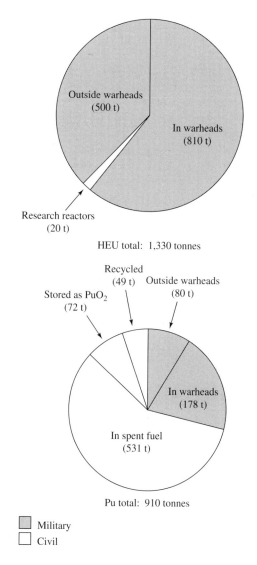

Figure 1.24   World inventories of plutonium
and *highly enriched uranium* (HEU) in 1990.
Source: F. Berkhout and H. Feiveson (1993).
Securing nuclear materials in a changing
world. *Annual Review of Energy and the
Environment* 18: 631–665. Copyright © 1993
by Annual Reviews Inc. Reproduced with
permission.

c. Nuclear waste disposal.   Even with safer reactor designs and an effective
means of combatting the spread of nuclear weapons, nuclear power remains potentially
hazardous because it produces radioactive wastes.

There are several points in the entire cycle of nuclear fuel where the dispersal of radioactive materials is an actual or potential problem (Figure 1.25). At the beginning of the cycle,
uranium mining is, itself, a hazardous occupation. Miners have a high risk of developing lung
cancer because they inhale dust and radon. Even more pressing in terms of public health are
uranium mine tailings, fine particles that remain after uranium ore is extracted from the rocks.
Although the levels of radioactivity are low, the amount of material is huge: 75 million tons
in the United States. Moreover, the material will pose a hazard for millenia since one of the

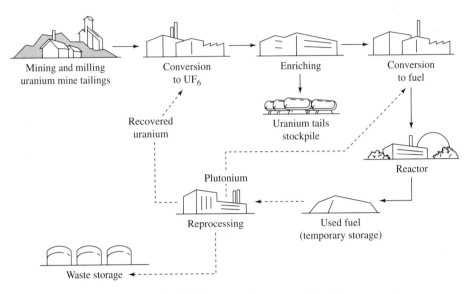

Figure 1.25   Fuel cycle for the light-water nuclear reactor; dashed lines are not yet
part of U.S. cycle.

important nuclides, thorium-230, has a half-life of 80,000 years. Because the radioactivity is
dilute, its danger went unrecognized for decades. The tailings were left in huge piles to be
dispersed by wind and leached by rain. Some material was even used as fill under buildings,
exposing the occupants of some buildings to levels of radioactivity exceeding the legal limits
for uranium miners. Safe disposal of tailings remains a major task that we have barely begun.

Further into the cycle, dispersal of radioactive materials is a hazard at the fuel reprocessing plant, where the spent fuel rods are dissolved in acid to separate the fission products
and the heavy element contaminants before the uranium or plutonium is sent back to fuel fab-
rication plants. Reprocessing produces high-level waste (HLW) that is highly radioactive and
of moderately high temperature. Since no method has been approved of disposing HLW permanently, it has been accumulating at various sites for almost half a century. One of the major
sites for storing HLW from military reactors is the 560-square-mile Hanford Nuclear
Reservation in Washington state, where millions of gallons of HLW sit in 177 tanks. Most
tanks have been in use for decades longer than planned, and every year several tanks begin
to leak. Cleanup of the Hanford site is complicated by some lax recording practices over
many years. Many of the tanks are filled with unknown mixtures of chemicals, some of which
are turning out to be highly explosive.

Reprocessing is no longer practiced in the United States and is not contemplated in the
near future. The spent fuel rods of U.S. reactors are being stored temporarily at the reactor
site, in storage tanks filled with water. Eventually these spent fuel rods must be reprocessed
or disposed of as waste themselves. As reactor operators fill their tanks to capacity, they are
forced to make more room either by rearranging the spent fuel rods or by transporting them
to other sites. The accumulation of spent fuel rods adds to the pressure of finding a long-term
solution to the problem of nuclear waste disposal.

But the problem is extraordinarily contentious because the waste must be isolated from the human environment for exceedingly long periods of time. The length of time for safety can be assessed by the rule of thumb that radiation sinks to negligible levels after ten half-lives: the total amount of radioactive material x, is reduced to $x/2^{10}$, or about 1/1,000. For fission products, this interval is a few hundred years, but plutonium and other heavy elements have half-lives of tens of thousands of years; ten half-lives is on the order of a quarter million years. Consequently, most strategies for disposal have searched for geological formations in which waste might be deposited for millenia. Few potential sites exist: most areas are too prone to disturbance by earthquakes and ground-water infiltration to be considered stable enough.

Current plans for disposal in the United States consist of immobilizing the waste, mainly in borosilicate glass, and then burying it either in salt beds in New Mexico or under Yucca Mountain in Nevada. The glass is relatively inert and impermeable to water, so it should resist leaching for a long time. Salt is very dry, and, moreover, tends to fill in spontaneously any cracks that might develop by movement of Earth. However, disposal at both these sites has been virtually blocked by continuing concerns about the long term stability of the buried wastes, and by resistance from local inhabitants. Understandably, no single region wishes to bear the burden of the nuclear wastes generated throughout the nation. As negotiations over sites continue within all levels of government, we may find waste disposal blocked more by politics than by technology.

6. Is Nuclear Power Part of the Future?

Even though nuclear power currently supplies about 21% of all electrical power in the United States, if current trends continue, U.S. nuclear power capacity will dwindle to virtually nothing by the year 2030. No new nuclear reactors have been ordered since 1979. Moreover, of the 135 plants planned in 1978, 85 have been cancelled or mothballed. Building a nuclear power plant requires an average of ten years, so by the turn of the century a maximum of 120 nuclear plants will be operating, with the newest plants online for an average lifetime of 30 years.

The decline of the nuclear industry is due to a combination of economics and public resistance. The nuclear industry was born amid projections that increasing demand for electrical power would outstrip the utilities' ability to produce electricity from other sources. That demand was much slower to materialize than planners thought, thanks to energy-saving measures set in motion by the 1973 oil crisis. During the same period, public resistance to nuclear power grew, fueled by heightened awareness of the hazards of nuclear power and the consequences of Three Mile Island and Chernobyl. Together, these trends made nuclear power unreasonably expensive. As resistance to nuclear power increased, so, too, did the costs of building, running, and insuring nuclear plants. But as demand for electricity softened, the opportunities for utilities to cover the cost of developing nuclear power decreased, until most utilities decided that building more nuclear power plants simply did not make good economic sense.

Is nuclear power dead, then? Certainly not yet. It is a vital source of energy for countries such as Japan and France; changing environmental assessments may yet revive it in the

United States. Demand for electricity continues to grow, as does our awareness of the full environmental cost of reliance on fossil fuels. Moreover, the accidents at Chernobyl and Three Mile Island have stimulated the design of newer reactors with more safety features; these designs may yet breathe new life into the nuclear industry and help prepare it for when the demand for electricity begins to outstrip its supply. Indeed, as shown in Figure 1.26, if the growth and decline of nuclear power were to follow the same trajectory as those for coal and oil, one would expect a significant increase well into the twenty-first century. Even so, the future of nuclear power is clouded by the continuing, unresolved problems of nuclear safeguards and of waste disposal.

7. Fusion

Nuclear energy can potentially be obtained not only from the fission of nuclei but from their fusion. As we saw in Figure 1.14, the most stable nuclei have intermediate masses; very small nuclei are less stable. In particular, helium nuclei (mass number = 4) are considerably more stable than hydrogen nuclei (mass number = 1 or 2). When two hydrogen nuclei fuse to form a helium nucleus, an enormous amount of energy is released. However, fusing the nuclei requires extreme reaction conditions in order to overcome the huge energy barrier caused by

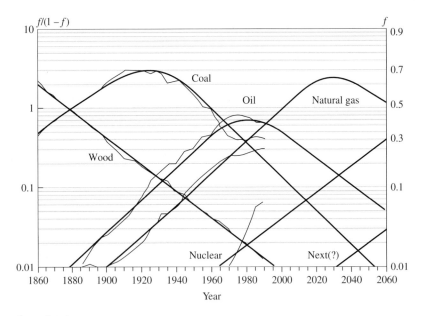

f = market share

Figure 1.26 Patterns of growth and decline in shares of global primary energy production (f signifies market share).   *Source:* Adapted from C. Marchetti and N. Nakicenovic (1979). *The Dynamics of Energy Systems and the Logistic Substitution Model,* report no. RR-79-13 (Laxenburg, Austria: International Institute for Applied Systems Analysis). Copyright © 1979 by International Institute for Applied Systems Analysis. Reprinted with permission.

the repulsion between the two protons. Such extreme conditions are found in the centers of stars, including our sun, whose energy outputs are due to fusion reactions. Similar conditions are also obtainable in hydrogen bombs, which use the force of a fission explosion as a trigger.

Fusion reactions are somewhat easier to carry out if the fusing nuclei are the heavy isotopes of hydrogen, deuterium (mass number = 2), and tritium (mass number = 3). Several different fusion reactions are diagrammed in Figure 1.27. All these reactions produce enormous amounts of energy; the fusion of one gram of deuterium and tritium produces as much energy as burning 8.4 tons of oil.

But for all these reactions, the *ignition temperatures* are on the order of 100 to 1,000 million °C. At such high temperatures all earthly materials vaporize; hence, one of the challenges in developing fusion technology has been to devise a way to contain the fusing nuclei. Two approaches appear promising: magnetic confinement, in which the fusing nuclei

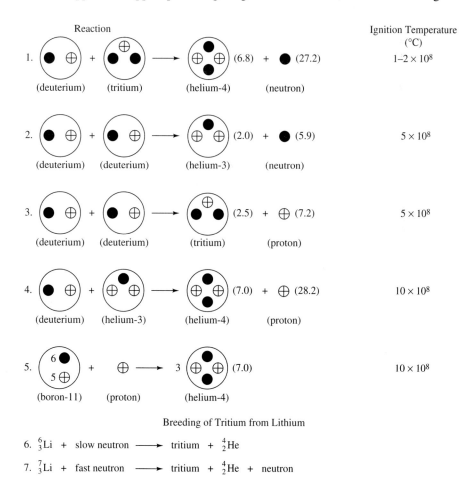

Figure 1.27 Energy-producing fusion reactions (numbers in parentheses are the energies of the product particles, in units of 10^7 kJ, obtained from fusion of one gram of reactants).

are suspended in a magnetic field, and inertial confinement, in which the fusing nuclei are forced together by the impact of high-power lasers.

a. Magnetic confinement: the Tokamak reactor.

In the magnetic confinement method, the fusion fuels are suspended in free space by using powerful magnets. At the high temperatures under consideration, atoms separate into their charged species, producing *plasmas*, or clouds of nuclei and electrons. Because the plasma is charged, it can be controlled with magnetic fields. The magnets keep the heated, charged plasma away from the walls of the reactor, preventing the particles from colliding with the walls and cooling down.

The most successful magnetic design is the Tokamak reactor shown in Figure 1.28. In this design, an electrical current is generated within the plasma; the current causes the magnetic field to bend around in a donut shape or *torus*, which confines the fusion reaction. The reactor walls are lined with light materials such as carbon, boron, or beryllium to minimize reactions of energetic particles that escape the plasma. Surrounding the walls is a blanket of lithium, which captures neutrons given off in the reaction and produces tritium for additional fuel (see reactions 6 and 7 in Figure 1.27).

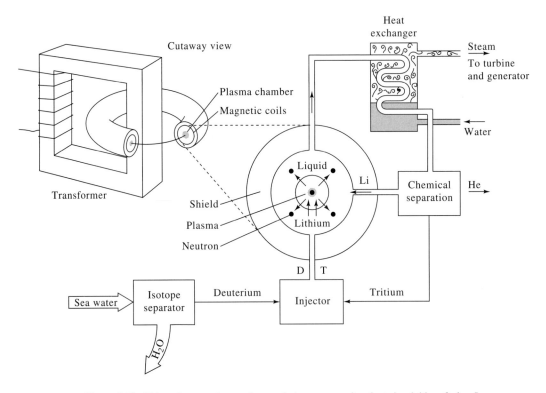

Figure 1.28 Tokamak magnetic-containment fusion reactor using deuterium/tritium fuel. *Source:* G. Gordon and W. Zoller (1975). *Chemistry in Modern Perspective* (Reading, Massachusetts: Addison-Wesley). Copyright © 1975 by the Addison-Wesley Publishing Company. Published with permission.

For fusion to generate power, the energy input needed to sustain the plasma conditions must be less than the energy output from the reactor. The break-even point where input and output are balanced has yet to be achieved in any experimental reactor. The break-even point depends upon three parameters: the density of the plasma (n), its temperature (T), and the confinement time (τ), or the length of time it would take the plasma to cool substantially. The target values of these three parameters for a Tokamak reactor are shown in Table 1.10. Different reactors configured in different ways have come close to or surpassed the individual target values; however, they have not been obtained simultaneously. The performance of Tokamak reactors over time is shown in Figure 1.29. Performance has improved steadily and is approaching the break-even point. Of course, the ratio of energy output to energy input must be

TABLE 1.10 COMPARISON OF TARGET VALUES AND EXPERIMENTAL RESULTS FOR TOKAMAK REACTORS

	Density, $n(10^{20}\text{ m}^{-3})$	Temperature T (keV)	Confinement time τ (s)
Target	1–2	15–30	2
Experiment	4–10	35	1.8

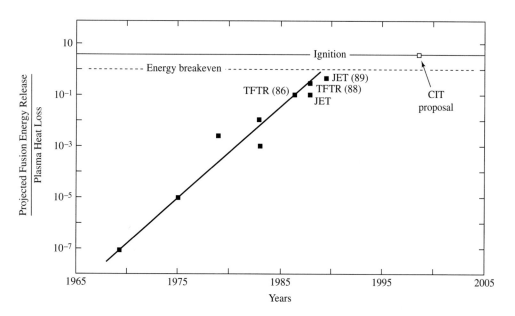

Figure 1.29 Progress in Tokamak reactors in achieving the conditions required for fusion power; TFTR = Tokamak fusion test reactor (U.S.), JET = joint European Tokamak (U.K.), CIT = compact ignition Tokamak (U.S. and international). *Source:* D. B. Montgomery (1991). The prospect for fusion energy in the 21st century. In *Energy and the Environment in the 21st Century,* J. W. Tester et al., eds. (Cambridge, Massachusetts: The MIT Press). Reprinted with permission from MIT Press.

considerably greater than 1:1, if fusion is to become truly useful. Hence, commercial development of a Tokamak-type reactor is still decades in the future.

b. Inertial confinement reactors. In the inertial confinement method, a target pellet containing deuterium and tritium is subjected to a very short and intense pulse of energy from lasers or particle beams. The intense energy vaporizes the outer suface of the particle, ejecting matter outward. The laws of physics require that the force of the outwardly moving material be matched by inwardly moving material; hence, the remaining part of the pellet is imploded by the laser, reaching very high densities and temperatures of up to 100 million °C, conditions sufficient for fusion. The deuterium-tritium pellet is confined by inertia for a fraction of a nanosecond (10^{-9} s), long enough to allow about 30% of the pellet to fuse before it comes apart.

The feasibility of laser confinement depends on advances in laser technology. If laser-directed fusion is to produce enough energy for power generation, it must deliver energy to the pellets in times well below one nanosecond, and by beams with precisely determined shapes. The energy-transfer efficiency of currently available lasers is very far from what will be required.

The pulsed nature of laser-directed fusion also creates problems in designing the reaction chamber. The systems needed to recover energy and breed fuel in the chamber would have to withstand explosions equivalent to 100 or more kilograms of TNT several times a second. At present, inertial confinement fusion is much less well-developed than magnetic confinement fusion.

c. Comparisons between different fusion reactions. Of the fusion reactions in Figure 1.27, the reaction between deuterium and tritium (D/T reaction) has the lowest ignition temperature, 100 to 200 million °C. This makes it by far the most easily achievable reaction. However, this reaction has the pronounced disadvantage that it requires tritium as a fuel. Tritium is a radioactive gas that leaks easily through metals and can contaminate water by exchanging with the water hydrogen atoms. It releases low-energy β particles that are not dangerous when emitted externally to the body but are dangerous enough when tritium is inhaled, or ingested in water.

Tritium has a half-life of 12.3 years, so there is no natural source of the isotope. It must be synthesized from lithium by reactions 6 and 7 shown in Figure 1.27. Lithium, in turn, is a relatively scarce element, and high-grade deposits are not plentiful. Tritium and lithium probably correspond to an energy resource comparable to the fossil fuels.

Another disadvantage of the D/T reaction is that it produces one high-energy neutron per nuclear fusion. Because neutrons are electrically neutral, they do not respond to the magnetic fields guiding the plasma. Instead, these neutrons fly out of the plasma at high speed, carrying with them 80% of the reaction's energy. As a result, the fuel quickly cools, and large quantities of heat must be pumped in to sustain the fusion reaction.

High-energy neutrons pose an additional problem for fusion reactors: They bombard the structural materials of the reactor, causing *neutron activation* and rendering the materials radioactive. These radioactive materials are a substantial problem for disposal; the volume of material to be discarded from the core of a fusion reactor is likely to be considerably greater than the volume of radioactive wastes generated by a fission reactor. Most fusion wastes will be relatively short-lived compared to fission wastes (see Figure 1.30), and some, such as silicon carbide, decay within a day to one-millionth the levels present in fission reactors. However, the most benign materials are expensive; less expensive ferritic steels require almost 100 years to decrease their levels of radioactivity significantly.

Fewer damaging neutrons are produced in the D/D reactions (numbered 2 and 3 in Figure 1.27); however, the ignition temperature is significantly higher, making these reactions more difficult to achieve. Neutrons are generated in 50% of the reactions, but they are of lower energy and pose fewer problems. Note, however, that 50% of the reactions generate tritium, some of which can react with deuterium via the D/T reaction to produce some high-energy neutrons. Thus, although the D/D reaction reduces radioactivity considerably, it does not represent a radical improvement.

Still better from an environmental perspective are the D/^{3}He and H/^{11}B reactions (4 and 5 in Figure 1.27), which produce no neutrons (although, if deuterium is present in the heated fuel, some of it will react with itself via the D/D reaction, producing some neutrons). However, both of these reactions require ignition temperatures that are 5 to 10 times greater

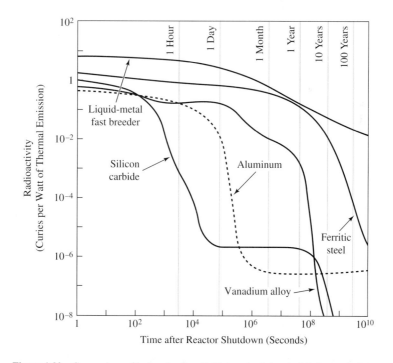

Figure 1.30 Comparison of induced radioactivity in potential materials from a fusion reactor with radioactive wastes from a liquid-metal fast breeder fission reactor. *Source:* R. W. Conn (1983). The engineering of magnetic fusion reactors. *Scientific American* 249(4): 60–71. Copyright © 1983 by Scientific American, Inc. All rights reserved.

than for the D/T reaction. It is questionable whether these reactions will ever be technologically feasible.

d. Is fusion the energy source of the future?   Fusion's greater appeal compared to nuclear fission rests mainly in fusion's greater safety. Unlike fission reactors, fusion reactors cannot sustain runaway reactions. Fuel must be continuously injected, and the amount contained within the reactor vessel at any given time can operate the reactor only for a matter of seconds. As a result, fusion reactors should require simpler post-shutdown or emergency cooling systems than fission reactors. If the materials are chosen carefully, the radioactivity created in the reactor should be considerably less hazardous than that of a fission reactor. Furthermore, the only radioactive material that could be released in an accident is tritium, because the other radioactive materials would be largely bound up in the structural elements.

The greater safety of fusion reactors is not without a stiff economic price. Although fusion fuels are relatively inexpensive, producing electricity from fusion will still entail great expense for building plants, maintaining them, and disposing of used radioactive materials. The plants must be enormous because of the low power densities of the reactor. Currently, economies in one area are likely to be offset by greater costs in other areas. For example, the costs of disposal depend greatly upon the materials used in making the first wall of the reactor, where bombardment by highly energetic neutrons is most frequent. If the walls were to be made of special materials such as vanadium and silicon carbide, the price of these materials would make the cost of constructing the walls very high. On the other hand, if less expensive materials such as steel were to be used, the walls would need replacement sooner and generate longer-lived radioactive materials, both factors that increase the cost of fusion power. Similarly, the costs of disposal depend upon the fuel; using fuels that do not produce neutrons would limit the amount of radioactive materials for disposal. However, the most environmentally benign fuels are also those that require the highest ignition temperatures, raising the projected costs of fusion energy.

D. RENEWABLE ENERGY

As we saw in Table 1.2, the major alternative to fossil and nuclear fuels is to harness renewable energy sources that, with the exception of geothermal energy, are derived directly or indirectly from sunlight. The annual energy deposited by sunlight on the continental United States is about 700 times the total annual U.S. energy consumption in 1990. Enough sunlight falls yearly on each square meter to equal the energy content of 190 kilograms of high-grade bituminous coal. The sun already provides us, in fact, with our most basic energy needs including heat, fresh water, and plant life. There is plenty left over for our other energy needs if we can learn how to use it. The difficulty is that sunlight is diffuse and intermittent. The technologies required to harness and store solar energy are currently expensive; it is cheaper to extract and consume fossil and nuclear fuels from the ground. However, the increasing environmental burdens associated with these fuels are sparking heightened interest in a variety of renewable energy technologies. A number of these are close to being commercially viable, and there is reason to think that with further technological improvements, renewable forms can play a substantial role in human energy utilization.

1. Solar Heating

The most convenient and straightforward application of solar energy is for space heating of residential and commercial buildings. This can be accomplished by passive solar design which maximizes the capture of direct solar energy to provide most of a building's heating needs. To achieve this, the building and its windows must be oriented so that solar radiation is admitted during the winter and avoided during the summer. Also, dense materials such as concrete and stone can provide thermal storage mass by absorbing solar radiation and storing it as a buffer against temperature fluctuations. Passive solar heating can be effective, however, only when buildings are well-insulated and sealed against air infiltration.

The U.S. Department of Energy has conducted extensive studies of heating efficiencies of buildings with solar passive design in different climate regions. The average indoor temperature of these buildings was 19.5°C (67°F), and the buildings required, on average, auxiliary heating of 49 kJ/°C-day-m^2. Conventional buildings typically use from 120 to 240 kJ/°C-day-m^2; the average for new constructions is 100 kJ/°C-day-m^2. Based on these results, auxiliary heating requirements in passive solar buildings appear to be about half that of conventional new construction, and less than one-fourth that of the existing housing stock.

Active solar energy systems are another alternative. They are termed *active* because, in contrast to *passive* systems, a source of energy other than solar energy is required to drive the system. One such a design is shown in Figure 1.31. Water circulates through the flat-plate collectors on the roof, where it is heated by the sun and pumped into a storage tank, which provides heat and hot water. Such buildings also function best when the demand for heat and hot water is minimized by proper insulation and water conservation. Many improvements in efficiency can be accomplished quite inexpensively, with short pay-back periods. For example, wrapping an insulating blanket ($125 to $25) around the tank of the water heater can reduce standby heat loss by 25% to 45%. Water-saving showerheads (under $10) use two to three gallons of water per minute, while conventional showerheads use five to eight gallons. Washing machines and dishwashers, the two major water-consuming appliances, are available in new designs that use only a fraction of the water of conventional models. Depending on the climate, a well-designed and properly sized solar water heater can provide up to two-thirds of a household's hot-water needs. It can save 50% to 85% of the hot-water portion of monthly electricity utility bills if the backup element is kept at 50°C (122°F). A disadvantage of solar water heaters is their high initial cost compared to conventional models.

Currently, space heating and hot water constitute 20–25% of U.S. energy needs. This consumption would be decreased significantly if solar heating systems were to gain a greater share of the market in servicing residential and commercial buildings.

2. Solar Electricity

A more demanding application is the generation of electricity from sunlight. It would take an area of 34,000 km^2, about the size of Maryland and Delaware, to supply the total electrical

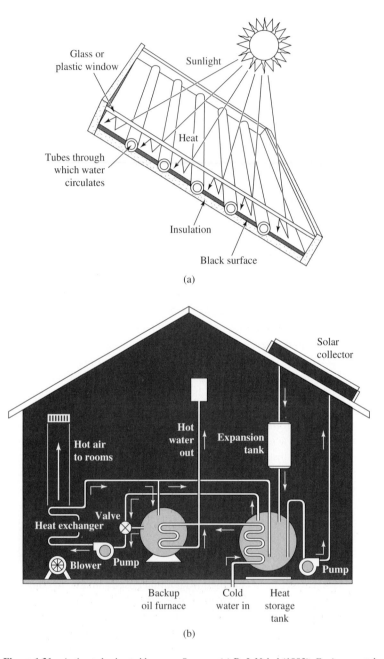

Figure 1.31 Active solar-heated house. *Sources:* (a) B. J. Nebel (1993). *Environmental Science: The Way the World Works* (Upper Saddle River, New Jersey: Prentice Hall). Copyright © 1993 by Prentice Hall. Reprinted by permission. (b) B. Anderson and M. Riordan (1976). *The Solar Home Book: Heating, Cooling and Designing with the Sun* (Harrisville, New Hampshire: Brick House Publishing). Copyright © 1976 by Brick House Publishers. Reprinted with permission.

energy needs of the United States if sunlight were converted to electricity with 12% efficiency, a feasible target efficiency. In the initial stages of development, solar electricity could complement existing power plants, especially in areas of plentiful sunshine. In this regard, it is an advantage that daylight hours coincide with high electricity demand, for heating, air-conditioning and many other purposes, so that solar generators are well-matched to the peak loads of electrical utilities.

a. Solar thermal electricity.   One approach to solar electricity is to collect the sun's heat and use it to run a steam generator. To achieve high-enough temperatures for efficient operation, the sunlight can be focused by parabolic mirrors to produce operating temperatures ranging from 100 to 400°C, as illustrated in Figure 1.32a. To achieve still higher temperatures, an array of mirrors can beam their reflected energy at a common receiver on a central tower. In this way, temperatures ranging from 500 to 1500°C can be achieved (Figure 1.32b). The higher the temperature, the more efficient the energy conversion (see pp. 77–80). A variety of solar thermal generators are undergoing large-scale tests, especially in Japan and in California, but none have yet been commercialized. Further development may lead to economic electricity via this route, however. One forecast for large central-receiver plants projects electricity costs by 2010 of less than 5 cents per kWh (kilowatt-hour), which is about the current production price for new coal-fired power plants. Inasmuch as the individual parabolic units are modular, they may lend themselves to more flexible and dispersed operations at an earlier stage.

b. Photovoltaic electricity.   Sunlight can be transformed directly into electricity via the photovoltaic (PV) effect. When light is absorbed in a PV material, positive and

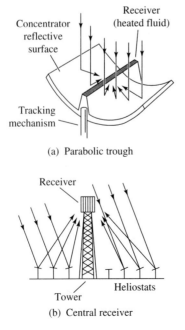

(a) Parabolic trough

(b) Central receiver

Figure 1.32   Two solar-thermal technologies: (a) the parabolic trough; (b) the central receiver.   *Source:* P. De Laquil et al. (1993). Solar-thermal electric technology. In *Renewable Energy, Sources for Fuels and Electricity,* T. B. Johansson et al., eds. (Washington, DC: Island Press). Copyright © 1993 by Island Press. Reprinted with permission.

negative charges are created, which can be collected by electrodes at either side, and passed to an external electric circuit. The most highly developed PV material, and the one most likely to succeed in large-scale applications, is silicon, the same material on which the electronics industry is based. The silicon solar cell was developed for the space program and has undergone extensive development over the past three decades. Electricity costs from solar cells have declined rapidly, from $60 per kWh in 1970 to $1 in 1980 and to 20–30 cents in 1994. Annual worldwide sales already exceed 40 megawatts of peak capacity. A variety of government–utility company collaborations around the world are testing out PV technologies and systems. For example, at the Rokko Island test facility in Japan, 100 residential PV systems have been connected to the power grid.

PV cells are modular and easy to use, and are already economical for power requirements that are remote from electricity grids. For example, the Georgia Power Company found that a $3,000 PV remote lighting system eliminated the need for a $35,000 electric grid extension. The governments of Italy and Spain have installed free-standing systems for remote areas. PV systems may become especially important in developing countries where rural electrification is embryonic. PV generators may well be more economical than extending power lines. On a lifetime basis, PV cells are now cost-effective compared to diesel generators at capacities below 20 kW. The market for such low-power generators is large. In India, for example, there are 4–5 million diesel power water pumps, each consuming about 3.5 kW.

1) Principles of the PV Cell. PV materials are semiconductors, solids with electrical properties between metals and insulators. In solids, the electronic levels spread into energy bands, owing to the mutual interactions of the orbitals on all the atoms in the solid lattice. The filled orbitals together produce the *valence* band, while the empty orbitals produce the *conduction* band, as illustrated in Figure 1.33. If an electron is injected into the conduction band of a solid, it can move freely throughout the lattice, since all the orbitals in the band are empty. Likewise, if an electron is removed from the valence band, the resulting *hole*, a center of positive charge, can move freely throughout the lattice via motions of the valence band electrons, to fill the successive vacancies. An electron in the conduction band and a hole in the valence band are called *mobile carriers* of electricity.

The energy difference between the top of the valence band and the bottom of the conduction band is called the *band gap*. Insulators have very large band gaps; the energy cost for promoting an electron from the valence to the conduction band is prohibitive. Consequently, an insulator will not carry a current. In a metal, the band gap is zero. There is no barrier to inhibit an electron from leaving the top of the valence band and entering the con-

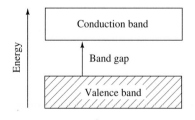

Figure 1.33   Valence and conduction bands in a semiconductor.

duction band, leaving a hole behind. Therefore, metals carry electric current when a potential gradient is applied.

The zero band gap of a metal results from delocalized bonding; the orbitals are partially filled, resulting in a continuous band, the lower part of which is filled and the upper part of which is empty. An example is lithium, in which each atom has one electron in a 2s valence orbital. Every atom is surrounded by twelve other atoms, in a close-packed array. The result is a sea of 2s orbitals, which are half-filled. An insulator, in contrast, has strongly localized bonds. The electrons are held in pairs between the atoms in bonding orbitals, which make up the valence band. The empty anti-bonding orbitals, which make up the conduction band, are much higher in energy. An example is diamond, in which each carbon atom is connected to four other carbon atoms by tetrahedrally directed bonds. The bonds reflect the mutual overlap of sp^3 hybrid orbitals, which are filled with electrons. The anti-bonding orbitals, in which the sp^3 atomic orbitals are combined with opposite phase, are very high in energy because of the strength of the bonds. The band gap of diamond is 5.2 ev, corresponding to light of 240 nm wavelength, in the deep ultraviolet region of the spectrum.

Semiconductors have relatively weak bonds and therefore, relatively low band gaps. Silicon has the same bonding arrangement as diamond does, since it lies directly below carbon in the periodic table, but Si has an extra shell of electrons and therefore has a larger radius than C. The mutual overlap of the valence sp^3 orbitals is less effective than in diamond, and the Si-Si bonds are substantially weaker than the C–C bonds. Likewise, the bonding-antibonding energy gap is lower. The silicon band gap is only 1.09 ev, corresponding to light of 1140 nm, in the infrared region of the spectrum.

This small band gap is one of the reasons that silicon is a good solar cell material. When light is absorbed by a semiconductor, an electron is promoted from the valence to the conduction band. This is the basis of the PV effect. To be absorbed, however, the light must have energy equal to or greater than the band gap. Light from the sun has a distribution of wavelengths, which peaks near 500 nm in the green-blue region of the spectrum (Figure 1.34). The distribution has a broad tail, extending far into the infrared region, but two-thirds of the photons have wavelengths shorter than 1140 nm. Consequently, silicon can capture most of the solar photons.

It is not enough, however, simply to shine sunlight on a piece of silicon. The electrons, which are promoted to the conduction band, have no reason to travel in any particular direction. After a short time, they fall back to the valence band, refilling the holes that had been created by the light. This process is called electron-hole *recombination*. In order to produce a flow of electrons in an external circuit, a PV material must have a built-in electrical potential, which is created by an intrinsic charge asymmetry. This charge asymmetry is produced by *doping* the semiconductor with foreign atoms that have either more valence electrons than the native atoms, or fewer valence electrons. As illustrated in Figure 1.35, silicon can be doped with arsenic, which has five valence electrons instead of four, or with gallium, which has only three valence electrons. The foreign atoms enter the Si lattice and are surrounded by four Si atoms. The four bonds utilize four valence electrons, however, leaving the As atom with an extra electron, or the Ga atom with a hole. Because of repulsion from the bonding electrons, the extra electron on the As atom has a high energy, close to that of the Si conduction band. It takes only a little energy for this electron to enter the conduction

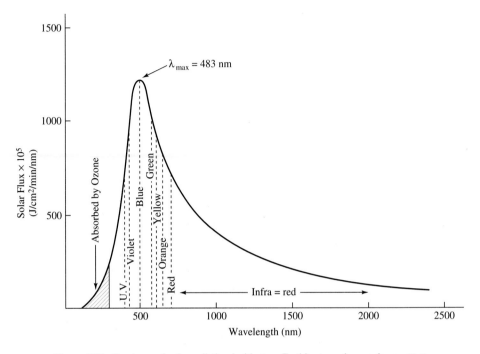

Figure 1.34 Spectrum of solar radiation incident on Earth's atmosphere; solar constant (area under curve) = 8.16 kJ/cm^2/min (energy of radiation increases with decreasing wavelength).

band and move freely through the lattice. A fixed positive charge is left on the As atom. Doping silicon with As therefore produces an *n-type* semiconductor, because *n*egative mobile carriers are created. Doping with Ga, on the other hand, produces a *p-type* semicon-

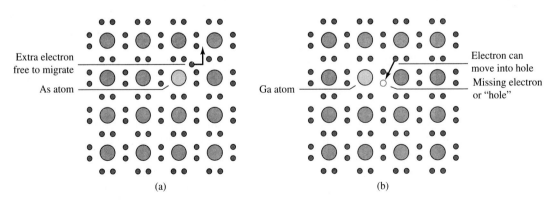

Figure 1.35 (a) An arsenic-doped *n*-type silicon semiconductor; (b) a gallium-doped *p*-type silicon semiconductor. *Source:* A. T. Schwarz et al. (1994). *Chemistry in Context* (Dubuque, Iowa: W. C. Brown Publishers). Copyright © 1994 by American Chemical Society. Reprinted with permission.

ductor, having *p*ositive mobile carriers. The missing electron on the Ga atom can be supplied by the surrounding Si lattice, creating a hole in the valence band, which becomes the mobile carrier. A fixed negative charge is now left behind on the Ga atom.

The charge asymmetry required by the PV cell is created by joining a *p*-type with an *n*-type semiconductor, forming a *p-n junction* (Figure 1.36). The mobile carriers migrate away from the junction because of repulsion from fixed charges of the same sign. Thus, the mobile electrons in the *n*-type semiconductor are repelled by the adjacent fixed negative charges on the Ga atoms of the *p*-type semiconductor, while the mobile holes in the *p*-type semiconductor are repelled by the adjacent positive charges on the As atoms of the *n*-type semiconductor. The result is an electric potential at the junction, created by the fixed negative charges on one side and the fixed positive charges on the other. If a photon is now absorbed in the *p*-type semiconductor, the electron that is injected into the conduction band will have a chance of being accelerated across the junction before it recombines with a hole. Likewise a photo-generated hole in the valence band of the *n*-type semiconductor can be accelerated in the opposite direction before it recombines with an electron. This directed movement of the photo-generated mobile charges produces an electric current in an external circuit.

The actual geometry used in a solar cell is shown in Figure 1.37. Which semiconductor is on top is immaterial, since the current is produced by both electrons and holes. But it is important to keep the top layer very thin, in order to avoid absorbing the photons before they reach the immediate region of the junction, where the charge separation occurs. The charges produced by photons absorbed outside this region will recombine before they can be accelerated across the junction.

Even in the absence of recombination, solar energy cannot be converted at 100% efficiency because of its spectral distribution. About a third of the photons have energies below the band gap and cannot be utilized. For the remainder, 1.09 ev is the maximum energy that can be extracted; energy increments above the band gap are wasted as heat. Consequently, only half the solar energy can be converted to photoelectrons. From this must be subtracted the efficiency losses due to recombination.

The recombination probability increases if there are imperfections in the silicon lattice. Consequently, the best solar cells are made of crystalline silicon. Efficiencies as high as 28% have been achieved with such cells. Unfortunately, crystalline silicon is expensive to manufacture in quantity; it must be cast in ingots and then cut into wafers. Interest has therefore

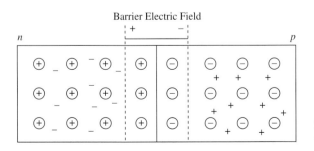

Figure 1.36   The *p-n* junction in a photovoltaic cell.

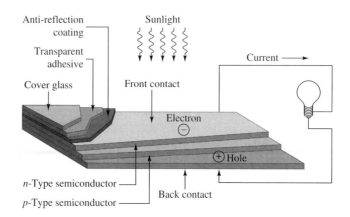

Figure 1.37   Schematic diagram of a photovoltaic cell.   *Source:* A. T. Schwarz et al. (1994). *Chemistry in Context* (Dubuque, Iowa: W. C. Brown Publishers). Copyright © 1994 by American Chemical Society. Reprinted with permission.

turned to amorphous silicon, which can be manufactured continuously in thin films at a much lower cost. Because of lattice imperfections, the conversion efficiency is reduced for amorphous silicon, but intensive development has raised the achievable efficiency to 16%. The lowered efficiency is more than made up by the lower costs and ease of manufacture.

2) Photosynthesis: Nature's Solar Cell.   Hundreds of millions of years before the invention of the *p-n* junction, nature solved the problem of harnessing photo-induced charge separation to produce useful energy. In photosynthesis, as in the silicon solar cell, electrons and holes are created by the absorption of solar photons, and are induced to travel in opposite directions. Instead of a *p-n* junction, nature has evolved a *photo-reaction center*, at which charge is separated across a biological membrane, as illustrated in Figure 1.38. The key components of the photo-reaction center are a collection of chlorophyll molecules (Figure 1.39) capable of absorbing sunlight and generating electrons and holes in their photo-excited states. The electrons hop from one molecule to another along a gradient of decreasing energy of the available empty orbitals. This gradient serves the same function as the charge asymmetry in a *p-n* junction, namely, to separate electrons and holes physically. Instead of an electric current, the charge separation leads to the production of energy-storing chemicals. Because the photo-reaction center is held in a membrane, the electron and hole are able to react separately with molecules that are located on opposite sides of the membrane. The electron is taken up in a series of biochemical steps that lead to the reduction of CO_2 to carbohydrates, while the hole is involved in another series of steps that oxidize water to O_2.

Because of its extended system of conjugated double bonds, chlorophyll absorbs light strongly in the 400–700 nm region of the spectrum (see the absorption spectrum in Figure 1.39). This region contains about 50% of the total energy in solar photons. A green plant absorbs about 80% of the incident photons in this range (the rest are lost to reflection, transmission, and absorption by other molecules), leaving about 40% of the total solar energy available for photosynthesis. Of this, 28% actually ends up in carbohydrates; the rest is

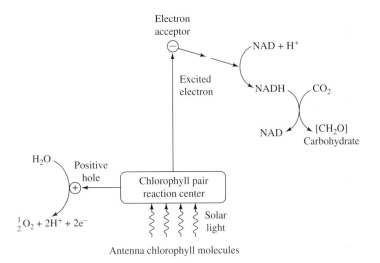

Figure 1.38   The photo-reaction center and energy storage in photosynthesis.

lost at the various electron transfer and chemical steps. Thus, the conversion efficiency from sunlight to carbohydrates is $0.5 \times 0.8 \times 0.28 = 0.11$. However, the plant uses about 40% of the energy for its own metabolic needs, leaving only $0.11 \times 0.6 = .067$ as the fraction of sunlight stored as photosynthetic energy. This maximum conversion efficiency applies to the so-called C_4 plants, in which the first product of photosynthesis is a 4-carbon sugar; these include corn, sorghum and sugar cane, and some other plants that grow best in hot climates. C_3 plants, in which the first photosynthetic product is a 3-carbon sugar, includes wheat, rice, soybeans, trees, and other plants that dominate temperate climates and account for 95% of global plant biomass. These plants are about half as efficient in photosynthesis as the C_4 plants. To these inefficiencies must be added the limitations on plant growth imposed by temperatures that are too low or too high, insufficient water and insufficient nutrients. These limitations and inefficiencies explain why only about 0.3% of the global insolation reaching Earth's surface is used by green plants and algae in photosynthesis.

3. Biomass

Humans have used plant matter for cooking and heating since the discovery of fire. Even today, wood is the dominant fuel for cooking and heating in developing countries. Where wood is scarce, crop residues and dung are used as fuel. Biomass accounts for 38% of energy consumption in developing countries. The burning of biomass is associated with environmental and health problems due to deforestation and air pollution. The gathering of biomass is labor-intensive, and the stoves and fireplaces in which it is burned are typically inefficient and smoky. Small wonder that, as development proceeds, biomass fuels are abandoned in favor of cleaner and more convenient forms of energy. In industrialized countries, biomass accounts for only 2.8% of energy consumption.

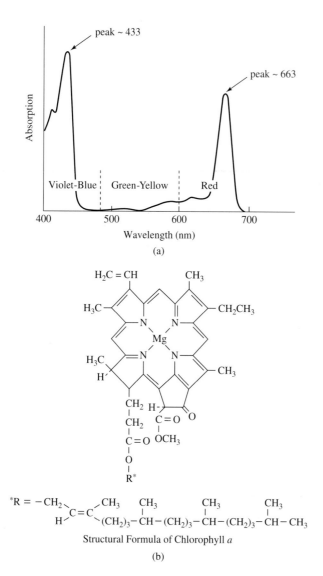

Structural Formula of Chlorophyll *a*

(b)

Figure 1.39 The chemical structure of chlorophyll *a* and its absorption spectrum.

Nevertheless, biomass represents a considerable energy resource, which could be converted by currently available technology to electricity, or alternatively, to liquid fuels such as alcohol (Figure 1.40). Synthesis of gaseous fuels is also possible. For example, biomass can be converted to fuel gas using the same chemistry discussed above for coal: a combination of reforming and hydrogenation to produce a mixture of methane, hydrogen, and carbon monoxide. In fact, the technology is more straightforward with biomass, because it is more reactive than coal and has low sulfur content, obviating the need for sulfur removal. Or, methane can be produced directly from biomass by digesting it in the absence of air

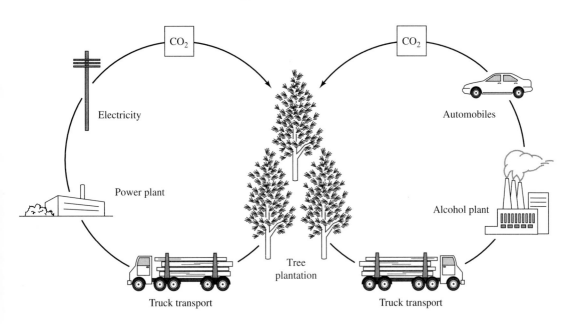

Figure 1.40   Alcohol fuel and electricity produced by biomass.   *Source:* C. J. Weinberg and R. H. Williams (1990). Energy from the sun. *Scientific American* 263(3): 146–155. Copyright © 1990 by Scientific American, Inc. All rights reserved.

with anaerobic bacteria. In either case, the fuel gas can be used for heating, or for power generation.

Alternatively, the gasification process could be adapted to the production of methanol, with the same technology used to convert natural gas or coal to methanol. Or the biomass can be converted directly to ethanol via a biochemical fermentation step. Indeed, Brazil already has half its automotive fleet running on ethanol produced from sugar cane. In the United States, surplus corn is likewise converted to ethanol and sold as "gasohol," a blend of 19% ethanol in gasoline. Either ethanol or methanol can be used as a fuel for the internal combustion engine, but methanol has the added potential of being a convenient fuel for fuel cell vehicles (see p. 84). Methanol can be reformed to hydrogen at low temperatures, whereas ethanol reforming requires considerably higher temperatures and cannot be carried out practically in an on-board reformer.

How is the biomass to be obtained for these purposes? Agricultural or forestry wastes are candidate sources. It has been estimated that these residues are equivalent in energy to about one-third the annual commercial energy use. Of course, some of the residue is desirable as fertilizer, and much of it is already being used as fuel, albeit at low efficiency. A great deal of it is simply burned in the open, contributing to air pollution. Much more energy could be produced by deliberately growing energy crops—fast growing trees or grasses. However, land area requirements for energy plantations are substantial. Photosynthesis is an inefficient process, and can be counted on to convert only about 1% of the sun's light to accumulated biomass; this rate compares poorly with photovoltaic cells, which are already more than 10%-efficient. On the other hand, growing plants is much cheaper than manufacturing solar cells.

In the industrialized world, substantial areas have been removed from agricultural production because of surpluses. If 7% of the world's land area, a target comparable to current reforestation goals, were devoted to biomass production for energy, the resulting supply would be equivalent to about four-fifths of worldwide commercial energy use. Especially for developing countries, with abundant sunshine, rainfall, degraded land, unemployed labor, and reliance on imported oil, the production of commercial fuels from biomass could provide a significant development option.

4. Hydroelectricity

Hydroelectric plants utilize part of the energy of the solar-driven hydrological cycle. The continents of the world receive more rain than the water they lose by evapotranspiration, and the excess runs off in rivers to the ocean. The running water can be used to turn a turbine and generate electricity. Although there are many small hydroelectric facilities that utilize river-flow directly, the larger installations, which account for most of the available hydropower, rely on dams to increase the hydraulic head (water pressure), and to even out the flow, thereby allowing the continuous production of electricity. Dams serve other purposes as well, including the provision of water for residential, industrial, and agricultural purposes; facilitating flood control and/or navigation; and providing recreational facilities. In fact, most dams do not generate electricity, although many could be retrofitted to do so.

The share of world electricity provided by hydropower is about 14%. This is a very small fraction of the available hydrological potential, but probably represents a third to a fourth of the potential sites that could be developed economically. Most of the undeveloped sites are in the former Soviet Union and in developing countries.

Like other forms of solar energy, hydroelectricity adds no CO_2 nor other emissions to the atmosphere, but it is not without environmental costs. The water backing up behind dams floods the shoreline, inundating human habitations, relics, and ecosystems, and the character of the river is permanently altered. The effects are particularly devastating on migratory fish, and are only partially remedied by fish ladders. The large dams on the Columbia River have contributed to the collapse of the salmon population in the northwest of the United States. Dams also hold back silt, which can have deleterious effects downstream. The most celebrated example is Egypt's Aswan dam, which stopped the annual flooding of the Nile valley, resulting in diminished nutrient input to crops, and salinization and subsidence of the Nile delta.

The silt behind the dam must eventually be flushed out before it engulfs the turbine intakes. The still waters behind the dam are subject to eutrophication (overfertilization—see pp. 228–231) and, especially in the tropics, can promote the spread of disease-carrying organisms. Most of these problems can be minimized by appropriate site selection and management, but public opposition now limits the number of available dam sites significantly. Small hydroelectric plants that work on river flow avoid most of these problems, but they are less efficient and are subject to the seasonal fluctuation of river levels.

5. Wind Power

The wind provides another form of solar power, since wind results from air-temperature differences associated with different rates of solar heating. A global circulation of air, the Hadley circulation, is created by moist hot air rising at the equator and being replaced by drier air flowing in from the region of 30 degrees north and south latitude. At higher latitudes, the air flows toward the poles and is deflected westward by Earth's rotation, creating a wavelike pattern known as the Rossby circulation. Regional variations in atmospheric temperature superimpose smaller circulation systems on the global pattern. Locally strong winds are created by sharp temperature differences between, for example, the land and the sea, and they can be channeled by mountains and valleys. Many regions have steady prevailing winds as a result of these conditions.

Wind technology is nearly as old as recorded history. Well over 2,000 years ago, windmills were used to pump water in China and to grind grain in Persia. Europe was introduced to windmills in the eleventh century by veterans of the Crusades; the windmills were improved in the succeeding centuries, especially in Holland and England. By the eighteenth century there were 10,000 windmills in Holland alone. In the United States, windmills were a vital asset of early settlers of the Great Plains, and there were 50,000 backyard windmills pumping water and generating electricity until 1950, after which they were phased out by the Rural Electricity Administration.

The 1973 oil crisis rekindled interest in wind energy and in other alternative energy sources. A number of development programs were set in motion, the most successful of which were in Denmark and California. These programs were based on the concept of windmill arrays, using advanced airfoils and turbines (see Figure 1.41), which feed electricity into the regional power grid. Because of this integration, the intermittence of the winds can be leveled out by the array and by other power sources on the grid. In both regions, the costs of wind electricity have decreased steadily since 1985 and are now nearly on a par with coal generation.

The energy potentially available in prevailing winds is huge. Current wind turbines are quite efficient and can convert up to 25% of the wind energy to electricity. Denmark and California now produce 2.5 and 1.1% of their electricity, respectively, with wind turbines. Denmark expects to increase this percentage to 10% by the year 2005. In the United States, the entire electricity demand could be met by assembling four million 500 kW wind turbines one half of a kilometer apart over 10% of the land cover where winds are most favorable (see Figure 1.42). Even if a small fraction of the potential wind energy sites are developed, the contribution to the world's electricity supply could be significant. An eventual contribution of 20% is considered to be not unrealistic. At this point, wind energy is the closest to commercialization of all renewable energy sources.

6. Ocean Energy

A great deal of energy is stored in the world's oceans in the form of tides and waves, and in gradients of temperature and salt concentration. But the oceans are vast and these energy

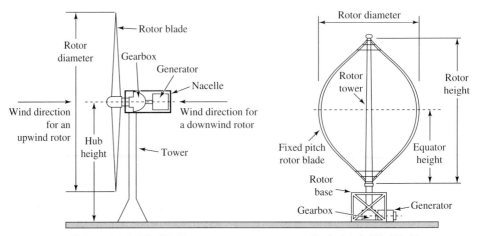

Figure 1.41 Two basic wind turbine configurations: the horizontal axis wind turbine and the vertical axis wind turbine. *Source:* A. J. Cavallo et al. (1993). Wind energy: technology and economics. In *Renewable Energy, Sources for Fuels and Electricity*, T. B. Johansson et al., eds. (Washington, DC: Island Press). Copyright © 1993 by Island Press. Reprinted with permission.

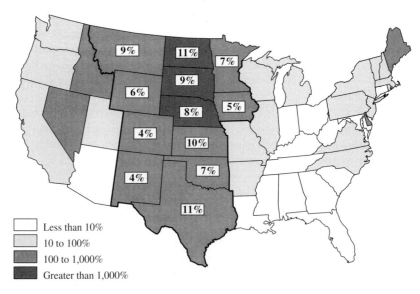

Figure 1.42 Wind energy potential in the United States (shading shows wind-electric potential as a percentage of in-state electricity demand; numbers refer to percentages of total U.S. wind-electric potential for the 12 states with the highest wind resources). *Source:* C. J. Weinberg and R. H. Williams (1990). Energy from the sun. *Scientific American* 263(3): 146–155. Copyright © 1990 by Scientific American, Inc. All rights reserved.

forms are highly dispersed. Tidal energy is the only form currently being exploited for commercial power generation at an installation on the Brittany coast in France. The tides result from the gravitational attraction of the moon and sun. In the open ocean, the waters rise and fall by about one meter, but at shorelines the tides are higher, and can reach several meters in estuaries, due to the funneling effect of the shore. Tidal power can be extracted much like power in falling rivers. A dam is built with sluices to admit the incoming tide to a reservoir, which is then emptied through turbines when the tide recedes. Although several tidal sites, especially in the United Kingdom and France, have potential for power generation (when the economic conditions become favorable), the number of such sites is insufficient for tidal energy to become an important part of the global energy supply.

Waves result from the action of wind on the waters and can carry significant amounts of energy. Where winds blow for long distances over open oceans, waves can be tens of meters high and displace tons of water. The west coasts of Europe and the United States and the coasts of Japan and New Zealand are particularly suitable for extracting the energy of waves. To accomplish this, a variety of mechanical devices have been proposed and are being researched. None are currently close to being practical, however.

The difference in salt concentration between the oceans and fresh water represents a large osmotic pressure, equivalent to a 240-meter head of water. But no practical method of harnessing the mixing energy has been devised.

More interesting are the possibilities for exploiting the energy stored in ocean thermal gradients. The sun heats the surface layers of the oceans, but the deeper layers remain cold. In tropical regions, the surface temperature is as high as 26°C, whereas at a depth of 1,000 meters the temperature is 5–6°C. Because of the expanse of the ocean, the total energy stored in the surface gradients is enormous, about two orders of magnitude greater than the energy of all the tides and waves. But capturing this energy is not easy. Not only is it widely spread, but the 20°C gradient limits the theoretical heat engine efficiency (see pp. 77–79) to under 7%. Consequently, a significant supply of electricity from this source requires a very large plant, and capital costs would be high. Moreover, transmission losses would limit such plants to operation near shores, although plants operating in the open ocean could, in principle, generate hydrogen electrolytically, for transport by ship (see discussion of the "hydrogen economy" in the "Energy Utilization" section that follows).

An OTEC (ocean thermal energy conversion) plant requires a large pipe to bring cold waters from depths of 600 to 1,000 meters to the surface. This cold water is the heat sink for a heat engine working in either a closed or open cycle (Figure 1.43). A closed cycle plant is like a refrigerator run backwards. A working fluid, such as ammonia, is evaporated by the warm surface waters; the vapors flow through a turbine, generating electricity, and are condensed by the cold waters from the deep. In an open cycle plant, the seawater itself is the working fluid. Warm seawater is evaporated at very low pressure; the water vapors run the turbine and are condensed by the cold water. This method requires extra pumps to de-gas the water and create the needed vacuum, but it produces freshwater as a byproduct.

Indeed, byproducts may be the key to successful operation of OTEC plants. In Hawaii, an experimental station is producing electricity and freshwater; in addition it circulates the pumped cold seawater for air-conditioning nearby buildings. Finally, the circulating sea water supports a variety of agricultural and maricultural enterprises. The deep waters of the

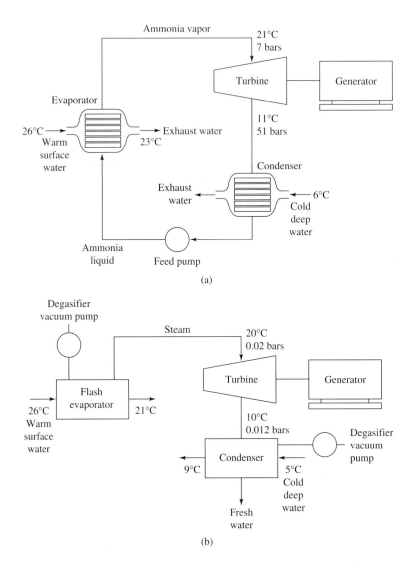

Figure 1.43 Schematic of an OTEC plant; (a) closed-cycle plant; (b) open-cycle plant. *Source:* J. E. Cavanagh et al. (1993). Ocean energy systems. In *Renewable Energy, Sources for Fuels and Electricity,* T. B. Johansson et al., eds. (Washington, DC: Island Press). Copyright © 1993 by Island Press. Reprinted with permission.

ocean are not only cold but also rich in nutrients from the falling remains of life forms at the surface. These nutrients can fertilize vegetable and fish farms in the vicinity of the OTEC plant. Similar opportunities for integrated power, water, air-conditioning, and farming activities may exist at many places on tropical shores.

7. Geothermal Energy

Geothermal energy is the one form of renewable energy that is unconnected with the sun. Earth itself generates heat from its molten core, and from the decay of naturally occurring radioisotopes. This heat source is thousands of times weaker than the sun's rays falling on Earth, but in some parts of Earth's crust, geological anomalies permit the concentrated up-welling of heat from below. Hot springs and volcanoes are familiar examples of these anomalies. Hot springs have been valued for bathing and for their curative powers for thousands of years, and the Romans used the hot water to heat their bath houses. A modern geothermal industry began in the nineteenth century, with the extraction of boric acid from hot springs near Laradello, Italy. By 1827, geothermal steam replaced firewood as a fuel for concentrating the boric acid, and the first geothermal power plant began operation in 1913, also in Italy. Since then, power plants have been installed at a number of sites around the world where geothermal steam is available. Such sites are limited in their capacity, however, and geothermal power currently supplies just 0.2% of the world's electricity; this share is projected to rise to 0.6% by the year 2000.

There is a much larger potential if geothermal sources other than readily available steam are tapped. There are extensive deep reservoirs of hot water under high pressure, and even more extensive regions of near-surface molten rock (magma) and of hot dry rock. In the United States alone, the pool of heat from these sources is 5.3×10^{20} kJ, thousands of times greater than annual U.S. primary energy consumption. Only a small fraction from this pool can be exploited, however, and technologies for doing so are still in the experimental stage. Geothermal steam plants operate like any other steam plant, except that special measures are needed to deal with the salts and gases, especially H_2S, entrained in the steam, in order to minimize corrosion and environmental contamination.

E. ENERGY UTILIZATION

Extraction of energy from the various available sources is one side of the human energy equation; utilizing the extracted energy is the other. We need energy for myriad human activities, but the amount of energy required depends strongly on the efficiency with which it is used. The efficiency of energy utilization can vary enormously. It depends on technology, on the integration of energy systems, and on our patterns of living.

1. Heat Engine Efficiencies: Entropy

The first law of thermodynamics tells us that energy is conserved. A joule is a joule, no matter what the form of the energy; the joules all add up. But the second law of thermodynamics tells us that energy can never be used with 100% efficiency; some fraction of the energy cannot be converted to useful work.

Consider a power plant, diagrammed in Figure 1.44a, that generates electricity by boiling water and forcing the steam through a turbine. The steam is then condensed and returned

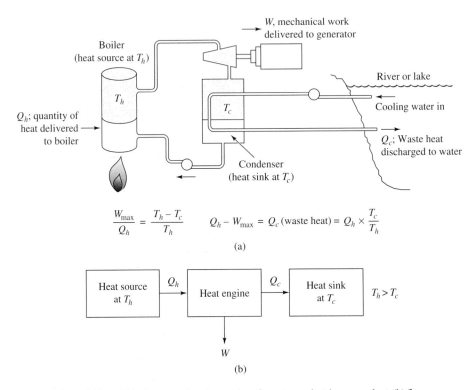

$$\frac{W_{max}}{Q_h} = \frac{T_h - T_c}{T_h} \qquad Q_h - W_{max} = Q_c \text{ (waste heat)} = Q_h \times \frac{T_c}{T_h}$$

(a)

(b)

Figure 1.44 (a) Maximum work and waste heat from steam electric power plant; (b) flow of energy in converting heat into work.

to the boiler. In this way, heat from the burning of fuel, whether fossil or nuclear fuel, is converted to electricity. Some heat has to be transferred to the environment (cooling water) to condense the steam, however. Since the total energy is conserved,

$$Q_h = W + Q_c \qquad (1.18)$$

where Q_h is the quantity of heat used to boil the water, Q_c is the quantity of heat transferred to the environment, and W is the quantity of work produced in the form of electricity. A diagram of the energy flow is given in Figure 1.44b.

The second law of thermodynamics puts a limit on the fraction of heat that can be converted to work:

$$W/Q_h < (T_h - T_c)/T_h \qquad (1.19)$$

where T_h is the absolute temperature of the heat source (boiler) and T_c is the absolute temperature of the heat sink (cooling water).

The physical basis for equation (1.19) is that the entropy of a system and its surroundings cannot decrease. The universe always tends toward greater disorder; this is a general statement of the second law. Entropy, S, which is a measure of disorder, is defined as Q/T. The entropy change for the power plant and its surroundings is:

$$\Delta S_{\text{total}} = \Delta S_{\text{source}} + \Delta S_{\text{sink}} \tag{1.20}$$

$$S_{\text{total}} = -Q_h/T_h + Q_c/T_c \tag{1.21}$$

where the plus and minus signs indicate a gain and loss in entropy, respectively. The overall change must be positive. Therefore

$$-Q_h/T_h + Q_c/T_c > 0 \tag{1.22}$$

Substituting

$$Q_c = Q_h - W \tag{1.23}$$

into this expression and rearranging it leads to equation (1.19).

This simple equation has very important implications for energy efficiency. High efficiency requires a high T_h or a low T_c, or both. If we could lower T_c to zero, 100% efficiency would always be available. But this is impossible. The temperatures in the formula are absolute temperatures, and absolute zero is $-273°C$. Since the temperature at the surface of Earth is, on the average, not far from $27°C$, we know that T_c in the formula cannot be lower than about 300 K (K stands for Kelvin, the absolute temperature scale obtained by adding 273 to the Centigrade temperature). Thus, the maximum efficiency depends on T_h.

The value of T_h depends on the design of the heat source and on the strength of materials in the boiler. In a modern coal-fired power plant, T_h is about $550°C$, or 823 K. Equation (1.19) then gives a theoretical efficiency of 64%. The actual efficiency is about 31%, however, because the boiler, the turbine, and the electrical generator are all less than 100% efficient in transmitting the energy; there are additional heat losses at each of these stages, beyond the minimum heat loss required by the second law of thermodynamics.

Other heat sources can have lower or higher efficiencies. A nuclear power plant has an overall efficiency of only about 30%, because T_h is restricted to lower values for reasons of safety. Substantially higher efficiencies can be achieved with gas turbine heat sources for power plants, which are beginning to be commercialized. In this technology, a turbine is driven directly by the gas combustion process, rather than by steam produced in a boiler. Based on advanced materials developed for jet engines, gas turbines can operate at very high temperatures, up to $1260°C$, and therefore offer very high theoretical efficiencies. In practice, the efficiency is limited by the fact that T_c is also quite high because the exhaust gas from the turbine is very hot, typically $500°C$. These values for T_h and T_c give a maximum efficiency of 50%; the actual efficiency is only about 33%, about the same as a nuclear plant. But the hot gases can themselves be used to run a steam turbine (Figure 1.45). This second stage operates at about the same conversion efficiency as a regular steam plant. By combining these two stages, one can extract close to 80% of the energy in the fuel.

The same thermodynamic considerations apply when energy is "consumed." The energy does not disappear, of course. Rather it is converted to "waste" heat, raising the entropy of the universe. En route to this destination, the energy can be utilized more or less efficiently. Heating a house provides an illustrative example. A gas furnace is more efficient for this

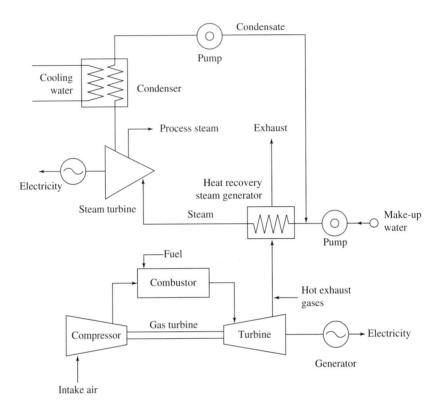

Figure 1.45 Schematic of gas turbine/steam turbine combined-cycle power plant. *Source:* R. H. Williams and E. D. Larson (1993). Advanced gasification-based biomass power generation. In *Renewable Energy, Sources for Fuels and Electricity*, T. B. Johansson et al., eds. (Washington, DC: Island Press). Copyright © 1993 by Island Press. Reprinted with permission.

purpose than is electrical heating, because most of the energy in the fossil fuel can be transferred directly to the house (assuming a highly efficient furnace), whereas with electrical heating, about two-thirds of the energy has been thrown away at the power plant to produce the electricity that is then converted back to house heat.

Electricity can be converted to heat in a much more efficient way, however, by using it to run a heat pump. This device (see Figure 1.46) is a small version of a power plant run in reverse. Mechanical work is used to condense a working fluid at the temperature of the heat sink. The fluid is then allowed to evaporate, thereby absorbing heat from the heat source. In this way heat can be pumped from lower to higher temperatures by the expenditure of work. This is how refrigerators and air-conditioners operate. A house can be heated in cold weather by reversing the direction of the air-conditioner.

The conversion of work to heat is governed by the same entropy considerations as the reverse process. The maximum degree of conversion is simply given by the inverse of equation (1.19):

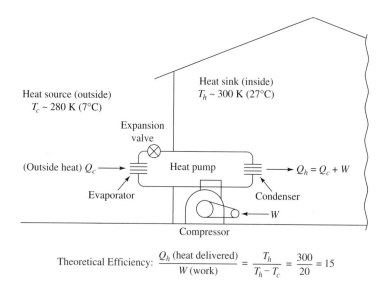

Theoretical Efficiency: $\dfrac{Q_h \text{ (heat delivered)}}{W \text{ (work)}} = \dfrac{T_h}{T_h - T_c} = \dfrac{300}{20} = 15$

Figure 1.46   The heat pump: an efficient means of residential space heating.

$$Q/W < T_h/(T_h - T_c) \qquad (1.24)$$

but T_h and T_c are reversed, since the heat source (T_c) is now colder than the heat sink (T_h). If the inside temperature is about 300 K, and the outside temperature is 20°C colder, then the amount of pumped heat is potentially 300/20 = 15 times greater than the amount of work. One kJ of electricity could transfer as much as 15 kJ of heat. Does this mean that we have violated the first law of thermodynamics? Not at all. It just means that we need to pay attention to the quality as well as the quantity of the energy. The higher the temperature, the higher the quality; high-grade heat can be converted to a larger quantity of low-grade heat, increasing the entropy in the process. Electricity is very high-quality energy, and this quality is largely wasted by converting it directly to heat at low temperature; the heat pump allows this quality to be more fully utilized. The high theoretical efficiency cannot be achieved in practice because of resistances to heat transfer, but a heat/work ratio of 2 is easily attainable, thereby recapturing a substantial fraction of the fossil fuel calories that are lost at the power plant.

2. Fuel Cells

Combustion is not the only way to extract useful energy from chemical fuels. Electricity can be obtained directly with the aid of a fuel cell. For example, hydrogen can be burned in oxygen, producing water and heat

$$H_2 + 1/2\, O_2 = H_2O \qquad (1.25)$$

or the same reaction can be carried out at two electrodes, with an electric current flowing between them, as illustrated in Figure 1.47. The electrode reactions are

$$H_2 = 2H^+ + 2e^- \tag{1.26}$$

$$2e^- + 1/2\ O_2 + 2H^+ = H_2O \tag{1.27}$$

Hydrogen is oxidized at the *anode*, where electrons are removed and passed through the external circuit to the *cathode*, where oxygen is reduced. The two electrode reactions together add up to reaction (1.25). But the energy is released mostly as electricity instead of as heat. Electricity production via the fuel cell is just the reverse of the familiar electrolysis process, in which hydrogen and oxygen are produced at a pair of electrodes immersed in water when an electric current is passed between them.

Because the reaction energy is transformed directly into electricity, the efficiency is not limited by the heat engine formula. The second law is still operative, however, and limits the efficiency to less than 100%, because the entropy of the molecules decreases in reaction (1.25). Less disorder is associated with the products than with the reactants, because three reactant molecules produce only two product molecules. In addition, the product is a liquid (if the reaction is carried out at less than 100°C and 1 atm. of pressure) and it occupies less volume than the reactants, which are gases. The decrease in entropy must be paid for by part of the reaction energy, decreasing the amount that can be converted to work. The thermodynamic expression for this fact is

$$\Delta G = \Delta H - T\Delta S \tag{1.28}$$

where ΔS is the change in entropy associated with the reaction (note that since $S = Q/T$, $T\Delta S$ has units of energy), ΔH is the change in the total energy, or *enthalpy*, and ΔG is the change in the *free energy,* the part of the total energy release that is available for work. The sign convention for this equation is that Δ refers to products minus reactants. ΔS is therefore negative, as are ΔG and ΔH; ΔG is less negative than ΔH because of the negative entropy change. (A theoretical efficiency greater than 100% is possible if ΔS is positive for the electrochemical reaction.)

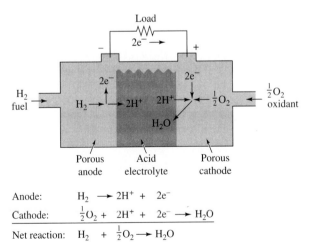

Figure 1.47   Schematic of a fuel cell using hydrogen as fuel. *Source:* D. L. Illman (1994). Auto-makers move toward new generation of "greener" vehicles. *Chemical and Engineering News* 72(31): 8–16. Copyright © 1994 by American Chemical Society. Reprinted with permission.

Since ΔG is the energy available for work, and ΔH is the total energy, the theoretical efficiency for the fuel cell reaction is $\Delta G/\Delta H$. At 300 K, $\Delta H = -295$ kJ/mol for reaction (1.25), while the energy lost to entropy, $T\Delta S$, is -58 kJ/mol. Consequently, $\Delta G = -237$ kJ/mol, and the theoretical efficiency is 80%. As the temperature goes up, $T\Delta S$ increases in proportion, and the theoretical efficiency goes down, in marked contrast to the efficiency of a heat engine. When water remains in the gas phase, however, the entropy change is diminished, as is ΔH, which includes the cohesive energy of water when it is a liquid. At 1000 K, for example, $\Delta H = -238$ kJ/mol (note that on a per mole O_2 basis, $\Delta H = 476$ kJ/mol, which is essentially the value obtained in our bond energy calculation in Table 1.4, which was based on the energies of isolated molecules), and $T\Delta S = -43$ kJ/mol, giving $\Delta G = -195$ kJ/mol, and a theoretical efficiency of 82%. Thus, high efficiency is, in principle, achievable over a wide temperature range, by converting the energy of reaction directly to electricity.

However, the theoretical efficiency can only be reached if no current is actually drawn from the fuel cell. When a current does flow through the external circuit, several resistances build up that reduce the efficiency, and increase the fraction of the energy that is converted to heat. These include 1) the electrical resistance of moving positive ions from the region of the anode to the region of the cathode to counterbalance the electron flow (electrolyte resistance); 2) the resistance to the movement of reactant molecules to the electrodes and product molecules away from the electrodes (mass transfer resistance); and 3) the resistance to the chemical reactions themselves due to slow reaction kinetics (significant activation barriers). These resistances can be minimized, but not eliminated, by optimizing the fuel cell design. Thus, the reaction kinetics can be speeded up with catalysts (platinum metal is most commonly used), mass transfer can be enhanced by increasing the surface area of the electrodes, for example, by making them porous, and electrical resistance can be minimized by connecting the electrode compartments with an electrolyte having a high conductance for cations.

All three resistances are reduced at elevated temperature, which increases rates for chemical reactions and for molecular and ion transport. Consequently, successful fuel-cell designs have tended to have elevated operating temperatures. The only fuel cell that was commercially available in 1995 is based on a phosphoric acid electrolyte and operates at 150–210°C. A fuel cell based on molten carbonate electrolyte, operating at 550–650°C, has been tested in the field. And there is much interest in a fuel cell whose electrolyte is a solid oxide ceramic, which operates at about 1000°C. These high-temperature fuel cells may soon find application as stationary power generators. Their efficiencies, currently in the 50–60% range, compare favorably with steam generators, and they lend themselves to modular construction and to reducing transmission costs by placement near the consumer. In addition, the waste heat, being generated at high temperature, can be used for industrial and space heating (a technique called "cogeneration"—see p. 88).

On the other hand, high-temperature fuel cells are not suitable for mobile power sources, which require light units and cold starts. The U.S. space program has utilized the alkaline fuel cell, whose electrolyte is potassium hydroxide. The alkaline fuel cell has a high energy density, can start cold, and operates in the range 70–250°C. It must use pure oxygen, rather than air, however, since the potassium hydroxide would react with the CO_2 in air, degrading the fuel cell performance. There is much current interest in the PEM (proton exchange membrane) fuel cell, in which the electrolyte is a polymer membrane (filled with ionic sulfonate groups) that can exchange cations. The membrane has a high conductance for protons. The PEM fuel cell can use air as the oxygen source, and is capable of operating from room temperature to 80–100°C. It holds promise for use in automotive transport, and an experimental bus powered with PEM cells is currently in operation in British Columbia.

Although in principle chemical fuels other than hydrogen could operate a fuel cell, none of the carbon fuels that are currently or potentially in use have been found to react at electrodes rapidly enough to sustain a significant electric current, with the catalysts tried so far. Although further research might lead to an adequate catalyst, it seems more likely that carbon fuels will be linked to fuel cells via the production of hydrogen, with the steam re-forming and water-gas shift reactions discussed above (p. 31) for coal (reaction 1.15 and 1.16). Steam reforming is not limited to coal, but can be utilized with other carbon fuels, such as methane:

$$CH_4 + H_2O = CO + 3H_2 \qquad\qquad (1.29)$$

In the case of methanol, the elements of water are built into the molecule, which can "reform" itself upon heating:

$$CH_3OH = CO + 2H_2 \qquad\qquad (1.30)$$

Adding steam converts the CO to CO_2 and produces another molecule of H_2 via the water-gas shift reaction. Methanol is reformed at sufficiently low temperature, that it is possible to design cars and trucks with a compact, onboard reforming unit in order to run a fuel cell with methanol. Since it is easier to fuel a vehicle with methanol than with hydrogen, this option could make fuel-cell-powered electric vehicles attractive before long. The solid oxide fuel cells are able to utilize liquid carbon-based fuels directly, because at their high operating temperature, the fuels are reformed to hydrogen inside the cell.

3. Electricity Storage: The Hydrogen Economy

Electricity storage and conversion is a critical issue for many power systems at both large and small scales. In many cases either the source of electricity or the need for it is intermittent, producing a mismatch between supply and demand. Efficient electricity storage is then needed to optimize the system. For example, the main drawback to the production of direct solar or wind electricity is the intermittent nature of the energy source. Electricity is produced only when the sun shines or the wind blows. On the other side of the ledger, power companies must cope with large fluctuations in the daily demand for electricity. Extra generating capacity is required to meet peak demands during daylight hours, particularly during the summer when air-conditioners draw a heavy load, and the extra capacity is left idle much of the time. To some extent these fluctuations can be coordinated, since the solar flux also peaks during daylight hours and during the summer, but there is still an important need for efficient energy storage. A few electric companies have developed water pump storage, in which excess electricity is used to pump water uphill to reservoirs and the water running back down-hill can then be used to run turbines to meet peak demands. Because of the continually fluctuating water levels, these reservoirs create ecological problems, and are especially contentious if they displace multiple-use ponds. Other energy storage schemes under consideration include storage of compressed air in caverns, mechanical energy storage in fly-wheels, and direct electrical storage in large superconducting magnets.

On the small-scale end of electricity utilization, electricity storage is also a critical issue for the development of the electric car. At present, electric cars run on current from the lead-acid storage battery. In this battery (Figure 1.48), electricity is stored in the chemical conversion of Pb^{2+} ions to metallic Pb at one electrode, and to PbO_2 at the other. The overall reaction,

$$2Pb^{2+} + 2H_2O = Pb + PbO_2 + 4H^+ \tag{1.31}$$

is energetically uphill. When current is drawn, both electrode reactions are reversed, and the reaction is allowed to run downhill. The lead-acid battery performs these energy conversion steps very efficiently, and can be charged and discharged many times before it is worn out through competing chemical processes. It is used in all automotive vehicles as a portable store of electricity for auxiliary needs. In electric vehicles, however, the need is not auxiliary; electricity is the source of power for locomotion itself. The lead-acid battery has two serious drawbacks as the main power source: 1) it is heavy and adds significantly to the weight of the vehicle, thereby lowering the efficiency and the driving range; 2) it takes several hours to charge, making "refueling" an inconvenient operation. Consequently, much effort has been devoted to developing alternative batteries that give more power with less weight and easier charging. Nickel-cadmium batteries, already in use for many appliances, offer greater power, energy density, and longer life than lead-acid batteries but are substantially more expensive. Promising alternatives that are currently under development include

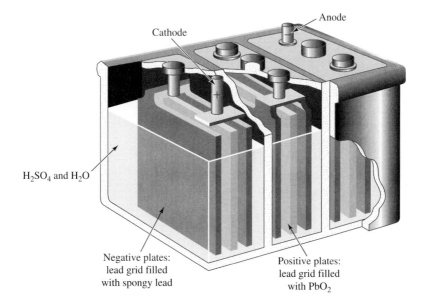

Figure 1.48   Cross-section of lead-acid storage battery.   *Source:* P. Buell and J. Gerard (1994). *Chemistry in Environmental Perspective* (Upper Saddle River, New Jersey: Prentice Hall).

nickel-metal hydride, sodium-sulfur, lithium-iron disulfide, and lithium-polymer batteries. But the reliable lead-acid battery, itself undergoing continuing improvements, will be hard to beat, despite its disadvantages.

In this situation, the hydrogen fuel cell offers an attractive alternative to any of the storage batteries. Indeed, it can itself be thought of as a battery, in which the storage medium is hydrogen and oxygen (Figure 1.47), instead of Pb and PbO_2. The same drivetrain can be utilized with a battery or a fuel cell. But the fuel cell is not required to run itself backward with external electricity in order to store its energy. A canister of hydrogen, or alternatively, a tank of methanol with an onboard reformer (see preceding section) is all that is required. Consequently, refueling is as fast as it is for gasoline-powered cars. And since hydrogen or methanol are much lighter than lead or most other metals, more energy can be stored for the same weight, giving a fuel-cell electric car a greater driving range than one with storage batteries.

Power companies could also use hydrogen for electricity storage, once appropriate fuel cells are commercialized. Electricity is readily converted to hydrogen through the electrolysis of water (the reverse of the fuel cell reaction), with efficiencies as high as 85%. Thus a combination of electrolysis and fuel cells, with tank storage of the hydrogen, could provide relatively efficient load-leveling capacity.

Moreover, hydrogen can be transported more efficiently than electricity. The cost of electrical transmission over large distances is high. Hydrogen transport by pipeline would be much more efficient and less expensive. The areas with the greatest amount of sunlight, where solar plants would be most efficient, are often far from centers of population. Transmission problems for ocean-based generating plants are even more severe. Instead of electricity, remote plants could generate hydrogen, which could be shipped or piped to urban

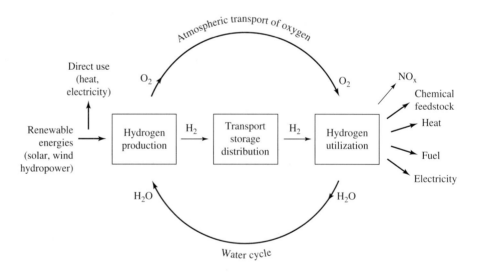

Figure 1.49 A solar-based hydrogen economy. *Source:* J. M. Ogden and J. Nitsch (1993). Solar hydrogen. In *Renewable Energy, Sources for Fuels and Electricity,* T. B. Johansson et al., eds. (Washington, DC: Island Press). Copyright © 1993 by Island Press. Reprinted with permission.

centers. These considerations have led to the concept of the hydrogen economy shown in Figure 1.49, in which hydrogen gas would become the main energy currency. It would be consumed directly for electrical generation and heating, either by combustion or by fuel cells. For transportation, it could also be used directly, via fuel-cell electric vehicles, or it could be used to synthesize liquid fuels by chemistry similar to that applied for coal conversion to methanol and liquid hydrocarbons (p. 31).

Hydrogen is often thought to be a particularly hazardous substance. Ever since the Hindenburg disaster of 1939, when a dirigible filled with hydrogen caught fire and crashed, killing all on board, hydrogen has haunted the popular imagination. But hydrogen is not particularly more dangerous than natural gas or gasoline. The lower flammability limit, the minimum percentage of fuel in air that can sustain a fire, is only a little lower for hydrogen, 4%, than it is for methane, 5%, and is substantially higher than for gasoline, 1%. Since it is much lighter than air, hydrogen disperses rapidly if there is a leak. Methane does also, but gasoline fumes, being heavier than air, tend to accumulate in the vicinity of a leak and are more likely to catch fire. There is actually a history of hydrogen use in home heating, because, before the widespread availability of natural gas, many utilities supplied "town-gas," manufactured from coal or wastes, which was rich in hydrogen. Process industries have long used hydrogen, and hydrogen pipelines several hundred kilometers long have operated safely in Germany, England and the United States.

Despite this experience in industrial settings, the widespread availability of hydrogen will no doubt be some time in coming, and the introduction of transfer facilities at vehicle filling stations will take much longer. One can imagine intermediate stages in the development of the hydrogen economy that would smooth the transition. There might initially be increased utilization of natural gas, as petroleum stocks dwindle and the pressure for utilization of clean fuels increases. The methane could be burned directly, in homes, power plants, and in new gas-powered vehicles, or it could be converted to methanol, which could serve as an intermediate fuel in fuel-cell vehicles with onboard reformers. Methanol from coal and from biomass could add to the liquid fuel supply. Eventually, the pipelines used for natural gas could be converted to hydrogen transport, as solar hydrogen becomes more feasible and demand for hydrogen increases.

4. Systems Efficiency: Transportation, Materials, and Recycling

To better understand the possibilities for increased energy efficiency, we need a more comprehensive view of the way energy is utilized in society. Figure 1.50 is a diagram of how energy actually flowed through the U.S. economy in 1990. On the left side are the inputs from coal, petroleum, natural gas, nuclear, and hydropower, while on the right side, the end uses are broken down into residential and commercial, industrial, and transportation categories. The units are $10^{14} \times kJ$. In 1990 the United States consumed 858.5 of these units of energy but produced only 684.4 from its own resources; the balance was made up mainly of petroleum imports, which accounted for 46% of the petroleum used. Electric utilities used 311.9 units, over one-third of total consumption, producing 97.3 units of electricity, which was split about 1:2 between industrial and residential plus commercial uses.

Nearly 70% of the energy consumed by the electric utilities are "lost" as waste heat, due in part to the inherent inefficiency of heat engines, as discussed above. But this is far

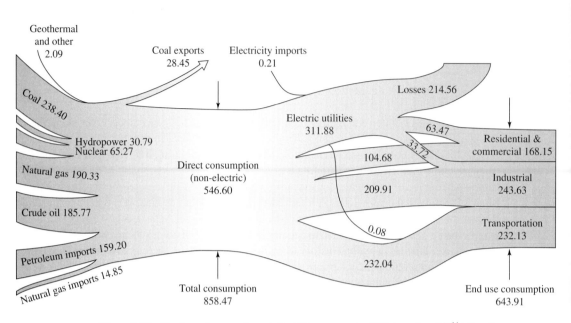

Figure 1.50 The flow of energy through the U.S. economy in 1990 (in units of 10^{14} kJ). *Source:* Energy Information Administration (1991). *Annual Energy Review 1990* (Washington, DC: U.S. Department of Energy).

from the only loss in the system. Much of the end-use energy is also lost, in the sense that it does not accomplish its intended use before ending up as waste heat. A liter of heating oil, for example, produces exactly its heating value, wherever it is burned, but the effect on the house being heated depends on the efficiency of the furnace and heating system, on the size of the house, and on how well it is insulated. Two houses of similar size, side by side, can require quite different amounts of oil or gas to achieve the same indoor temperature. Likewise, the amount of electricity used for heat, light, and appliances depends on the efficiency of electricity conversion to the intended use. For example, compact fluorescent light bulbs are now available, which provide light equivalent to a traditional tungsten bulb with a fraction of the electric current requirement, and sensors are available that turn lights off when no one is in a room. Clearly, the magnitude of the end-use energy requirement for residential purposes could be less than it now appears to be, if the residents adopted energy-saving measures. Much has, in fact, been accomplished in this direction, with improved building codes and enhanced efficiencies in appliance designs, for example, but there is still considerable room for improvement. Likewise, there are many opportunities in the industrial and transportation sectors to improve energy efficiency and reduce the energy utilization rate. A useful strategy is "cogeneration," in which the "waste" heat from electricity generation is put to use in industrial or residential heating. Since heat cannot be transported efficiently over long distances, cogeneration is often not practical for large central power stations, but it could become more feasible with modular distributed power, based on gas turbine, solar thermal, or fuel cell sources, for example.

All of the energy flowing through the economy eventually ends up as waste heat. The question is what fraction of it actually accomplishes human ends on its way to its entropic destination. This fraction determines just how much energy we actually need.

a. Transportation.   Transportation is an especially important energy sector, not only because of its high rate of energy consumption (as seen in Figure 1.50, it accounts for more than one-third of end-use consumption), but also because of the international economics and politics of oil. About half of the world's oil production goes into transportation, while about 40% is used in space heating and industrial processing (including production of petrochemicals), and about 10% in electricity production.

Table 1.11 lists the common forms of transport and their energy intensities (the amount of energy required to transport a passenger or a ton of freight, for a given distance). For freight, there are large disparities in the energy intensity. Transport by ship, rail, or pipeline takes far less energy per ton of material per kilometer of travel than does transport by truck, which in turn takes far less energy than transport by airplane. Of course, these different transport modes are appropriate to different kinds of goods of different value and perishability,

TABLE 1.11   EFFICIENCIES OF FREIGHT AND PASSENGER TRANSPORT

		Passenger (1990)	
Mode	Activity unit	Activity (billions)	Energy intensity (10^2 kJ/unit)
Bus	PKT	35.6	21.6*
Rail	PKT	30.9	25.6
Auto	VKT	2013	39.3
Light truck	VKT	822	54.4
Airplane	PKT	826	37.4[†]
High speed rail	PKT	0	9.8[†]

		Freight (1990)	
Mode	Activity unit	Activity (billions)	Energy intensity (10^2 kJ/unit)
Truck	TK	2564	20.3
Rail	TK	1213	3.2
Ship	TK	1243	2.9
Airplane	TK	11.4	135.8
Pipeline	TK	1236	1.9

PKT = passenger kilometers traveled per year;   VKT = vehicle kilometers traveled per year

*assumes average load for urban transit bus is 17

[†]assumes average load factor is 0.61

TK = ton kilometers per year

Source: Union of Concerned Scientists et al. (1992). *America's Energy Choices: Investing in a Strong Economy and a Clean Environment, Technical Appendixes,* Appendixes C and D (Cambridge, Massachusetts: The Union of Concerned Scientists).

and are not freely interchangeable. Nevertheless, the growth of trucking, and especially of air freight, is a significant factor in the transportation energy demand. For passenger transport, the disparities are not as pronounced as for freight. It takes nearly twice as much energy per kilometer traveled to carry people by airplane than by bus, at average loadings. For trains, the energy intensity is about the same as buses, but drops to half as much for high-speed rail. For autos and light trucks, the intensity depends greatly on the number of passengers. The values in Table 1.11, which exceed the intensity for airplanes, are for vehicle kilometers traveled. In 1990, the average number of passengers in U.S. autos was 1.6, a loading that brings the energy intensity for autos into the range for rail and buses. Therefore, the choice of transport mode makes less difference to total energy consumption than is often supposed. What matters most is level of occupany, and how far and how often people travel. However, this comparison leaves out of account systemwide effects. For example, an effective mass-transport system might significantly reduce traffic congestion, thereby improving the energy efficiency of all forms of transport in the region.

The amount that people in the United States traveled in the different transport modes is plotted in Figure 1.51 for 1970–1987. The share of bus and rail travel (indicated as *mass transit*), already low by international standards, fell from 2.4% in 1973 to 1.9% in 1987, while the share of travel by air and by light truck increased substantially, accounting for the 2.2% annual increase in total passenger travel during the period. Travel by car stayed roughly constant, but it still dominates all forms of travel, accounting for 65% of the total. In the world

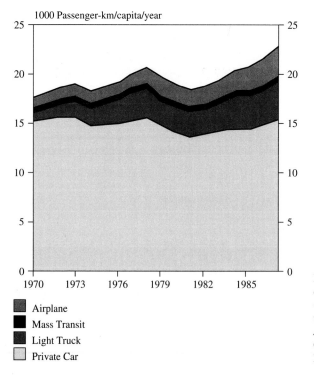

Figure 1.51   Passenger transportation per capita by mode.   *Source:* L. Schipper et al. (1990). United States energy use from 1973 to 1987: the impact of improved efficiency. *Annual Review of Energy* 15: 455–504. Copyright © 1990 by Annual Reviews Inc. Reproduced with permission.

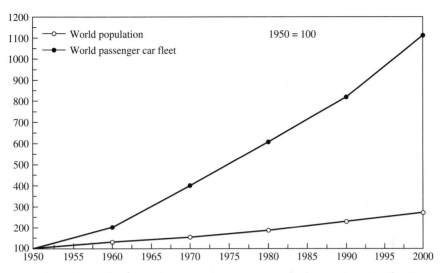

Figure 1.52   Comparison between population growth and passenger car growth globally since 1950.   *Source:* M. K. Tolba et al., eds. (1992). *The World Environment, 1972–1992, Two Decades of Challenge* (London: Chapman & Hall on behalf of The United Nations Environment Programme).

overall, the automobile is also becoming the dominant travel mode. Figure 1.52 shows that the world car fleet increased by more than a factor of eight between 1950 (when about 55 million cars were in use) and 1990, outstripping the growth in population by a factor of four.

Consequently, the energy efficiency of the automobile is a critical issue of global dimensions. Figure 1.53 shows that the energy efficiency or, equivalently, the gasoline use per distance traveled, has significantly improved for the United States following the 1973 Arab oil embargo, although this improvement has since leveled off (not shown in the figure), as oil prices have continued their decline from the mid-'80s. Average car efficiency is still lower than in Japan or Europe. In these regions the efficiency improvement has been less marked because they were more efficient from the start. In general, today's cars perform much better than those of twenty years ago, thanks to continuing improvements in design and materials.

Still, there is substantial room for further improvements. Figure 1.54 is a diagram of the energy losses in a typical car between the fuel tank and the wheels. We see that the fraction of the fuel energy actually delivered to the driveline is only 26% on highways, and 18% in urban driving, while the energy delivered to the wheels is even lower, 20% and 13%, respectively. Although engine design is already well-advanced, these losses could be cut substantially with improved materials that permit higher-temperature operation. Alternatively, electric motors offer greatly improved energy conversion efficiency, by factors of two or three, once storage batteries and/or fuel cells become viable. In addition, there is room for improvement in cutting the losses due to aerodynamic drag and to braking. But the biggest effect on fuel efficiency is in the weight of the car. Most of the improved efficiency in U.S. cars came from reducing their weight while maintaining adequate interior space, by using lighter and stronger materials.

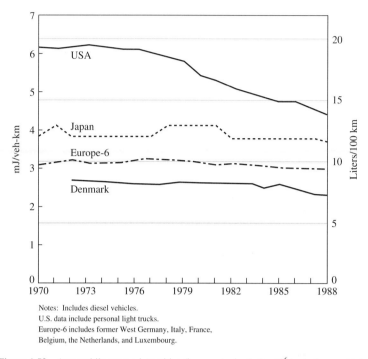

Notes: Includes diesel vehicles.
U.S. data include personal light trucks.
Europe-6 includes former West Germany, Italy, France,
Belgium, the Netherlands, and Luxembourg.

Figure 1.53   Automobile energy intensities (in *megajoules* (mJ = 10^6 joules) per vehicle-kilometer), and fuel economy (in liters per 100 kilometers) from 1970 to 1988.   *Source:* L. Schipper and S. Meyers (1992). *Energy Efficiency and Human Activity: Past Trends, Future Prospects.* (Stockholm: Stockholm Environmental Institute). Reprinted with permission from Stockholm Environmental Institute.

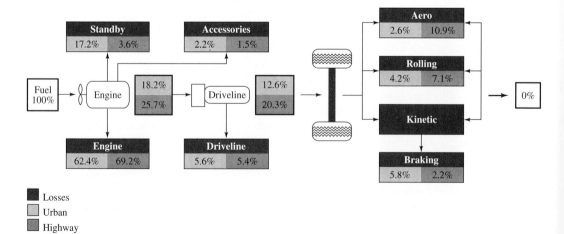

Figure 1.54   Energy losses in automobile transport point to strategies for improving fuel efficiencies in automobiles.   *Source:* D. L. Illman (1994). Auto-makers move toward new generation of "greener" vehicles. *Chemical and Engineering News* 72(31): 8–16. Copyright ©1994 by American Chemical Society. Reprinted with permission.

b. Materials.   The development of stronger, lighter, more durable materials has an enormous impact on the efficiency of energy utilization. In the case of cars, the substitution of strong but lightweight plastics and composites (mixtures of different structural elements such as glass or carbon fibers with resins, to increase strength) for metal body parts has decreased the weight required for the same carrying capacity. New materials have decreased the weight of many consumer and industrial products, thereby decreasing the energy costs for their transportation, and frequently for their manufacture as well. This trend is augmented by miniaturization of many products, also made possible by advanced materials. More durable materials also mean longer life for the products and a lower production rate. These trends are sometimes called the "dematerialization" of industrial societies, as advanced materials and better information make it possible to make do with less stuff.

In addition, materials able to withstand high temperatures have a direct impact on energy efficiency by improving the performance of heat engines. The most notable example is the steady improvement of jet engines for aircraft, through the development of alloys and ceramics that permit the engines to operate at higher temperatures. These engines have improved aircraft fuel economy substantially. The same technology, in the form of gas turbines, is now being introduced to improve power plant efficiencies (pp. 79–80).

The choice of materials can also affect energy efficiency in more mundane ways. For example, some communities have banned disposable styrofoam containers because of concerns that, being nonbiodegradable, discarded styrofoam leaves unsightly trash and fills up overburdened landfills. Whether styrofoam is actually inferior to other disposable materials, principally paper, is a matter of debate, since biodegradation of paper can be quite slow, especially in sanitary landfills, and since styrofoam is probably more amenable to recycling, although not much of it is actually being recycled at the moment. But an important consideration was left out of the early debates on paper versus styrofoam, namely the environmental costs of producing the two materials. Table 1.12 compares energy and water use as well as emissions for manufacturing paper and styrofoam cups. When the values are compared on a *per cup* basis, styrofoam is found to have considerably lower environmental impacts (see problem 23, Part I). The main reason for the difference is that a styrofoam cup of a given capacity weighs less than a sixth as much, on average, as a paper cup. Styrofoam is a stronger material than paper, especially because, being a hydrocarbon material, it is not wet by aqueous liquids. In contrast, paper, which is made of cellulose, a molecule covered with hydroxyl groups, interacts with water via hydrogen bonds, and is gradually dissolved (see discussion of water and hydrogen bonding, pp. 199–200). Consequently, the paper cup requires more material to maintain its integrity while in use, and its production has a much larger impact on energy use and the environment.

c. Recycling.   The recyclability of paper versus polystyrene is just one of the many complex issues around recycling. For most people, recycling arises in the context of solid waste disposal. The more the trash is recycled, the less there is to dump in landfills, which are rapidly being filled, and the less pressure there is to build incinerators as an alternative disposal method. But there are many barriers to recycling—political, economic, and technical. The technical barriers can be thought of in terms of the second law of thermodynamics. When materials are mixed together in products, and then mixed again when discarded products are mingled in trash, entropy is increased. Un-mixing the materials requires a decrease

TABLE 1.12   RAW MATERIAL, UTILITY, AND ENVIRONMENTAL SUMMARY FOR HOT DRINK CONTAINERS

Item	Paper cup*	Polyfoam cup†
Per cup:		
Raw materials		
Wood and bark	25 to 27 g	0 g
Petroleum fractions	1.5 to 2.9 g	3.4 g
Other chemicals	1.1 to 1.7 g	0.07 to 0.12 g
Finished weight	10.1 g	1.5 g
Per metric ton of material:		
Utilities		
Steam	9000 to 12,000 kg	5500 to 7000 kg
Power	980 kWh	260 to 300 kWh
Cooling water	50 m^3	130 to 140 m^3
Water effluent		
Volume	50 to 190 m^3	1 to 4 m^3
Suspended solids	4 to 16 kg	0.4 to 0.6 kg
BOD	2 to 20 kg	0.2 kg
Organochlorines	2 to 4 kg	0 kg
Metal salts	40 to 80 kg	10 to 20 kg
Air emissions		
Chlorine	0.2 kg	0 kg
Chlorine dioxide	0.2 kg	0 kg
Reduced sulfides	1 to 2 kg	0 kg
Particulates	2 to 3 kg	0.3 to 0.5 kg
Chlorofluorocarbons	0	0‡
Pentane	0 kg	35 to 50 kg
Sulfur dioxide	~ 10 kg	3 to 4 kg
Recycle potential:		
To primary user	Possible. Washing can destroy.	Easy. Negligible water uptake.
After use	Possible. Hot melt adhesive or coating difficulties.	Good. Resin reuse in other applications.
Ultimate disposal:		
Proper incineration	Clean	Clean
Heat recovery	20 MJ/kg	40 MJ/kg
Mass to landfill	10.1 g/cup	1.5 g/cup
Biodegradable	Yes, BOD to leachate, methane to air.	No. Essentially inert.

*Uncoated fully bleached kraft paper cup.

†Molded polystyrene foam bead (seamless) cup.

‡Many producers of foamable beads have never used CFCs.

Source: Updated and adapted by M. B. Hocking, from original article in M. B. Hocking (1991). Paper versus polystyrene: a complex choice. *Science* 251: 504–505.

in the entropy of the materials, and this requires the input of energy. The trash has to be collected and sorted (the degree of cooperation of the populace in presorting the trash is a key variable), and the materials in the sorted products may have to be separated mechanically or chemically. The difficulty depends on the product and on the material. For example, lead-acid batteries are recycled at a high rate. They are easy to collect, and the lead is readily extracted. Aluminum cans are also recycled at a high rate. They are easily separated from trash and contain little except aluminum; they can be said to have relatively low entropy. Plastic consumer products are another matter. Although pure plastics can be readily recycled, many different plastics are mixed in trash; often a given product contains more than one plastic. Separating them completely may be prohibitively costly, and often they cannot be processed together. For example, a little polyvinylchloride, the ingredient of plastic films, can ruin the recyclability of polyesters, the ingredient of soft-drink bottles. Another recycling issue is the quality of the recycled product. Paper is recycled in substantial quantities, but the cellulose fibers are degraded in the process and lose strength; they can be recycled no more than four times, before dissolving completely.

Despite these problems, recycling has a significant impact on energy efficiency. Although energy is required to restore materials that have been dispersed, this energy is likely to be substantially less than the energy required to produce the materials in the first place. The bar graph in Figure 1.55 shows the amount of energy required to produce a metric ton of steel, paper, and aluminum from primary materials, compared with the energy required for production from recycled materials. We see that aluminum production from ore is particularly energy-intensive. Recycled aluminum requires only 5% as much energy to process as primary aluminum ore. Even so, enough aluminum is thrown away in the United States

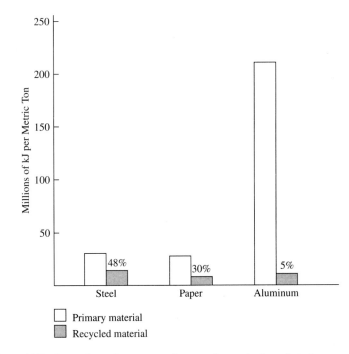

Figure 1.55 Comparison of energy requirements for production of steel, paper, and aluminum from primary and recycled materials.

to rebuild its commercial aircraft fleet every three months. The savings from using steel scrap and paper waste are not as high—52% and 70%, respectively. Nevertheless, the potential for energy conservation is substantial when we consider the volume of these materials produced annually. The production of steel, aluminum, and paper consumes more than 20% of the total industrial energy expenditure in the United States.

5. Energy and Well-Being

It is useful at this point to ask how much energy we actually need. The general assumption has been that increasing energy consumption means an increased standard of living. To evaluate this assumption, we can examine Figure 1.56, which is a plot of per capita energy consumption against per capita gross national product for many countries. There is a rough correlation between the two, with poor countries clustered near the bottom on both scales. At higher GNP, however, the graph fans out. Canada and the United States use twice as much energy per capita as does Denmark, although all three have the same high GNP per capita, ~$20,000. Japan and Switzerland have even higher GNP per capita, yet use energy at

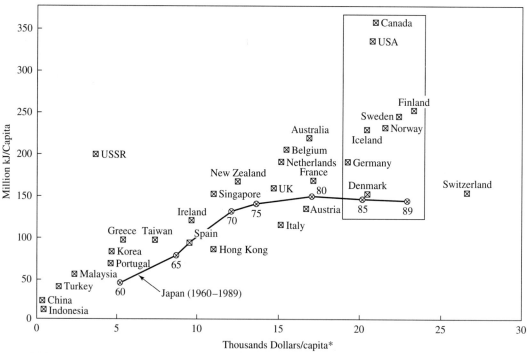

* Calculated using 1989 prices and 1989 exchange rates for U.S. dollars

Figure 1.56 Energy use per capita versus gross national product per capita. *Sources:* Organization for Economic Cooperation and Development (OECD) (1991). *National Accounts, Main Aggregates, vol. 1, 1960–1989* (Paris: OECD); International Energy Agency (1991). *Energy Balances of OECD Countries, 1980–1989* (Paris: OECD).

the same per capita rate as Denmark. For highly developed countries, it is apparent that more GNP does not necessarily require more energy. This point is illustrated also by the history of Japanese energy use, shown as the solid curve in the figure. From 1960 to 1970 Japanese energy utilization increased steadily in proportion to the GNP. But after 1973, energy use leveled off while GNP kept on growing. Thus the Japanese energy *intensity*, defined as energy consumption per unit of GNP, has been decreasing since 1973. This has been the experience of other economically advanced countries as well. Even the U.S. energy intensity is lower now than it was in 1973, although it is still much higher than it is for Denmark, Japan, and Switzerland.

The discussion in the preceding sections suggests that there are many opportunities for decreasing energy intensity even further. Just how far can one expect to go in this direction? No one knows the answer to that question, but it is possible to sketch some energy scenarios for a variety of economic and technological assumptions. Figure 1.57 shows the result of one such exercise for the United States, carried out by a group of conservation-minded

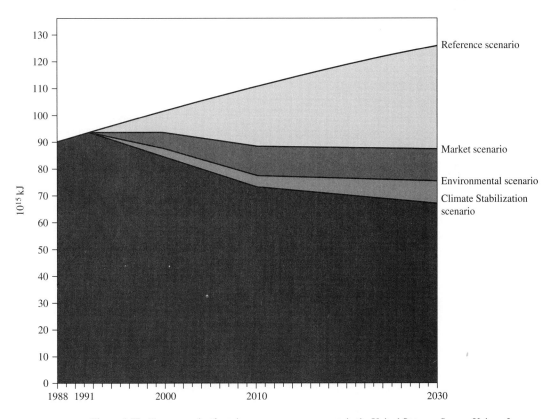

Figure 1.57   Four scenarios for primary energy requrements in the United States.   *Source:* Union of Concerned Scientists et al. (1991). *America's Energy Choices: Investing in a Strong Economy and a Clean Environment* (Cambridge, Massachusetts: Union of Concerned Scientists). Reprinted with permission from Union of Concerned Scientists.

organizations. The "reference" scenario, which anticipates a one-third energy consumption increase by the year 2030, is based on current policies, practices, and trends adapted from U.S. government analyses. Against this are set three alternative scenarios that anticipate a decrease in current energy utilization, despite the assumption of steadily increasing GNP (about 2.1% annual growth in all four scenarios). The "market" scenario anticipates the introduction of cost-effective energy efficiency and renewable energy technologies, while the "environmental" scenario adds the environmental and security costs of fossil fuels in establishing cost-effectiveness for new technologies. Finally, the "climate stabilization" scenario is designed to achieve a 25% reduction in CO_2 emissions by 2005 and a 50% reduction by 2030. In addition to the lowered total energy requirement, the three alternative scenarios project substantially greater adoption of renewable energy than does the reference scenario (Figure 1.58), thereby achieving greater improvement in emissions reduction and climate stabilization.

Are these scenarios realistic? The authors of the report forecast substantial savings from all three alternative scenarios after subtracting the initial investment required for the in-

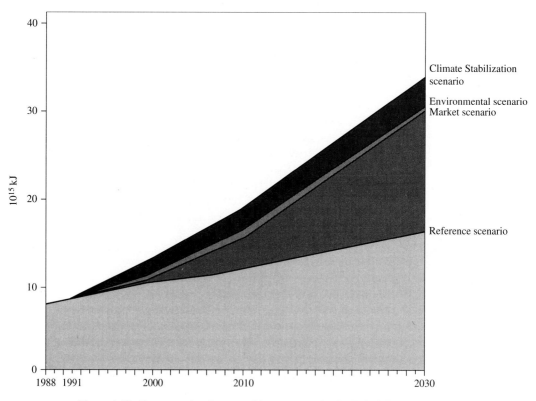

Figure 1.58 Four scenarios for renewable energy use in the United States. *Source:* Union of Concerned Scientists et al. (1991). *America's Energy Choices: Investing in a Strong Economy and a Clean Environment* (Cambridge, Massachusetts: Union of Concerned Scientists). Reprinted with permission from Union of Concerned Scientists.

troduction of new technologies. They recommend a series of policies to promote these scenarios, such as providing tax credits and eliminating regulatory barriers to the introduction of new technologies, taxing pollution, setting standards for vehicle and building efficiency, and promoting least-cost planning on the part of utility commissions and regulatory bodies.

Others may find fault with the numbers and assumptions in the analysis. In any event, the standard economic argument is that if savings are as evident as the analysis suggests, then investment will follow automatically without intervention or planning. But markets do not work ideally without full dissemination of information and when there are pre-existing regulatory and tax disincentives, as there are in the energy field. Moreover, despite much study and discussion, there is no agreed-on method of including environmental costs in profit and loss analyses. Thus, energy policy will remain in the political arena for a long time to come, and it will be necessary for citizens to stay abreast of developments in energy technology and policy.

F. SUMMARY

Our survey of sources has shown that energy is supplied to us in abundance by the sun, and this steady energy flow is sufficient to maintain the human population at a steady state, which we must, in any event, achieve over a period of time. Capturing the sun's energy and putting it to useful work are challenges to science and technology. At the moment it is much more expensive to do this than to tap the solar energy that has been stored over the millennia in fossil fuels, which we have used to develop our industrial civilization. Today we are completely dependent on fossil fuels and particularly on petroleum and natural gas. These fuel stocks are being drawn down and the easily accessible deposits will be depleted in a matter of decades for oil and gas and perhaps in a few centuries for coal. In the meantime, the development of atomic science has provided the key for unlocking the very large amounts of energy stored in the nuclei of uranium atoms and potentially in those of hydrogen atoms. The development of nuclear power on a large scale, however, poses unprecedented problems for humanity in terms of the dangers of nuclear weapons proliferation and the long-term hazards associated with massive amounts of radioisotopes.

On the supply side of the energy equation, the key questions are to what extent coal and nuclear fuels can be used safely with acceptable environmental costs, and how long it will take to introduce alternative technologies involving solar power and possibly the fusion reaction. Equally important questions exist on the demand side of the energy equation. A close examination of how energy is actually used shows that significant savings are possible in situations where energy consumption has been predicated historically on the availability of a cheap and plentiful supply. Energy savings have a greater impact on the equation than energy supplies because of the low efficiency with which fuels are used. Each joule of electricity that is not used represents a saving of 3 joules of oil, coal, or nuclear fuels. Savings on the demand side of the energy equation are desirable because they increase the range of choices with respect to the energy supply. The lower the projected energy requirements, the greater the flexibility for providing energy from a variety of sources, and the more time there is to develop the safest, most efficient, and least environmentally harmful technologies.

PROBLEM SET I: ENERGY

1. Given that evaporation of 1 ml of water requires 2.46 kJ of energy at abient temperatures, how much energy is released in a 2-cm rainfall over an area 10 by 10 km? If 1 ton of TNT releases 4.18 $\times 10^6$ kJ of energy, how many tons of TNT would be equivalent to the energy released in the rainfall?

2. How efficient is photosynthesis? Annual photosynthesis is estimated to produce an average 320 g (dry weight) of plant matter per m^2, 50% of which is carbon.

   (a) Calculate the total grams of carbon "fixed" each year as plant matter by an area 1000 m^2.

   (b) The photosynthetic reaction can be represented by the production of glucose:

$$6CO_2 + 6H_2O \rightarrow C_6H_{12}O_6 + 6O_2 \tag{1}$$

   Of the glucose produced, 25% is used by the plant to fuel respiration; the rest is converted to plant matter. Given the answer to part (a), calculate the total moles of glucose produced annually in our reference 1000 m^2 area through reaction (1).

   (c) Each mole of glucose produced represents the absorption of 2,803 kJ of solar energy. If the average energy available from sunlight is 1.527 kJ/cm^2 per day, what percentage of the incident solar energy is converted to chemical energy in our 1000 m^2 area?

   (d) Would this area produce more energy via biomass, or via solar collectors, assuming a 15% collector efficiency?

3. If world energy consumption continues to increase at the present annual rate of 2.8%, how long will it take to double? At this rate of increase, how long would it be before human energy consumption equals 1% of the solar energy absorbed at Earth's surface (equal to 25.1 $\times 10^{20}$ kJ/yr)? Assume that present world energy consumption is 36.8 $\times 10^{16}$ kJ/yr.

4. Some people worry that we will use up Earth's oxygen supply if we burn all the fossil fuel. Are they right to worry? To decide this question calculate how many moles of O_2 would be used up, given that the energy available in recoverable fossil fuels is estimated to be 5.3 $\times 10^{19}$ kJ, and that about 335 kJ are released per mole of O_2, on average, when fossil fuel is burned. Now calculate what fraction of the atmosphere's O_2 this represents from the following facts: the atmosphere weighs 1000 g for each square centimeter of Earth's surface, and is 22% O_2 by weight. The radius of Earth (r) is 6.4 $\times 10^6$ m.

5. The major component of natural gas used by utility companies is methane, CH_4; the gas used in gas barbecues, etc., is propane, C_3H_8. Assuming that each of these gases burns completely, compare the amount of energy released by each a) in terms of kJ/mole of CO_2 produced, and b) in terms of kJ/g of fuel. Do the calculations based on chemical bonds broken and formed in the combustion reactions. Use Table 1.3 to obtain bond energies.

6. (a) Compare the costs per unit of energy for electricity and gasoline. Assume current prices are $0.05 per kWh for the former and $1.50 per gallon for the latter. Gasoline weighs 5.51 lb/gal, and its energy release on combustion is 19,000 Btu (British thermal units) per pound. Energy units: 1 calorie = 4.18 joules = 1.16 $\times 10^{-6}$ kWh = 3.97 $\times 10^{-3}$ Btu.

   (b) Taking into account that gasoline engines are 20–25% efficient and that electric cars are 50–80% efficient, which is more economical for transport with respect to fuel costs?

7. (a) Consider that in the southwestern United States the average insolation is 270 W per m^2. Calculate the solar energy per m^2 per year in kWh. Given that the energy in a kilogram of coal is 8.14 kWh, the solar energy per m^2 per year is equivalent to how many kg of coal? Calculate the electrical energy that can be produced from a 1-meter flat-plate photovoltaic cell, assuming an ef-

ficiency of 15%. Calculate the electrical energy produced in a coal-fired power plant from the coal equivalent to the solar energy per m^2 per year, assuming an efficiency of 33%.

(b) Assume that the coal contains 72% carbon and 2% sulfur. After conversion to electrical energy via insolation and coal combustion, calculate and compare the residual flows of mass and energy from the two sources. (Base calculations on m^2 per year of solar energy, and its coal equivalent.)

(c) Comment on the relative merits of the two sources of electricity.

8. The half-life of ^{14}C is 5730 years.

(a) How old is a wooden bowl whose ^{14}C activity is one-fourth of the activity of a contemporary piece of wood?

(b) At Stonehenge, a charcoal sample is dug up, presumably the remains of a fire; its ^{14}C activity is 9.65 disintegrations per minute per gram of carbon. Living tissues have a ^{14}C activity of 15.3 disintegrations per minute per gram of carbon. When were the fires of Stonehenge ignited?

9. Describe three major environmental and security problems associated with nuclear power.

10. Plutonium is very damaging when inhaled as small particles because of the ionization of tissue by its emitted alpha particles. One microgram of plutonium is known to produce cancer in experimental animals. From its atomic weight (239) and half-life (24,360 years), calculate how many alpha particles are emitted by a microgram of plutonium over the course of a year. About how many ionizations do the particles produce?

11. A neutron generated by fission typically possesses a kinetic energy of 2 Mev. When such a neutron collides with a hydrogen atom 18 times, its energy is reduced to its thermal energy of 0.025 ev, that is, the kinetic energy it would possess by virtue of the temperature of its surroundings. The same neutron would have to collide with a sodium atom more than 200 times to reduce its energy by the same amount. Explain how these characteristics make water a suitable coolant in the pressurized light-water reactor that utilizes U-235 as the fuel, whereas sodium is the suitable coolant in the breeder reactor.

12. How does a breeder reactor extend the supply of nuclear fuel? What are the problems associated with the breeder design?

13. From the standpoint of weapons proliferation, why is it more dangerous to fuel reactors with plutonium than with uranium in which U-235 is enriched to 2–3%?

14. Considering the ^{238}U decay scheme, which of the daughters of uranium are likely to be most abundant in uranium-bearing soil, and why?

15. Any large-scale nuclear process produces radioactive waste. Compare the problems of waste from uranium fission reactors and from tritium fusion reactors.

16. Should you install a solar water heater? The average home has a 200-liter (50–60 gallon) hot-water tank, which is effectively drained and replenished three times a day. Assume that the entering tap water is 15°C and is heated to 55°C.

(a) Given an average energy from sunlight of 1.53 kJ/cm^2 per day, how large would the collection area of a solar water heater need to be if its efficiency is 30%?

(b) Assume that the price of a solar collector is $375/m^2. How much would it cost to install the hot-water system in (a)?

(c) If the price of oil remains at $0.75/liter for 20 years, how much would the solar collector save? Assume that the heating content of oil is 2.51×10^4 kJ/liter and it can be burned with 90% efficiency.

17. **(a)** Assume a wind turbine with a hub 50 meters above the ground, a rotor diameter of 50 meters, and a wind-conversion efficiency of 25%. The turbine operates in an area with an annual average wind-power density of 500 watts/m^2 at 50 meters altitude. How much electrical energy (in kWh) can the turbine generate per year?

(b) Wind densities equal to or greater than 500 watts/m^2 at an altitude of 50 meters are exploitable with today's technologies; about 1.2% of the land area of the contiguous United States possesses such wind densities. If, on average, windfarms contain eight turbines per km^2, what is the U.S. potential for electrical-energy production from wind power? (Assume a uniform power density of 500 watts/m^2, and the same specifications as for the turbine in part (a).) In 1990, 2,745 TWh of electricity were produced in the United States. What percentage of U.S. electricity consumption could be met by wind power?

(c) Continuing technical advances will allow wind-power generation on lands where the wind power density is 300 watts/m^2 at 50 meters. In this case, wind power could be harvested on 21% of U.S. land. Assuming that one-third of the land were covered by windfarms (again with a density of eight turbines per km^2), how much electricity would be generated? What percentage of the U.S. demand could be met? (Assume a uniform power density of 300 watts/m^2, and wind turbines with the same design as in part (a).) (The area of the contiguous U.S. is 7,827,989 km^2; 1 TWh = 10^9 kWh; 1 kWh = 3.6×10^3 kJ.)

18. Assume that due to dwindling supplies of crude oil and natural gas, the United States embarks on a plan to implement the "hydrogen economy," for which hydrogen will be generated by electrolysis of water. The source of electricity is from photovoltaic conversion of sunlight in the southwestern United States.

    (a) Assume that a fixed, flat-plate PV system is used, with a solar conversion efficiency of 15%, and a hydrogen-production efficiency of 80%. Assume that the average annual insolation in the southwest is 270 watts per m^2. Calculate the annual electrical energy produced per m^2 in kWh and in kJ. Calculate the energy content of the H$_2$ produced per m^2, as well as the number of moles, and the weight of H$_2$.

    (b) The United States consumed 34.5×10^{15} kJ of petroleum in 1990. How many square meters of PV collectors would be needed to supply the equivalent amount of energy? The area of the southwest (the states of New Mexico, Arizona, Colorado, Utah, and Nevada) is 1,386,370 km^2. What percent of the land area would be covered by PV collectors?

    (c) How much water per year would be required to produce the hydrogen? Assume a conversion efficiency of 80%. What percentage of total national water use would this correspond to, considering that in 1990, water use in the United States was 4.7×10^{14} liters.

19. An electric hot-water heater has a standard efficiency rating of 90% (that is, for every 10 kJ of electricity consumed, 9 go into heating the water). If the heater normally heats the water from the ambient temperature, 20°C (68°F), to 80°C (176°F), what is the *second law* efficiency (that is, what is the ratio of the amount of energy that an ideal heat pump would use to do the same job, to the amount of energy actually consumed)?

20. Calculate the theoretical efficiency of a heat pump when the room temperature is 20°C (68°F) and the outside temperature is –20°C (–4°F). If the heat pump were drawing heat from a water tank at 5°C (41°F), what would be the theoretical efficiency? Describe how solar energy in conjunction with a heat pump can provide an efficient means of heating a house.

21. Petroleum consumption in the United States in 1990 supplied about 34.5×10^{15} kJ of energy. Of this quantity, about 15.9×10^{15} kJ (46%) was supplied by foreign sources. The transportation sector is the major user of petroleum, accounting for 23.2×10^{15} kJ of consumption (about 67% of the total). Overall automobile fuel economy in the United States was about 21 miles per gallon (mpg) in 1990. Automobile efficiencies in Europe and Japan were about 27 mpg. If the U.S. fleet had averaged 27 mpg in 1990, how much less petroleum would have had to be imported? Prototype cars exist with efficiencies of 100 mpg or more. If the U.S. auto fleet had averaged 100 mpg in 1990, how much less petroleum would have been used?

22. A large power plant is being constructed that will produce 6.7×10^{10} kJ/day of electrical energy. For every kJ of energy produced, 2 kJ of waste heat is discharged. If a plant draws 2×10^9 liters/day of river water at 20°C into its cooling condensers, calculate the rise in temperature of the cooling water. If the river has a flow rate of 10^{10} liters/day, estimate the rise in temperature of the water downstream from the plant.

23. **(a)** As presented in Table 1.12, making a metric ton of paper requires 980 kWh of energy, whereas making a metric ton of polystyrene requires about 300 kWh. Given that an average 8 oz. paper cup weighs 10.1 g and an average polystyrene cup weighs 1.5 g, what is the ratio of the power requirements on a per-cup basis?

(b) The table also lists the amount of heat recovered from incinerating each type of cup; 20 MJ/kg (megajoule per kilogram) for paper; 40 MJ/kg for polystyrene. The heat could be converted to electricity at a power plant with an efficiency of about 30%. Compare the amount of electrical power available from incinerating discarded paper and polystyrene cups with the amount of energy needed to produce them. (1 kWh = 3.6×10^6 joules; 1 MJ = 10^6 joules.)

(c) A fast-food restaurant on a busy street uses polystyrene cups to serve coffee at a rate of 2.5 gross/day (1 gross = 12 dozen = 144). Reacting to pressure from an environmental advocacy group, the restaurant switches over to using paper cups. What is the effect of this decision in terms of kWh of power per week, if the town has no incinerator? What is the effect if the town sorts and incinerates paper and plastic for electricity production?

(d) Write a paragraph explaining to the proprietors of the restaurant in question (c) what they might do if they want to be as environmentally responsible as possible.

SUGGESTED READINGS I: ENERGY

Energy for Planet Earth (September, 1990). Special edition of *Scientific American* 263(3).

M. K. TOLBA, O. A. EL-KHOLY, E. EL-HINNAWI, M. W. HOLDGATE, D. F. MCMICHAEL, and R. E. MUNN, eds. (1992). *The World Environment 1972–1992, Two Decades of Challenge* (London: Chapman & Hall on behalf of The United Nations Environment Programme).

R. H. SOCOLOW, D. ANDERSON, and J. HARTE, eds. (annual reports). *Annual Review of Energy and the Environment* (Palo Alto, California: Annual Reviews Inc.).

Union of Concerned Scientists, Alliance to Save Energy, American Council for an Energy-Efficient Economy, Natural Resources Defense Council (1991). *America's Energy Choices: Investing in a Strong Economy and a Clean Environment* (Cambridge, Massachusetts: Union of Concerned Scientists).

National Energy Strategy: Powerful Ideas for America (1st ed. 1991/1992) (Washington, DC: U.S. Government Printing Office; or Springfield, Virginia: National Technical Information Service, U.S. Department of Commerce).

Energy Information Administration (annual reports). *Annual Energy Review* (Washington, DC: U.S. Department of Energy).

J. W. TESTER, D. O. WOOD, and N. A. FERRARI, eds. (1991). *Energy and the Environment in the 21st Century* (Cambridge, Massachusetts: The MIT Press).

W. M. BURNETT and S. D. BAN (1989). Changing prospects for natural gas in the United States. *Science* 244: 305–310.

A. NERO (1989). Earth, air, radon, and home. *Physics Today,* April: 32–39.

F. Berkhout and H. Feiveson (1993). Securing nuclear materials in a changing world. *Annual Review of Energy and the Environment* 18: 631–665.

T. B. Johansson, H. Kelly, A. K. N. Reddy, and R. H. Williams, eds. (1993). *Renewable Energy, Sources for Fuels and Electricity* (Washington, DC: Island Press).

L. Schipper, R. B. Howarth, and H. Geller (1990). United States energy use from 1973 to 1987: the impact of improved efficiency. *Annual Review of Energy* 15: 455–504.

A. H. Rosenfeld and D. Hafemeister (1988). Energy-efficient buildings. *Scientific American* 258(4): 78–85.

D. L. Illman (1994). Auto-makers move toward new generation of "greener" vehicles. *Chemical and Engineering News* 72(31): 8–16.

Part II

Atmosphere

We now turn our attention to environmental issues associated with Earth's atmosphere, a topic that flows naturally from our previous contemplation of energy sources and uses, because the burning of fuels has a large impact on the atmosphere. Air pollution is a major problem in most of the world's cities, and it often takes on regional dimensions. The atmosphere is a repository for emissions from combustion and from many other human activities; the air can be cleansed by natural mechanisms, but these can be overwhelmed by the amounts of pollutants being produced. On a global scale, we are carrying out vast inadvertent experiments on the atmosphere. Human activities are increasing the atmospheric concentration of carbon dioxide and other "greenhouse" gases, thereby altering the way the sun's heat is distributed on Earth's surface and in the atmosphere. In addition, the stratospheric ozone shield, which protects us from the sun's ultraviolet rays, is threatened by the emission of ozone-destroying chemicals. Although some of the world's leading scientists and the most powerful computers are trying to predict how these experiments will turn out, all scenarios for the future are riddled with uncertainties.

A. CLIMATE

1. Radiation Balance

As we emphasized in Part I, the sun provides Earth with an enormous input of energy every day. Earth rids itself of energy at the same rate and thereby maintains a steady state, with a constant average temperature. It loses energy by radiating light. Of course, Earth does not glow the way the sun does. A hot body gives off radiation with a range of wavelengths.* The distribution of the radiation shifts toward shorter wavelengths with increasing temperature. That is why a piece of iron heated in a furnace glows red and then white as its temperature increases. The wavelengths of Earth's rays are too long to be detected by our eyes.

The spectral distribution of radiation from the sun and from Earth is shown in Figure 2.1. The curves are somewhat idealized; they are the expected "black body" radiation of objects with the temperature of the sun and of Earth. The spectrum of a black body is smooth, whereas the actual spectra of the sun and Earth are somewhat bumpy because specific atomic and molecular transitions contribute to the emissions. For a black body, the peak wavelength of the radiation is inversely proportional to the absolute temperature (Wein's law):

$$\lambda_{peak} \text{ (nm)} = 2.9 \times 10^{6} \text{ (nm} \cdot \text{K)/T\}(K) \tag{2.1}$$

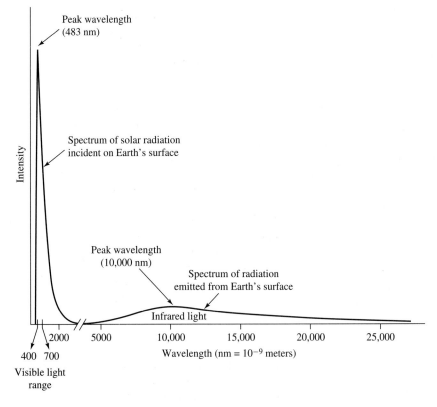

Figure 2.1   Spectral distribution of solar and terrestrial radiation.

*Wavelength (λ) is inversely related to energy by the equation $E = hc/\lambda$, where h is Planck's constant and c is the speed of light. Hence, the shorter the wavelength of a given photon, the greater is its energy.

The sun is a very hot body; its peak wavelength is 483 nm, corresponding to a temperature of 6,000 K. Most of its rays fall between 400 and 700 nm, in the region of visible light; these wavelengths are visible because our eyes have evolved in response to sunlight. Earth emits radiation with a peak wavelength of about 10,000 nm, corresponding to an average temperature of 288 K. Thus, while Earth absorbs radiation mainly in the visible region, characteristic of the high temperature at the surface of the sun, it gives off radiation in the infrared region, which corresponds to the much longer wavelengths characteristic of Earth's cooler surface temperature.

All the energy that Earth absorbs from the sun must eventually be re-emitted, so we can calculate Earth's steady-state temperature by setting its radiation rate equal to the rate at which Earth absorbs energy from the sun. The flux of solar energy directed at Earth, S_0, is 1370 watts/m^2.* It strikes Earth as if Earth were a disk of area πr^2, r being Earth's radius. Since Earth radiates from its entire surface, whose area is $4\pi r^2$, we need to divide S_0 by 4. Moreover, not all of the sun's rays are absorbed; the fraction reflected back to space, the *albedo* (*a*), is close to 0.3. The rate at which solar radiation is absorbed and re-emitted is therefore

$$S = (1 - a)S_0/4 = 240 \text{ watts/m}^2 \qquad (2.2)$$

This rate can be used to calculate the temperature according to the Stefan-Boltzmann law, which states that the rate at which a black body radiates energy, S, is proportional to the fourth power of its absolute temperature

$$S = kT^4 \qquad (2.3)$$

where k is the Stefan-Boltzmann constant, 5.67×10^{-8} watts/m^2 × K^4. According to this calculation, the temperature of the surface of Earth determined from the rate of solar emission is 255 K. But this is colder by 33 K than the average temperature at the surface of Earth. What is the source of this discrepancy? The atmosphere provides the answer: it traps much of the heat emanating from Earth's surface and radiates it back, raising the surface temperature.

This atmospheric trapping of infrared radiation is the *greenhouse effect*. Despite the name's negative associations, the greenhouse effect makes our planet habitable. At the much colder temperature that would prevail in its absence, all of Earth's water would freeze. The concern about the greenhouse effect is that it may become too much of a good thing, if the heat-trapping efficiency of the atmosphere increases further as a result of increasing concentrations of CO_2 and other greenhouse gases. (See section on Greenhouse Effect, pp. 118–127.) For now we note that because of the heating from below, the atmosphere grows colder with increasing elevation above Earth's surface. The number calculated with equation (2.3), 255 K, is the average temperature that prevails at an altitude of about 5 km. The Earth-air system acts as if it radiates from somewhere in the middle of the atmosphere.

Can the energy balance be affected by human energy consumption? In principle, if we keep increasing the rate of fossil and nuclear fuel burning, the global heat load might become significant. From equation (2.3) we can calculate how much the energy input to Earth would

*The watt is a unit of *power* defined as the rate at which work is done; 1 watt = 1 joule sec^{-1}.

need to rise in order to increase the average temperature by 1 K. For such a small change, we can differentiate equation (2.3) and divide by the total flux, to obtain

$$dS/S = 4\, dT/T \tag{2.4}$$

In other words, the fractional increase in the energy budget is four times the fractional increase in the temperature. For a 1 K rise from 255 K, $dT/T = 0.00392$, and $dS/S = 0.0157$. Thus, human energy utilization would have to equal 1.6% of the solar input to produce a 1 K rise in the average temperature. Currently, the ratio of human energy consumption to solar energy is only 0.01% (see Table 1.1, p. 5), leaving a comfortable margin. How long until we reach the 1.6% ratio? Currently, world energy consumption is increasing annually by 2.8%, which corresponds to a doubling time of 25 years; to reach 1.6% of the sun's energy would take about 185 years. But for the reasons discussed in Part I, this rate of increase is unlikely to be sustained that far into the future. Moreover, renewable energy does not count in the balance because renewable energy sources simply divert the current solar flux. To the extent that an increasing fraction of total human energy is extracted from renewable sources, Earth's heat load will be ameliorated.* Thus, direct heating of the planet through our increasing use of energy is not likely to become a serious problem (although a "heat-island" effect can raise the temperature of urban areas by several degrees relative to the surrounding countryside). More serious is the potential for altering Earth's temperature indirectly through changes induced by human activity in either the albedo or the greenhouse effect. These matters are considered in the following sections.

The actual flows of energy through the atmosphere are quite complicated. Figure 2.2 shows the planet's inputs and outputs of energy in units of 10^{20} kJ per year. About 54.4 units of solar energy impinge on Earth and its atmosphere, but about 16.3 units (30%) are reflected to space, exerting no influence on Earth's heat balance. Most of this light is reflected by clouds and atmosphere; a smaller amount (2.2 units) is reflected by Earth's surface. The remaining 38.1 units (70%) are absorbed, 13.0 units (24%) by the atmosphere and clouds, and 25.1 units (46%) by Earth's surface.

To maintain the heat balance, Earth's surface must lose an amount of energy equal to the absorbed solar energy. This loss occurs via three processes. About half of the energy (12.5 units) is lost as *latent heat,* which drives the hydrological cycle through the evaporation of water. A smaller amount (3.8 units) is lost as *sensible heat,* in which the atmosphere is heated from the surface by updrafts.

The third process, *radiation* of infrared light from Earth's surface, accounts for the remaining 8.8 units that need to be lost to complete the energy balance. Actually, a much larger flow of energy, 62.7 units, radiates from the surface (corresponding to a surface temperature of 288 K, consistent with the Stefan-Boltzmann law). A small amount of this radiation, 3.2 units, is emitted directly to space through the so-called *atmospheric window* (see pp. 120–122). Most of it, 59.5 units, is absorbed by clouds and the atmosphere, which re-radiate 53.9 units back to the surface, for a net radiative surface loss of 5.6 units. It is this exchange

---

*For example, whether a given amount of biomass is burned as fuel or undergoes microbial degradation to CO_2 and H_2O, the amount of energy released is the same; rates of release, however, will differ since the latter process is far slower than the former.

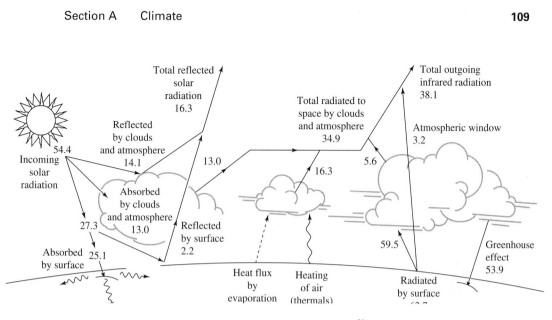

Figure 2.2   Heat balance of Earth (in units of 10^{20} kJ/yr).

of long-wave radiation between the surface and the atmosphere that precisely defines the greenhouse effect. Adding together the surface's radiative losses (3.2 and 5.6 units) provides the 8.8 units necessary for the balance.

The picture is not complete, however, because inputs and outputs of energy to the atmosphere must also be in equilibrium. As noted above, the atmosphere and clouds absorb 13.0 and 59.5 units of energy from absorption of solar and surface radiation, respectively, for a total input of 72.5 units. Radiative losses from the clouds and atmosphere are 34.9 units to outer space, and 53.9 units to Earth's surface, which add up to 88.8 units. Thus, radiative inputs are 16.3 units lower than the outputs, corresponding to a net radiative cooling of the atmosphere and clouds (as well as a net radiative heating of Earth's surface by the same amount). The equilibrium is restored by the transfer of latent and sensible from the surface to the atmosphere.

The above discussion refers to global averages. In fact, the heat flow pattern is not uniform over Earth's surface. Most of the sun's rays are absorbed in the tropics, while the outgoing radiation of Earth is more uniform with latitude (Figure 2.3). Thus, there is a constant movement of energy from the equator toward the poles, through the atmosphere and the oceans. These complexities do not alter the fundamental fact that the Earth-atmosphere system, taken as a whole, receives its energy from the sun and re-radiates it to outer space. The radiation balance has to be maintained.

2. Albedo: Particles and Clouds

a. Clouds.   The reflectivity of solar radiation varies greatly from place to place. The effect on the overall planetary albedo for a specified surface type depends not only on its reflectivity but also on its spatial coverage. The albedo of Earth's surface varies considerably

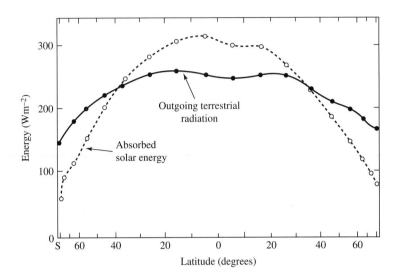

Figure 2.3   Absorbed solar radiation and outgoing terrestrial radiation as a function of latitude.   *Source:* T. H. Von der Haar and V. E. Suomi (1971). Measurement of Earth's radiation budget from satellites during a five-year period. Part I. Extended time and space means. *Journal of the Atmospheric Sciences,* 28: 305–314. Copyright © 1971, American Meteorological Society. Reprinted with permission.

(Figure 2.4). The darkest regions (with the lowest albedos) are the oceans, which constitute about 70% of Earth's total area. Albedos of the oceans range from 6–10% in the low latitudes to 15–20% near the poles due to the low solar elevation. Sea ice with overlying snow has an albedo of 40–60%. The darkest land surfaces are the tropical forests and cultivated land (10–15%). The brightest parts of the globe are the snow-covered polar areas, with albedos as high as 80%. The major deserts have albedos of 25–40%.

The surface albedo can be affected by human activities. For example, the local albedo can be increased by clearing forest land for agriculture, followed by erosion and desertification. However, the dominant factor in the global albedo is clouds. Peak cloud reflectivity occurs over the mid- and high-latitude oceans and in the tropical cirrus systems. As a global average, cloud albedo is around 35–40%. Since the annual average global cloud cover is around 54%, the total sunlight reflected by clouds is around 20%, two thirds of the global albedo. The remaining third is divided between backscattering from air molecules (6% of incident sunlight) and by Earth's surface (only 4%).

As noted in the above discussion, clouds absorb both solar and long-wave radiation. The high albedo of clouds tends to cool Earth's surface by reflecting sunlight to space, but the absorption of outgoing radiation works to warm the surface by the greenhouse effect. The overall role of clouds in determining Earth's heat balance depends on the relative strengths of these two contrasting processes, which is currently the subject of intensive research. Even small shifts in global cloud coverage might contribute to significant changes in the heat balance.

Clouds are a natural part of the hydrological cycle, but the extent of cloudiness is extremely hard to predict. The sun's heat evaporates water at Earth's surface; as the moist air

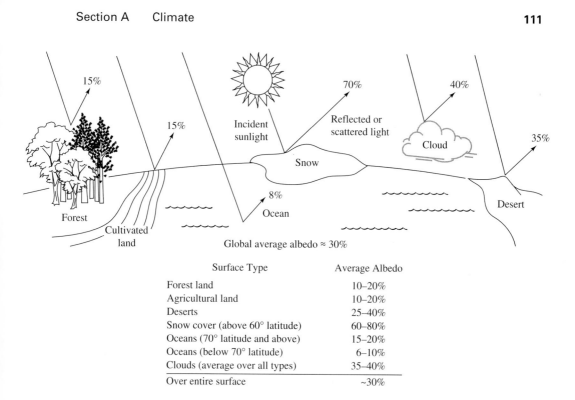

Surface Type	Average Albedo
Forest land	10–20%
Agricultural land	10–20%
Deserts	25–40%
Snow cover (above 60° latitude)	60–80%
Oceans (70° latitude and above)	15–20%
Oceans (below 70° latitude)	6–10%
Clouds (average over all types)	35–40%
Over entire surface	~30%

Figure 2.4   Variation in albedo with cloud cover and type of surface on Earth.

rises and cools, the water condenses out as droplets, forming the clouds. If we consider only the vapor/liquid equilibrium

$$H_2O(g) = H_2O(l) \tag{2.5}$$

water condenses from the gas phase when its partial pressure, p, exceeds the equilibrium vapor pressure of liquid water, p_0. This is expressed in the free energy difference, ΔG, for the phase change:

$$\Delta G = -RT \ln p/p_0 \tag{2.6}$$

where R is the gas constant (8.314 J/K) and T the absolute temperature. As the relative humidity, p/p_0, exceeds unity (100%), ΔG becomes negative, and liquid water forms spontaneously at equilibrium.

In pure air, however, the water molecules must first find a way to get together to form raindrops. Due to their large surface tension, very small drops evaporate quickly, even at relative humidities higher than 100%. In the condensation of a small number of molecules,

$$nH_2O = (H_2O)_n \tag{2.7}$$

the surface free energy of the droplet is a significant part of the free energy change,

$$\Delta G = -nRT \ln p/p_0 + 4\pi r^2 \gamma \tag{2.8}$$

where γ is the surface tension (72.8 dynes*/cm at 20°C), r is the droplet radius, and n is the number of moles of water contained in the droplet. n is also related to the radius of the droplet, via

$$n = (4\pi/3)r^3 \rho/M \qquad (2.9)$$

where $(4\pi/3)r^3$ is the volume of the drop, ρ is its density (1 g/cm^3), and M is the gram molecular weight (18 g). Thus ΔG is the result of two opposing terms that have a different dependence on r. Figure 2.5 is a plot of ΔG against r for a given value of p/p_0 (1.001, or 100.1% relative humidity). The curve goes through a maximum, which defines a critical radius, $r_c = 1$ μm. Droplets larger than this radius will accumulate more water molecules and become stable; droplets smaller than 1 μm will evaporate. A 1-μm drop contains 0.23×10^{-12} mole of water [see equation (2.9)] or 1.38×10^{11} molecules. It is very improbable that this many molecules can come together simultaneously to form a growing droplet; consequently with respect to precipitation, water vapor in pure air at 100.1% humidity is stable indefinitely. The critical droplet radius depends on the extent to which p/p_0 exceeds unity, that is, the extent to which the air is "supersaturated" with water. The dependence [obtained by differentiating equation (2.8) and setting $d(\Delta G)/dr$ equal to zero] is given by

$$r_c = 2M\gamma/[\rho RT \ln (p/p_0)] \qquad (2.10)$$

Thus, r_c decreases slowly as p/p_0 increases. Pure air can be supersaturated to a high degree without precipitation.

In nature, however, condensation actually does occur in the range of 100.1–101% relative humidity, corresponding to $r_c = 1$–0.1 μm (see problem 5, end of Part II). This is be-

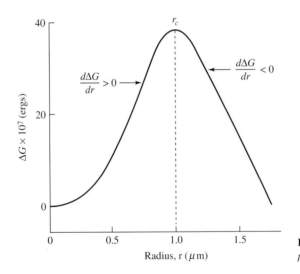

Figure 2.5   Variation of ΔG with drop size at $p/p_0 = 1.001$ ($T = 20°C$).

*The *dyne* is a unit of force, which, multiplied by distance (in centimeters), yields work in *ergs;* 10^7 *ergs* = 1 *joule.*

cause natural air has many suspended particles that can serve as condensation "nuclei" by providing a surface for the accumulation of water molecules. A film of water surrounding the particles has sufficiently low surface tension to permit droplet growth rather than evaporation. The principle of rainmaking by "cloud-seeding" is to inject particles that are effective in nucleating raindrops into supersaturated vapor.

There is a trade-off between droplet size and the number of condensation nuclei. A given amount of water vapor can form a small number of large drops or a large number of small ones. An excess of condensation nuclei can produce droplets that are too small to fall as rain. The fogs that often hover over cities probably reflect the large number of condensation nuclei in polluted air. An increase in the number of atmospheric particles is likely to increase the cloud cover and therefore the albedo.

b. Aerosol particles.   Total suspended particulate matter in air varies from less than 1 μg m^{-3} over polar ice caps and in mid-ocean, to as much as 30,000 μg m^{-3} in desert dust storms or forest fires. In a typical sample of urban air, mineral dust, sulfuric acid, ammonium sulfate, organic material, and soot may be found both as pure and as mixed particles (solid or liquid) in concentrations of around 100 μg m^{-3}. The effect of atmospheric particles on the heat flux of the atmosphere depends less on total concentration than on particle size and composition. Large, dark particles tend to absorb light, thus warming Earth's atmosphere. The most important of such particles is soot, arising from incomplete combustion of carbonaceous fuel and the burning of savannas and forests. In contrast, very small particles, regardless of color and composition, tend to scatter light, thus increasing the albedo of the atmosphere. The light-scattering effect seems to prevail at most latitudes, but absorption can dominate at high latitudes, especially over highly reflective snow- or ice-covered surfaces.

Two major natural sources of light-scattering aerosols appear to be 1) sulfate generated from biogenic gaseous sulfur emission in the deep ocean, and 2) organic carbon from partial oxidation of biogenic organic compounds such as terpenes emitted from forests. In polluted atmospheres, light-scattering particles are produced by reactions with sulfur-, nitrogen-, and carbon-containing gases generated mostly from combustion processes. As a global average, anthropogenic sources are estimated to comprise between 25% to 50% of total aerosols.

The amount of light-scattering by small particles or molecules is given by

$$s = [128\,\pi^5 r^6 / 3\lambda^4] \times (m^2 - 1)/(m^2 + 1) \qquad (2.11)$$

where r and m are the radius and refractive index of the particle, respectively, and λ is the wavelength of the incident light. Because of the $1/\lambda^4$ dependence, blue light is scattered more strongly than red. That is why the sky, which is seen in scattered light, is blue, while sunsets, which are seen in transmitted light, are red. The blue haze over the Great Smoky Mountains of the eastern United States is due to light-scattering from small particles formed by the oxidation of volatile terpenes emitted by the sap of coniferous trees.

The particles in the atmosphere are known collectively as the atmospheric *aerosol*. Their size distribution is very wide, but as seen in Figure 2.6, the aerosol is dominated by the smallest particles. Dust and sea spray are blown about by the wind in great quantity, but these particles are large and settle out rapidly. The submicrometer particles are formed mainly by oxidation of sulfur-, nitrogen- and carbon-containing gases in the atmosphere.

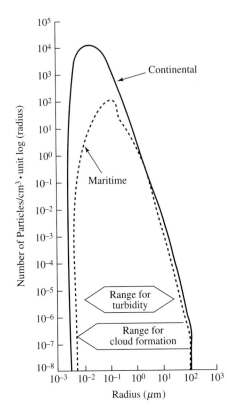

Figure 2.6 Size distribution of particulate matter in the lower atmosphere.

These particles stay aloft much longer, with lifetimes of days or weeks. Gradually, they grow to larger ones via collisions and aggregation, and settle out. No long-term accumulation in the lower atmosphere or troposphere is thus possible.

According to recent research, the particulates in the atmosphere that are most important in influencing the global heat balance are the sulfate aerosols, due to their unique optical and chemical properties. Not only do they efficiently scatter the sun's rays directly, they also exert an indirect effect in their role as a major source of cloud condensation nuclei. They increase the concentration of cloud droplets, resulting in an increase in the scattering surface of clouds. Also, rain may be inhibited, since the mean droplet size decreases with increasing numbers of droplets, leading to a further increase in the cloud cover. On the other hand, as shown in Figure 2.2, clouds also absorb solar and terrestrial radiation; heat gained from this absorption will counter the heat lost by the scattering effect of clouds. Although the relative strengths of these two processes are not known with precision, evidence accumulated thus far indicates that the scattering effect dominates, with sulfates and cloud formation resulting in a net cooling of the planet.

Large volcanic eruptions provide dramatic evidence of the ability of sulfate aerosols to affect global climate. The volcanic aerosols which dominate the effect on climate are micrometer-sized droplets of sulfuric acid, formed from SO_2 emission injected directly into the stratosphere from the force of the blast. Due to their small size and the quiescence of the stratosphere (see pp. 142–145), the aerosols may persist there for a year or more after the eruption. The Mount Pinatubo eruption in the Philippines in June 1991, one of the largest blasts in the past 170 years, emitted an estimated 10 Tg (teragrams, 10^{12}g) of sulfur; it was followed by a perceptible decline in the mean global temperature over the next two years (Figure 2.7).

c. Sulfur cycle. The global sulfur cycle is diagrammed in Figure 2.8. Sulfur is abundant in Earth's crust in sulfide minerals and in calcium and magnesium sulfates. Considerable amounts of sulfur are processed throughout the microbial world. Some photosynthetic bacteria use reduced sulfur compounds as hydrogen donors, while other organisms, the *chemoautotrophic* (living on energy derived from chemicals, rather than from the sun) sulfur bacteria obtain energy from the oxidation of sulfide or sulfur to sulfate. Still other organisms are sulfate reducers, deriving their energy from the anaerobic oxidation of organic compounds, with concomitant reduction of sulfate to H_2S. These bacteria are responsible for the sulfur smell and black coloration, due to iron sulfides, of some aquatic habitats.

Microbes and plants release many volatile sulfur compounds, including hydrogen sulfide (H_2S), methylmercaptan (CH_3SH), dimethylsulfide (CH_3SCH_3), dimethyldisulfide (CH_3SSCH_3), carbon disulfide (CS_2), and carbonylsulfide (OCS). The main form of terrestrial sulfur emissions is H_2S; over the open ocean, the main sulfur compound is CH_3SCH_3, which is produced in plankton by the enzymatic cleavage of dimethylsulfonopropionate, a compound that may help plankton achieve osmotic balance in the salty ocean water. It is estimated that some 65 Tg of sulfur are emitted to the atmosphere annually in volatile sulfur

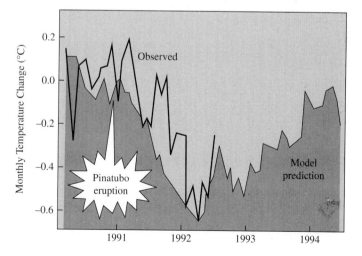

Figure 2.7 Observed versus predicted changes in global temperatures after the Mt. Pinatubo eruption. *Source:* F. Pearce (1993). Pinatubo points to vulnerable climate. *New Scientist* 138 (1878): 7. (Based on model of J. E. Hansen, Goddard Institute of Space Studies, National Aeronautics and Space Administration (NASA), New York, New York). Copyright © 1993 by New Scientist. Reprinted with permission.

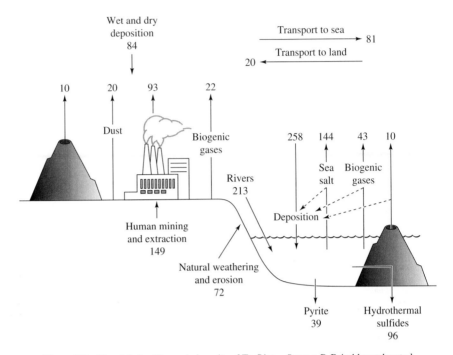

Figure 2.8   The global sulfur cycle in units of Tg S/yr.   *Source:* P. Brimblecombe et al. (1989). Human influence on the sulphur cycle. In *Evolution of the Global Biogeochemical Sulphur Cycle,* P. Brimblecombe and A. Y. Lein, eds. (Chichester, UK: John Wiley & Sons). Copyright © 1989, Scientific Committee on Problems of the Environment (SCOPE), Paris. Reprinted with permission.

compounds from biogenic sources, of which CH_3SCH_3 accounts for 43 Tg. The sulfur in these compounds is oxidized in the atmosphere to SO_2. In addition, as noted above, volcanoes occasionally spew large amounts of H_2S and SO_2 into the atmosphere.

The SO_2 from all sources is subsequently oxidized to sulfuric acid (under some conditions, the reduced sulfur compounds may be oxidized directly to H_2SO_4, without an intermediate SO_2 stage):

$$2SO_2 + O_2 + 2H_2O = 2H_2SO_4 \tag{2.12}$$

Direct reaction of SO_2 with O_2 is very slow, and the oxidation is actually carried out by more reactive species, particularly the hydroxyl radical and hydrogen peroxide (see Oxygen Chemistry, pp. 138–139). The average lifetime for SO_2 is only about a day or so before it is oxidized further. Some of the sulfuric acid in the atmosphere is neutralized (see discussion of acid rain, Part III, p. 224) by ammonia or by calcium carbonate particles. The sulfuric acid and the sulfate salts, being very hygroscopic, form particles that serve as cloud condensation nuclei. Alternatively, the SO_2 is oxidized in the raindrops themselves. In either case, the sulfur cycle is completed by rainout (Figure 2.8).

This natural cycle is now being strongly perturbed by human activity. Large quantities of SO_2 are emitted into the atmosphere by the burning of fossil fuels, especially coal, and by the smelting of sulfide minerals. The amount of sulfur emitted from anthropogenic sources is estimated to be 93 Tg/year, somewhat larger than the biogenic emission rate, and it is rising. This extra SO_2 contributes significantly to acid rain; it also increases the concentration of cloud condensation nuclei. The effect is concentrated in industrial regions because the sulfate is washed out of the air within a few days and can travel no further than about 1000 km, on average. Contour lines of sulfate precipitation in the United States are shown in Figure 2.9. The highest values are over the industrial Northeast, and the lowest are in the West. The average concentration, ~1 mg/l, is about ten times the concentration over remote oceans.

How important is the sulfur contribution to the overall energy budget of Earth? It has been calculated that sulfate aerosol scatters enough sunlight to reduce the absorbed radiation by 1.3 watts/m^2. In addition, comparison of clouds and aerosol in the northern and southern hemispheres suggests that sulfate aerosol increases cloudiness by about 15%, reducing solar heating by an additional 1 watt/m^2. The 2.3 watts/m^2 combined reduction is 1% of the absorbed radiation [see equation (2.2)]; on the basis of the Stefan-Boltzmann relation [equation (2.4)], it is expected to cool Earth by 0.6 K. This negative "radiative forcing"* just about balances the positive forcing calculated for the current level of anthropogenic greenhouse gases (see next section). Thus, it seems that fossil fuel burning may be having two opposite

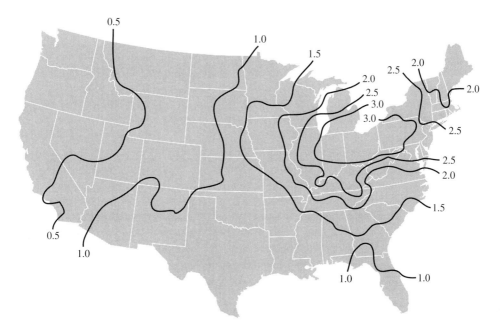

Figure 2.9 Precipitation-weighted concentration of sulfate (mg/l) over the United States, 1987. *Source:* R. J. Charlson et al. (1992). *Global Biogeochemical Cycles* (London: Academic Press). Copyright © 1992, Academic Press. Reprinted with permission.

---

*Due to the complexity of the global climate system and its numerous feedback mechanisms, it is not currently possible to estimate precisely how each of these factors affects the climate. For this reason, scientists estimate "radiative forcing," rather than possible climatic responses. "Radiative forcing," expressed in Wm^{-2}, is the calculated change in the heat balance of Earth.

effects on Earth's climate: greenhouse warming due to the emitted CO_2, and albedo cooling due to the emitted SO_2. The climate effect is complex, however, because the greenhouse gases are spread around the globe, whereas the sulfate aerosol is concentrated in industrial regions. There is currently a substantial effort to reduce SO_2 emissions in order to reduce air pollution and acid rain. An ironic and unintended consequence may be to reduce the cooling due to the sulfate aerosol, thereby augmenting the greenhouse warming.

3. Greenhouse Effect

   a. Infrared absorption and molecular vibrations.   As mentioned above, the greenhouse effect is the trapping of heat from below by the atmosphere. Earth's atmosphere admits the visible rays from the sun but traps the infrared rays emanating from Earth's surface. Some of Earth's heat is carried from the surface by air currents or by the evaporation of water (Figure 2.2) but most of it is radiated to the atmosphere, which traps it and radiates much of it back, releasing the rest to outer space. But how is this trapping accomplished? Why do we worry about CO_2 and other minor constituents of the atmosphere, when it is made up almost entirely of N_2 and O_2 (see Table 2.1)?

   The answer is that the major atmospheric gases are unable to absorb infrared light. They do not meet the two fundamental requirements for the absorption of electromagnetic radiation:

   1) When radiation is absorbed by a molecule, the molecule undergoes a quantum transition, involving the movement of either its electrons or its nuclei; the energy of the radiation must therefore match the energy of the molecular transition. In the infrared region of the spectrum, the available transitions involve movement of the nuclei in molecular vibrations. That is why argon, the third most abundant atmospheric constituent (0.9%) is transparent to infrared radiation. Since argon is monatomic, it has no vibrations.

   2) Because radiation is electromagnetic, its absorption requires that the transition change the electric field within the molecule, that is, the transition must alter the molecule's dipole moment (the vector sum of atomic charges times their distances from the molecule's center of mass). This second requirement is the reason that N_2 and O_2 are unable to absorb Earth's infrared radiation. Although their nuclei do vibrate along the bond joining them, and the energy of the vibration is in the infrared region, the vibration does not change the dipole moment, because the molecule remains symmetrical. The vibration is infrared-*inactive*. This is true for all *homonuclear* diatomic molecules. The dipole moment is altered by vibrations of *heteronuclear* diatomic molecules, such as CO, NO, and HCl, since these nuclei have different partial charges. However, these molecules do not contribute significantly to the greenhouse effect because their concentration in the atmosphere is too low and their absorption too weak.

   In contrast, polyatomic molecules have numerous vibrations ($3n - 6$ for nonlinear molecules, where n is the number of atoms, or $3n - 5$ for linear molecules); at least some of these vibrations change the dipole moment and are infrared-active. All the gases that contribute significantly to the greenhouse effect are polyatomic. The two most important greenhouse molecules are water and carbon dioxide. Their vibrations are illustrated in Figure 2.10. All three vibrations of water change its dipole moment. For carbon dioxide, the symmetric stretching motion of the two O atoms leaves the dipole moment unchanged: the dipoles that each O atom generates relative to the carbon atom cancel one another due to the linear geometry. However, the net dipole moment is altered by the asymmetric stretch and by the bending vibration.

Both water and carbon dioxide contribute to the greenhouse effect, but water is not listed in Table 2.1 because its content in the atmosphere varies greatly from place to place and time to time; on average, water molecules make up 0.4% of the atmosphere. The total amount of water is an order of magnitude greater than the amount of carbon dioxide. If Earth's temperature rises as a result of increasing carbon dioxide and other greenhouse gases, the water vapor pressure will also rise, and the amount of water in the atmosphere is expected to increase. This is an example of *positive feedback:* the greater the concentration of greenhouse gases, the higher the surface temperature; higher temperatures lead to greater amounts of atmospheric water, which in turn amplifies the greenhouse effect.

When greenhouse molecules absorb the terrestrial infrared radiation, they re-radiate it in all directions as illustrated in Figure 2.11. Absorption and re-radiation continue repeatedly with increasing altitude, leaving less than 10% of the absorbed infrared radiation available near the top of the atmosphere (Figure 2.2). As a result, the lower atmosphere is warmer, but the higher atmosphere is colder than it would be in the absence of infrared absorbers.

It might appear that water and carbon dioxide would not be effective in heat trapping because Earth's emissions cover a wide spectrum of wavelengths, whereas the molecular vibrations correspond to specific energies. However, the molecules can undergo not only vibrations but also rotations. For each molecular vibration, infrared photons can induce transitions to many different rotational levels (rates of rotation). Consequently, each vibration

TABLE 2.1 COMPOSITION OF DRY AIR AT GROUND LEVEL IN REMOTE CONTINENTAL AREAS

Constituent	Formula	Concentration (by volume, ppm)
Nitrogen	N_2	780,900
Oxygen	O_2	209,400
Argon	Ar	9,300
Carbon dioxide	CO_2	350
Neon	Ne	18
Helium	He	5.2
Methane	CH_4	1.7
Krypton	Kr	1.1
Hydrogen	H_2	0.5
Nitrous oxide	N_2O	0.3
Xenon	Xe	0.08
Carbon monoxide	CO	0.04–0.08
Organic vapors		0.02
Ozone	O_3	0.01–0.04

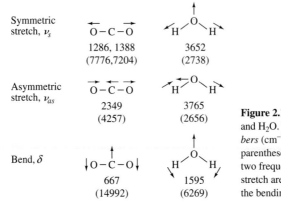

Figure 2.10 Molecular vibrations of CO_2 and H_2O. The top numbers are *wave numbers* (cm^{-1}) of the vibrations; the numbers in parentheses are the *wavelengths* (nm).* The two frequencies for the CO_2 symmetric stretch are due to a resonant interaction with the bending mode overtone (2×667 cm^{-1}).

has a broad absorption band. These bands are shown for water and carbon dioxide in Figure 2.12. In the top panel of the figure, the absorptions for these two molecules are added up and superimposed on Earth's emission spectrum. The combined absorption bands can be seen to block most of the terrestrial radiation. There is, however, a relatively unobstructed region of the spectrum between 8,000 and 12,000 nm through which radiation can escape. This region is called the *atmospheric window*.

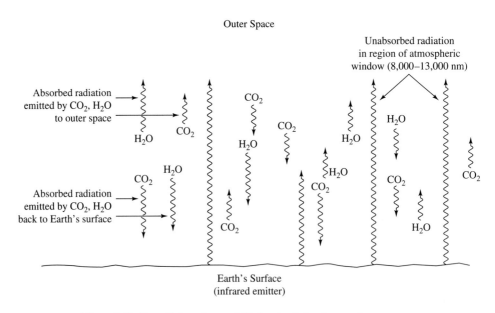

Figure 2.11   Re-radiation of terrestrial infrared radiation by greenhouse gases.

*The frequency of vibration, v, can be expressed in *wave numbers*, $\bar{v}$(cm^{-1}), where $\bar{v} = 1/\lambda$ and λ is the *wavelength*. $\bar{v}$ can be converted to λ (in nm) from $\lambda = 1 \times 10^7/\bar{v}$. See Appendix A for further explanation.

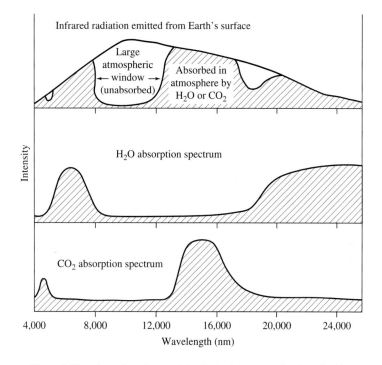

Infrared radiation emitted from Earth's surface

Large
atmospheric
← window →
(unabsorbed)

Absorbed in
atmosphere by
H_2O or CO_2

Intensity

H_2O absorption spectrum

CO_2 absorption spectrum

4,000   8,000   12,000   16,000   20,000   24,000

Wavelength (nm)

Figure 2.12   Absorption of terrestrial radiation by water and carbon dioxide.

This window can be filled by other polyatomic molecules such as the chlorofluoro-carbons (CFCs) (Figure 2.13), methane (CH_4) and nitrous oxide (N_2O). The CFCs are of major concern as destroyers of stratospheric ozone (see pp. 152–153), but they are also important greenhouse gases. Their impact is significant even though their concentration is about five orders of magnitude lower than that of carbon dioxide. At the current CO_2 concentration, the CO_2 absorptions are nearly "saturated," that is, most of the radiation emitted within the absorption bands is already absorbed. As a result, each extra CO_2 molecule contributes only a relatively small amount to the total absorption. (The same is true for each additional water molecule.) In contrast, even though the CFCs are very dilute and absorb only a small fraction of the radiation, since they absorb radiation in the window region, the contribution to overall absorption is relatively high for each extra CFC molecule. In fact, one CFC molecule contributes over 10,000 times more to the greenhouse effect than each extra CO_2 molecule. The relative radiative forcing is compared in Table 2.2 for the two most widely produced CFCs, as well as for the two other major greenhouse gases, methane and nitrous oxide. This forcing depends on the lifetime of the gas, as well as on its effectiveness as an infrared absorber. Thus, the N_2O contribution per molecule is much larger than that of CH_4, mainly because it is much longer lived. CH_4 is destroyed by reaction with hydroxyl radicals in the atmosphere (p. 138) whereas N_2O is only removed by drifting into the stratosphere, where it is photolyzed by ultraviolet rays (pp. 153–154).

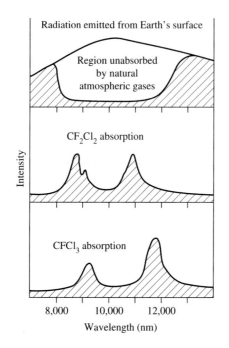

Figure 2.13 Absorption spectra of chloro-fluoromethanes (CF_2Cl_2 and $CFCl_3$) and their coincidence with the atmospheric window (8,000 to 13,000 nm).

TABLE 2.2 SUMMARY OF PROPERTIES OF GREENHOUSE GASES AFFECTED BY HUMAN ACTIVITIES

	CO_2	CH_4	CFC-11	CFC-12	N_2O
Atmospheric concentration	ppmv	ppmv	pptv	pptv	ppbv
Preindustrial (1750–1800)	280	0.8	0	0	288
Current (1990)	353	1.72	280	484	310
Current rate of change/year	1.8	0.015	9.5	17	0.8
(% Increase/year)	(0.5%)	(0.9%)	(4%)	(4%)	(0.25%)
Atmospheric lifetime (years)	(50–200)	10	65	130	150
Per molecule ratio of radiative forcing (ΔF (GHG)/ΔF(CO_2))	1	21	12,400	15,800	206
Major removal mechanism	a	b	c	c	c

ppmv, ppbv, and pptv are parts per million, parts per billion, and parts per trillion by volume, respectively.

a—slow exchange of carbon between surface waters and deeper layers of ocean.

b—reaction with hydroxyl radical in troposphere.

c—photolysis in stratosphere.

Source: Drawn from data and tables given in J. T. Houghton et al., eds. (1990). *Climate Change: The IPCC Scientific Assessment* (Cambridge, UK: Cambridge University Press).

b. Greenhouse gas trends.   Since the introduction of CFC compounds for a variety of products in the 1950s, the CFC contribution to greenhouse warming has been increasing rapidly (Figures 2.14 and 2.15, and Table 2.2). Although the CFCs are being phased out of production due to international agreements to protect stratospheric ozone, those already produced will be around for a long time. Their lifetimes are on the order of 100 years (Table 2.2); like N_2O they are removed from the atmosphere only after transport to the stratosphere and encountering energetic ultraviolet radiation (see pp. 152–154).

N_2O is a side-product of the microbial process of *denitrification,* in which NO_3^- is the oxidant for energy production by anaerobic bacteria, and N_2 is the major product (see pp. 226–229); the ratio of N_2O to N_2 is about 1:16 but can vary depending on conditions (for example pH, O_2 concentration). There are also indications that N_2O is a side-product of *nitrification,* in which ammonia is oxidized to nitrate. Natural sources are thought to account for 4–8 Tg of N per year as N_2O, while the anthropogenic contribution is anywhere from 0.1 to 3 Tg/year (although actual emissions could well be twice as high; see note in Table 2.3). Most of the anthropogenic contribution derives from nitrogen fertilizer use, which increases both nitrification and denitrification (Table 2.3). Quite possibly, intensive agriculture may be largely responsible for the observed increase in atmospheric N_2O, currently estimated to be 0.25% annually. Nevertheless, there may also be significant industrial sources. For example, it was recently found that nitric acid oxidation of cyclohexanol to adipic acid, a precursor in nylon production, was producing copious quantities of N_2O, contributing as much as 10% to the estimated annual N_2O increase;[*] the process has since been altered to minimize N_2O emission.

Methane is the second most important greenhouse contributor after CO_2. Although it has a shorter lifetime (10 years) than CFCs or N_2O, its atmospheric concentration is higher, and it has a substantial rate of increase, 0.9% annually (Table 2.2). Some of the methane

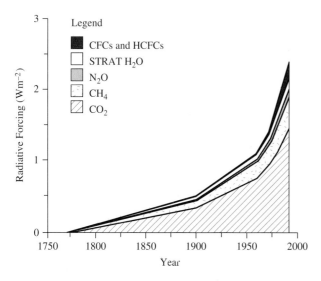

Figure 2.14   Changes in radiative forcing due to increases in greenhouse gas concentrations between 1765 and 1990. *Source:* J. T. Houghton et al., eds. (1990). *Climate Change: The IPCC Scientific Assessment* (Cambridge, UK: Cambridge University Press).

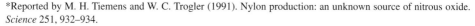

[*]Reported by M. H. Tiemens and W. C. Trogler (1991). Nylon production: an unknown source of nitrous oxide. *Science* 251, 932–934.

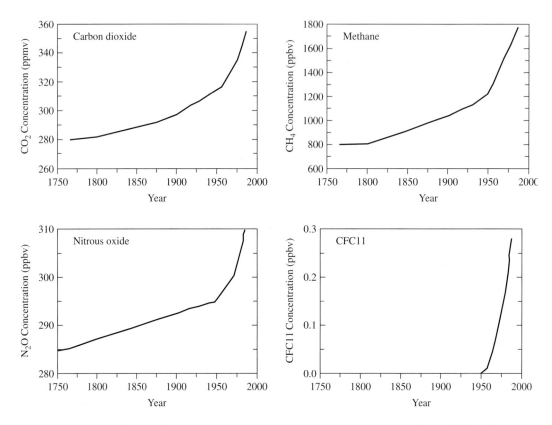

Figure 2.15   Estimated historical concentrations of major greenhouse gases   *Source:* J. T. Houghton et al., eds. (1990). *Climate Change: The IPCC Scientific Assessment* (Cambridge, UK: Cambridge University Press).

comes from natural gas deposits, from leaks in the gas distribution system, and from coal mines (Table 2.3). Most of it, however, results from the action of anaerobic bacteria, the *methanogens,* which produce methane as an end-product of their metabolism. These bacteria are abundant in wetlands, rice paddies, and landfills, and also in the fore-stomachs of ruminant animals and in termites. Natural and anthropogenic sources are estimated to be of the same order of magnitude, roughly 235 and 290 Tg CH_4 per year, respectively. About half of the natural contribution is from wetlands, and most of the rest is from wild animals and termites. Rice paddies account for about 40% of the anthropogenic contribution, with another 30% from gas and coal extraction. Domestic animals, mostly cattle, account for an estimated 14%. Rice cultivation, gas and coal consumption, and cattle raising have all increased substantially since the 1950s and may account for the marked upturn in the methane emission curve since that time (Figure 2.15). But there may also be significant natural variations as well.

Although the relative share of the radiative forcing due to CFCs, N_2O, and CH_4 has been increasing (Figure 2.14), the largest effect is still due to CO_2. Anthropogenic produc-

TABLE 2.3 SOURCES OF ATOMOSPHERIC EMISSIONS OF GREENHOUSE GASES
(DECADE OF THE 1980s)

Greenhouse gas	Natural sources	Anthropogenic sources
Carbon dioxide	(a)	*Total:* 6,800 Tg C/yr Burning of fossil fuels* (77%); Net destruction of vegetation (deforestation) (23%)

Methane	*Total:* 235 Tg CH_4/yr
	Naural wetlands (49%);
	Enteric fermentation (wild
	   animals) (17%);
	Termites (17%)
	Biomass burning (8%);
	Oceans/freshwaters (6%);
	Others (2%)
		Total: 290 Tg CH_4/yr
		Rice paddies (38%);
		Gas drilling, transmission (16%);
		Landfills (14%); Enteric fermentation
		   (domestic animals) (14%);
		Coal mining (12%);
		Biomass burning (7%)

Halocarbons	None (b)	*Total:* 1.1 Tg CFC/yr Synthetic chemicals (100%)
Nitrous oxide[†]	*Total:* 4.3–7.9 Tg N/yr Oceans (33%); Soils (67%)	*Total:* 0.1–2.6 Tg N/yr Fertilizer (85%); Combustion (11%); Biomass burning (4%)[‡]

*Does not include other minor industrial processes that produce CO_2 such as cement production.

[†]Known sources of N_2O do not balance estimates based on the stratospheric destruction of N_2O plus its increasing concentration in the atmosphere, which in total suggest emissions of 10–17.5 Tg N/yr.

[‡]Percentages given only for upper limit of total emissions.

(a) Respiration of vegetation and soil detritus causes an annual atmospheric emission of 100 to 120 billion tons of CO_2, but this is balanced by photosynthesis which absorbs CO_2 from the atmosphere.

(b) Does not include methyl chloride (CH_3Cl), which may have a significant natural source.

Source: J. T. Houghton et al., eds. (1990). *Climate Change: The IPCC Scientific Assessment* (Cambridge, UK: Cambridge University Press).

tion of CO_2 from burning fossil fuels is 5,000 Tg C/year, which dwarfs the other greenhouse gases. Deforestation is estimated to add another 1,800 Tg C/year. Of course, the natural flux of CO_2 to the atmosphere due to the constant respiration of the biosphere is much greater, but this flux is in balance with photosynthesis. The balance is nicely illustrated by the historical record of CO_2 concentration at the monitoring station in Mauna Loa, Hawaii (Figure 2.16). Every year the CO_2 declines to a minimum concentration in the summer when photosynthesis in the fields and forests of the northern hemisphere converts CO_2 to biomass; it rises to a maximum in winter when the dead vegetation decays, releasing its stored carbon as CO_2. The oscillatory pattern is regular from year to year, but it is superimposed on a rising background of average CO_2 concentration, which increased by 12.7% between 1958 and 1992 due to the continued extra CO_2 production from human activities.

Of the total 6,800 Tg C/year emitted by fossil fuels and deforestation, only about 3,000 Tg C/year stay in the air. Where does the remaining CO_2 go? The oceans are an obvious possibility. Because seawater is alkaline and CO_2 is acidic, the oceans are a vast reservoir of

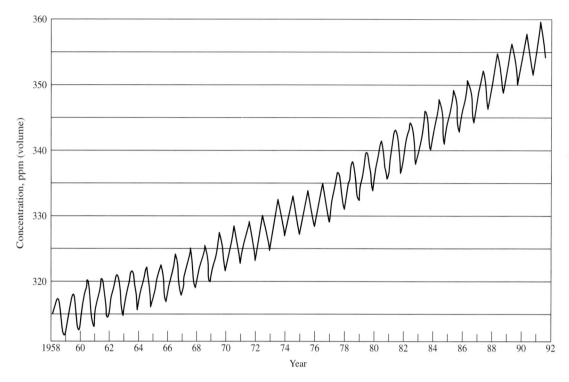

Figure 2.16   Increase in the atmospheric concentration of carbon dioxide between 1958 and 1992 at Mauna Loa, Hawaii.   *Source:* C. D. Keeling et al., Scripps Institution of Oceanography, La Jolla, California; National Oceanic and Atmospheric Administration, Washington, DC.

CO_2. However, only the surface layer of the ocean, the top 75 meters, is in equilibrium with the atmosphere, and its capacity for absorbing CO_2 is limited. Exchange of the surface layer with the deep oceans takes hundreds of years. It is estimated that the surface layer absorption capacity is 1,600 Tg of CO_2 carbon per year. This is 2,200 Tg C/year less than the total 3,800 Tg C/year not found in the atmosphere.

The location of the "missing" carbon has been a subject of considerable debate, but it is now generally accepted that the biosphere itself is the reservoir. A recent study of $^{13}C/^{12}C$ isotope ratios (which provide a measure of biosphere CO_2 flux since plant photosynthesis discriminates against ^{13}C) has established a temperate latitude Northern Hemisphere CO_2 sink that is equivalent in magnitude to roughly half of the global fossil fuel emissions in 1992 and 1993.* Some of the carbon increase may be attributed to forests that have regenerated on former agricultural land. Moreover, there is some evidence that in recent years northern forest biomass has increased, perhaps because the rising CO_2 level in the atmosphere is stimulating plant growth. Such stimulation is known to occur in laboratory settings although, in natural ecosystems, plant growth is usually limited by other factors such as temperature,

*P. Ciaeis et al. (1995). A large northern hemisphere terrestrial CO_2 sink indicated by the $^{13}C/^{12}C$ ratio of atmospheric CO_2. *Science* 269: 1098–1102.

water, or other nutrients. The limiting nutrient is frequently nitrogen (see discussion starting on p. 253), and another possibility is that forest growth has been promoted by increased precipitation of nitrates derived from the nitrogen oxide pollutants produced by burning fossil (as well as biomass) fuels.

Figure 2.17 illustrates the global carbon cycle with estimates for the amounts of carbon residing in the air, soil, biomass, and ocean reservoirs, as well as the annual fluxes. These fluxes are very large; it is not surprising that we find it hard to account fully for the fate of the relatively small anthropogenic contribution. Yet it is this contribution that is driving the accumulation of CO_2 in the atmosphere and, with it, the greenhouse warming. Evaluating the mechanisms for carbon storage is important if we are to anticipate just how much CO_2 is likely to be left in the atmosphere in the future.

4. Climate Modeling

The mean global temperature has risen about $0.5°C$ or more over the past century (Figure 2.18). Is this rise due to human activity? Or is it part of the natural variation? If Earth's temperature is tracked back over longer periods of time, using several geophysical techniques as well as fossil records, the temperature is found to have varied greatly. These

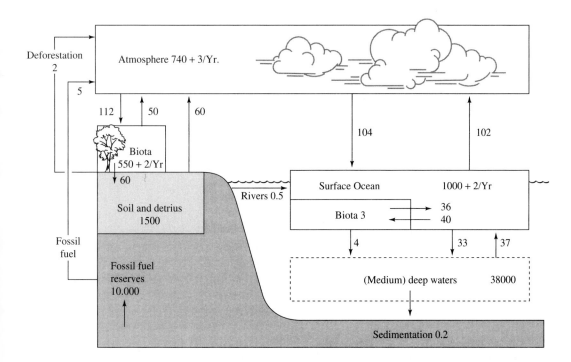

Figure 2.17   The global carbon cycle (in units of 10^3 Tg).   *Sources:* Adapted from global model of the National Aeronautics and Space Administration (NASA), Washington, DC, and J. T. Houghton et al., eds. (1990). *Climate Change: The IPCC Scientific Assessment* (Cambridge, UK: Cambridge University Press).

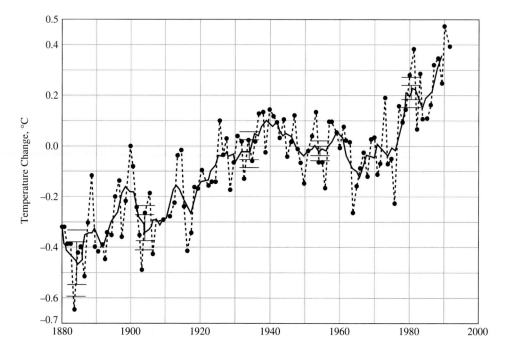

Figure 2.18   Global average temperatures from 1880 to 1991 (relative to the average for 1951–1980). *Source:* H. Wilson and J. E. Hansen, Goddard Institute for Space Studies, National Aeronautics and Space Administration (NASA), New York, New York.

variations are illustrated in Figure 2.19. The graph on the left shows that over the past several hundred years, the temperature has fluctuated more than 1°C. Temperatures dipped down from about 1400 to 1850, a period known as the Little Ice Age in Europe. Much larger fluctuations occurred over a span of hundreds of thousands of years (right graph). These reflect the global ice ages, the last one of which reached its lowest temperature about 20,000 years ago.

What caused these temperature swings? The matter is still under intense debate, but a role for the greenhouse effect is strongly suggested by the variation in CO_2 and CH_4 content found in the air bubbles trapped in the Antarctic ice sheet. The record now extends back over the past 150,000 years; as shown in Figure 2.20 the temperature is remarkably well-correlated with the CO_2 and CH_4 levels. Peaks and troughs in the records of temperature and of greenhouse gases line up very well. Of course, a correlation does not establish cause and effect. It is possible that high temperatures induced higher CO_2 and CH_4 levels, rather than vice versa, or that some other factor induced similar variations in all three variables.

Given this variability in past climates, predicting the effects of human activities on future climates is a hazardous undertaking. It is also an important challenge, one that atmospheric scientists are trying to meet by calculating climate changes using global circulation models (GCMs). The general idea behind these models is to apply the laws of physics to the atmosphere and oceans, and to use the inputs of energy from the sun to predict the dynamics of air parcels around the globe. The system is extremely complicated and requires

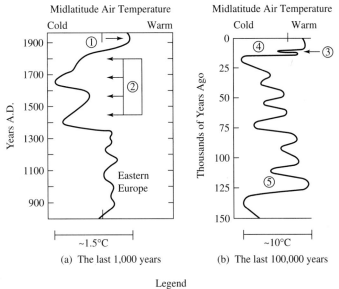

Midlatitude Air Temperature

Midlatitude Air Temperature

(a) The last 1,000 years

(b) The last 100,000 years

Legend

1. Thermal maximum of 1940s
2. Little Ice Age
3. Cold interval
4. Present interglacial (Holocene)
5. Last previous interglacial (Eemian)

Figure 2.19   Long-term natural fluctuations in Earth's temperature.   *Source:* National Research Council (1975). *Understanding Climatic Change, a Program for Action* (Washington, DC: National Academy Press).

many approximations, even with the most powerful supercomputers. There are also conceptual difficulties such as determining the role of aerosols and clouds, including the myriad chemical reactions in the atmosphere, and describing chemical exchanges with the ocean and with the biosphere. In an effort to gain consensus based on the best scientific information, the *Intergovernmental Panel on Climate Change (IPCC)* was established in 1988 jointly by the World Meteorological Organization (WMO) and the United Nations Environment Programme (UNEP). Over 300 scientists from 26 nations have participated in the scientific assessment of climate change. The first report of the IPCC in 1990 projected temperature changes based on different energy development scenarios to the year 2100. The findings show that the *equivalent CO_2 concentration** could rise between two to four times the preindustrial level by the year 2100, with a corresponding global temperature rise of between 2 and 4.2°C (Figure 2.21).

Such large temperature changes have not been experienced on earth since the ice ages, and their effects would be profound. Due to both the expansion of water upon warming and the melting of mountain glaciers, the sea level would rise, flooding low-lying coastal zones

---

*The *equivalent CO_2 concentration* expresses the total radiative forcing of all greenhouse gases in terms of the amount of CO_2 that would give that forcing.

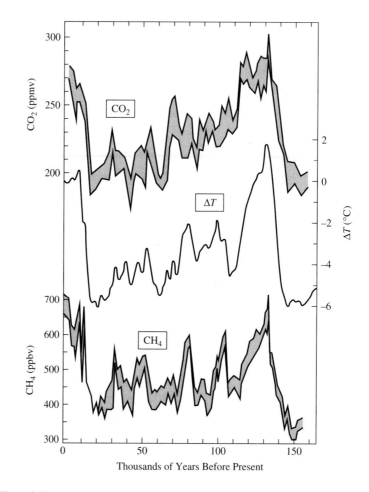

Figure 2.20 Antarctic ice core records of local atmospheric temperature, and correspond-
ing air concentrations of carbon dioxide and methane for the past 160,000   *Source:* J. T.
Houghton et al., eds. (1990). *Climate Change: The IPCC Scientific Assessment* (Cambridge,
UK: Cambridge University Press).

where a substantial fraction of the human population currently lives. Computer models suggest about a half-meter rise over the next century. (It is worth noting that melting of Arctic
ice would not affect sea level because most of it is floating on the water. The much thicker
Antarctic ice sheet rests on land but is considered unlikely to melt unless the temperature
rises even higher than currently predicted.) The increasing temperature would also probably
dry out the interior of continents, while increasing the precipitation rate in regions close to
open water. The resulting gradients of temperature are likely to increase the severity of
storms. Additionally, as the average temperature rose, the climatic zones for flora and fauna
would shift toward the poles. These shifts would force major readjustments of agriculture.
There is also concern that tropical diseases would spread, as insect vectors increase their
range. Not only is the magnitude of the projected change larger than experienced in Earth's

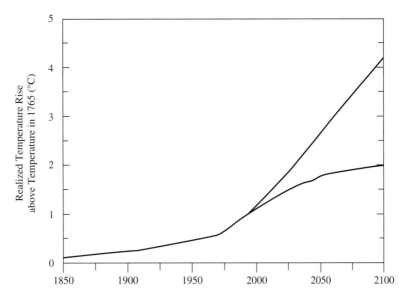

Figure 2.21   Simulations of the increase in global mean temperature from 1850 to 2100. The higher projection assumes a coal-intensive energy supply, only modest gains in energy efficiency, complete depletion of tropical forests, and little effort to implement further controls on CH_4, N_2O, and CFCs. A quadrupling of the equivalent CO_2 concentration, relative to the pre-industrial level (280 ppm), is anticipated in this scenario. The lower projection is based on a shift to renewable energy sources and nuclear energy in the first half of the next century, complete phase-out of CFCs, and major gains in energy efficiency, reforestation, and reduction of agricultural emissions. Even with these measures, a doubling of the equivalent CO_2 concentration is anticipated.   *Source:* J. T. Houghton et al., eds. (1990). *Climate Change: The IPCC Scientific Assessment* (Cambridge, UK: Cambridge University Press).

recent history, but the rate of change is unprecedented. Forests could be devastated if the rate of climate change outpaced the rate at which forest species could migrate.

Of course, the computer projections could be wrong. Because the causes of climate change are not well understood, it is impossible to be sure about the natural variation upon which the anthropogenic effects are superimposed. But the uncertainty cuts both ways. It is as likely that the models underestimate as overestimate the future climate change. What is clear is that the rising greenhouse gas concentrations are producing a change in the radiative forcing that is very significant, on the scale of effects that are capable of changing global climate.

Recognizing the seriousness of the problem, the industrialized countries committed themselves, as part of the international climate-change treaty signed at the 1992 Earth Summit in Rio de Janeiro, to reducing emissions of greenhouse gases to 1990 levels by the year 2000. There has been little agreement, however, on what practical measures will be put into effect. Moreover, even if the target is achieved by the year 2000, additional measures would have to be considered. The lifetimes of the major greenhouse gases are long, and the ability of the oceans and biosphere to absorb CO_2 quickly is limited. To prevent the greenhouse gas radiative forcing from rising further would require very large reductions in the emission rates (Table 2.4). For example, stabilizing CO_2 concentrations at their current levels requires

TABLE 2.4 1990 EMISSIONS OF GREENHOUSE GASES AND REDUCTIONS REQUIRED TO STABILIZE RADIATIVE FORCING

Gas	1990 Anthropogenic emissions (Tg)*	Relative contribution over 100 yr (%)[†]	Stabilization reduction (%)[‡]
CO_2	6,800	61	>60
CH_4	290	15	15–20
N_2O	14[**]	4	70–80
CFCs	1.1	11	70–85
HCFC-22[§]	0.1	0.5	40–50

*Emissions given as Tg C in CO_2 and Tg N in N_2O. Emissions of other gases are expressed in molecular weights; e.g., Tg CH_4 in methane.

[†]Not mentioned here are other minor gases, whose total contribution is 8.5%.

[‡]These are the percentage reductions in the rates of anthropogenic emissions that would be required to stabilize concentrations of greenhouse gases at present-day levels.

[§]HCFC is an abbreviation for a partially hydrogenated CFC. HCFC-22 is a designation for $CHClF_2$, chlorodifluoromethane.

[**]Emission of N_2O is somewhat higher than the range presented in Table 2.3; here it is assumed that known sources of emission underestimate actual emission (see footnote to Table 2.3, p. 125).

Source: J. T. Houghton, et al., eds. (1990). *Climate Change: The IPCC Scientific Assessment* (Cambridge, UK: Cambridge University Press).

reducing emissions by more than 60%, which implies dramatic reductions in the rates of fossil fuel burning and deforestation. Thus our inadvertent experiment in global warming is already under way; its outcome will depend greatly on just how high the greenhouse gas emissions become.

B. OXYGEN CHEMISTRY

The reactivity of oxygen is the controlling factor in the chemistry of the atmosphere and of Earth's crust. Because oxygen has such a high electronegativity, second only to fluorine, oxygen is avid for electrons and combines with most other elements to form stable oxides (Figure 2.22). Solid oxides of iron, aluminum, magnesium, calcium, carbon, and silicon make up Earth's crust, while the oceans are filled with the oxide of hydrogen. The atmosphere contains volatile oxides, especially carbon dioxide and sulfur dioxide. The only common element stable in the presence of oxygen is nitrogen. The bond connecting the two nitrogen atoms is so strong that converting N_2 to any of the nitrogen oxides costs considerable energy. Nevertheless, nitrogen oxides, N_2O, NO, and NO_2, are found in the atmosphere because once formed, they are only slowly converted back to N_2 and O_2.

1. Kinetics and Thermodynamics

The nitrogen oxides illustrate the complex interplay between reaction rates and equilibria in atmospheric chemistry. Due to the strength of the N–N and O–O bonds relative to N–O

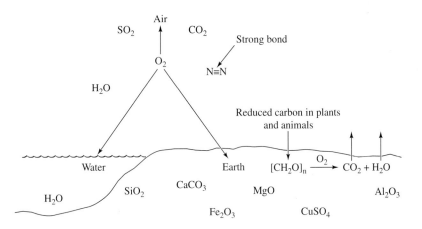

Figure 2.22   Most elements, except nitrogen, are stable as oxides.

bonds (Table 1.3, p. 18), the reaction

$$O_2 + N_2 = 2NO \qquad \Delta H = 180.8 \text{ kJ} \qquad (2.13)$$

is uphill energetically (Figure 2.23); at ambient temperatures, it does not proceed spontaneously. (The enthalpy change, ΔH, for this reaction is just twice the standard enthalpy of
NO, since the standard enthalpies of the elements in their stable forms are defined as zero;
see Table 2.5 for the standard enthalpies and free energies of some atmospheric compounds.)
But at high temperatures, the reaction does proceed to a significant extent because the kinetic
energy of the hot molecules can make up the energy deficit. In general, high temperature
shifts the position of equilibrium toward reaction products for an endothermic (energy-
absorbing) reaction, and toward the reactants for an exothermic (energy-releasing) reaction.

The controlling thermodynamic relationship is

$$\Delta G = \Delta H - T\Delta S \qquad (2.14)$$

where ΔG is the change in free energy associated with the reaction, and ΔS is the change in

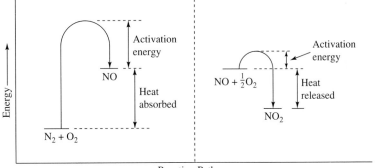

Figure 2.23   Energy profiles of nitrogen-oxygen reactions.

TABLE 2.5   STANDARD (25°C) ENTHALPIES AND FREE
ENERGIES OF FORMATION FOR SOME ATMOSPHERIC
COMPOUNDS (kJ)

	ΔH_o	ΔG_o
O_3	142.2	163.4
CO_2	−393.4	−394.3
CO	−110.5	−137.2
NO	90.4	86.7
NO_2	33.8	51.8
HNO_3 (aq)*	−206.5	−110.5
SO_2	−296.8	−300.3
SO_3	−395.1	−370.3
H_2SO_4 (aq)*	−907.3	−741.8
$H_2O^{\dagger}$	−241.8	−228.6
H_2O (l)‡	−285.7	−237.2

*Values for aqueous solutions
†Gas phase
‡Liquid phase

entropy. Also

$$\Delta G = -RT \ln K \tag{2.15}$$

where K is the equilibrium constant for the reaction, and R is the gas constant. Therefore, the equilibrium constant is related to the enthalpy and entropy by

$$\ln K = -\Delta H/RT + \Delta S/R \tag{2.16}$$

As the temperature increases, $\ln K$ becomes less negative (favoring products) if ΔH is positive, and less positive (favoring reactants) if ΔH is negative. If K is known at a standard temperature, T_0 (usually 25°C), it can then be calculated at any other temperature from the enthalpy change,

$$\ln K - \ln K_0 = (1/T_0 - 1/T)\Delta H/R \tag{2.17}$$

assuming that ΔH does not depend on the temperature (a good assumption for gas-phase reactions, which are not subject to solvation changes).

Because of the positive ΔH for reaction (2.13), NO formation becomes significant whenever air is heated sufficiently, as in combustion processes, or in the path of lightning bolts. Lightning and forest fires are major sources of atmospheric NO, and the anthropogenic contributions from fuel combustion are comparable to the natural sources (see Figure 4.1, p. 254).

When the temperature is lowered, the reverse of reaction (2.13) should proceed spontaneously. Once outside the combustion zone, NO is unstable with respect to conversion back to N_2 and O_2. But the reaction has a high activation energy (left side of Figure 2.23) and is slow. It is so slow that another reaction intervenes (right side of Figure 2.23), namely

$$2NO + O_2 = 2NO_2 \qquad \Delta H = -113.2 \text{ kJ} \tag{2.18}$$

Since ΔG is negative (see Table 2.5), this reaction likewise proceeds spontaneously at low temperatures. It is much faster than the reverse of reaction (2.13) (despite the larger enthalpy change for the latter) because there is a more efficient mechanism, involving oxygen free radicals (see p. 138), which lowers the activation energy.

The NO_2 is in turn converted to HNO_3

$$4NO_2 + O_2 + 2H_2O = 4HNO_3 \qquad \Delta H = -388.6 \text{ kJ} \qquad (2.19)$$

The nitric acid molecules, being hygroscopic, are absorbed in raindrops and rained out of the atmosphere. (Correspondingly, the enthalpy change for reaction (2.19) is determined from the standard enthalpies for liquid water and aqueous nitric acid, in Table 2.5.) This is the mechanism that removes nitrogen oxides from the atmosphere.

The nitrogen oxides illustrate the relative importance of kinetics and thermodynamics in atmospheric chemistry: the lack of an efficient mechanism for converting NO to N_2 and O_2 allows the steady-state concentration of the nitrogen oxides to exceed their concentrations at equilibrium. From reaction (2.13) we can obtain the equilibrium expression

$$[NO]^2/[N_2][O_2] = K_{eq} = 10^{-\Delta G/(2.303)RT} \qquad (2.20)$$

(the factor 2.303 converts natural to base-ten logarithms; also $R = 8.314$ J/K). From the standard free-energy change (Table 2.5), we can obtain $K_0(25°C) = 10^{-30.3}$. Using the enthalpy change, we can calculate [via equation (2.17)] that at an ambient temperature of 288 K (average global surface temperature), $K_{eq} = 10^{-31.4}$, while at a high temperature of, say, 2000 K, $K_{eq} = 10^{-3.4}$. Since, by volume, the atmosphere is 78% N_2 and 21% O_2, $[N_2] = 0.78$ atm, and $[O_2] = 0.21$ atm at sea level. Then, using equation (2.20), we calculate that $[NO] = 10^{-16.1}$ atm at 288 K, and $10^{-2.1}$ atm at 2000 K, an enormous difference. The average atmospheric concentration of NO is roughly 10^{-4} ppm, or 10^{-10} atm at sea level, about six orders of magnitude higher than the equilibrium value at 288 K.

2. Free Radical Chain Reactions

Gas-phase reactions such as those described above are generally slow because there is no low-energy pathway for the direct reaction of neutral molecules. For example, a mixture of H_2 and Cl_2 is stable indefinitely, even though their reaction to form HCl is highly exothermic

$$H_2 + Cl_2 = 2HCl \qquad \Delta H = -184.6 \text{ kJ} \qquad (2.21)$$

The exothermicity results because the H–Cl bonds are stronger than the H–H and Cl–Cl bonds; but the mixture is stable because there is no easy way to rearrange the latter into the former. One way would be to break the H–H and Cl–Cl bonds simultaneously and recombine the atoms, but at ambient temperatures the energy for bond breaking is unavailable. Another way would be to transfer electrons from the less electronegative (H_2) molecule to the more electronegative (Cl_2) molecule, thereby weakening the bonds. However, the appropriate molecular orbitals do not have the right energies: the highest occupied molecular orbital (HOMO) on H_2 (the σ orbital on the left side of Figure 2.24) is too low in energy, and the lowest unoccupied molecular orbital (LUMO) on Cl_2 (the σ* orbital on the right side of Figure 2.24) is too high in energy to allow electron transfer.

Under certain conditions, the reaction occurs rapidly. If the H_2/Cl_2 mixture is irradiated with blue light, it reacts explosively. Blue light triggers the reaction because the blue

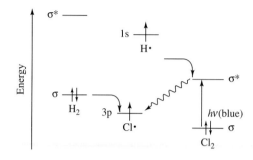

Figure 2.24 Schematic energy level diagram for the HOMO and LUMO of H_2 and Cl_2, and of H· and Cl·. A direct reaction path is unavailable energetically, but Cl· and H· atoms can react via electron transfer to or from their half-filled valence orbitals. A radical chain reaction is initiated by a Cl· atom, generated by photodissociation of Cl_2.

photons are absorbed by the Cl_2, which then dissociates into Cl· atoms:

$$Cl_2 + h\nu(\text{blue}) = 2Cl· \tag{2.22}$$

The LUMO of the Cl· atom (the 3p orbital in the center of Figure 2.24) is at a low enough energy to allow interaction with the H_2 HOMO, leading to the formation of an H–Cl bond at the expense of the H–H bond:

$$Cl· + H_2 = HCl + H· \tag{2.23}$$

The H· atom left behind can then attack a Cl_2 molecule in the same way:

$$H· + Cl_2 = HCl + Cl· \tag{2.24}$$

Reactions (2.23) and (2.24) together constitute a *radical chain reaction*. The Cl· and H· atoms are *free radicals,* having a single electron in a valence orbital. Each reaction produces a product molecule and a new free radical, which can then continue the chain. The free radicals are the *chain carriers.* The chain reaction is *terminated* when radicals react with each other, for example

$$H· + Cl· = HCl \tag{2.25}$$

producing no new free radical. But chain termination is an improbable event because the concentration of the chain carriers is much lower than that of the starting molecules, H_2 and Cl_2. Therefore, the chain reaction continues for many cycles and constitutes an efficient mechanism for the overall reaction, (2.21).

3. Oxygen Radicals

Reactions of O_2 molecules in the gas phase are slow for an additional reason, having to do with the electronic structure of oxygen. The highest occupied orbitals of O_2 are a pair of degenerate π^* orbitals (Figure 2.25), each of which contains an unpaired electron; O_2 is a triplet molecule. Even though these orbitals are sufficiently low in energy to attract electrons from donor molecules (a reflection of oxygen's high electronegativity), if the donor electrons are paired, the reactions cannot take place; they are "spin-forbidden." Electron pair donors do react rapidly with "singlet oxygen," O_2 molecules in which the two valence electrons are paired up in one of the π^* anti-bonding orbitals, leaving a two-electron vacancy in the other (Figure 2.26). But singlet oxygen is an excited state of O_2, lying 97.9 kJ/mol above the

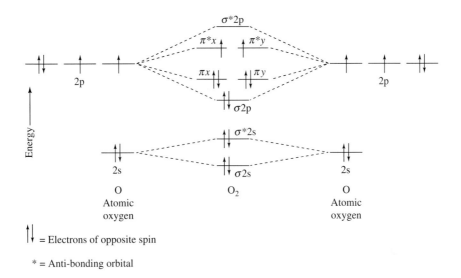

= Electrons of opposite spin

* = Anti-bonding orbital

Figure 2.25   Molecular orbital diagram showing the electronic structure of O_2. (π^* is an anti-bonding orbital. Electrons occupying these orbitals weaken the O–O bond. See problems 13 and 14, end of Part II.)

ground state. Under ambient conditions, the population of singlet O_2 molecules is negligible. It is for this reason that organic molecules, even electron-rich ones, are stable in air. The special electronic structure of O_2 allowed life to evolve in the presence of an oxidizing atmosphere.

Triplet O_2 *does* react rapidly if it encounters electron donors with unpaired electrons. These can be free radicals, or they can be transition metal ions with partially filled *d* orbitals. In general, free radicals control the gas-phase reactivity of O_2, while transition metal ions control its reactivity in aqueous solution and in biological tissues.

Upon reaction with a free radical such as the *alkyl* radical, $RCH_2\cdot$ (an alkyl radical is used for illustration, but the process is quite general), one of the O_2 electrons becomes paired with that of the radical, producing a new R–O bond:

$$RCH_2\cdot + O_2 = RCH_2O_2\cdot \qquad (2.26)$$

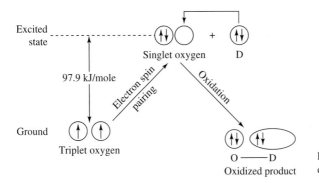

Figure 2.26   Slow reaction of oxygen with donor (D) containing two paired electrons.

The other O_2 valence electron remains unpaired, so the product is also a radical, an *alkylperoxyl* radical. Because in $RCH_2O_2\cdot$ one of the O_2 π^* orbitals is fully occupied, the O–O bond is weakened. Consequently, alkylperoxyl radicals are good O-atom donors to species that are avid for oxygen:

$$RCH_2O_2\cdot + X = XO + RCH_2O\cdot \tag{2.27}$$

X can be a variety of entities, including NO and SO_2, or organic molecules with C=C double bonds; in fact, this reaction with peroxyl radicals is the primary mechanism for the oxidation of these molecules in the atmosphere.

If R is an alkyl group, the $RCH_2O\cdot$ species remaining after the O-atom donation is an *alkoxyl* radical. It quickly reacts with O_2, donating an H atom

$$RCH_2O\cdot + O_2 = RCHO + HO_2\cdot \tag{2.28}$$

to produce an aldehyde and the *hydroperoxyl* radical. Like alkylperoxyl radicals, the hydroperoxyl radical readily donates an O atom to an acceptor, generating the *hydroxyl* radical:

$$HO_2\cdot + X = XO + HO\cdot \tag{2.29}$$

4. Hydroxyl Radical

The hydroxyl radical plays a special role in atmospheric chemistry. It readily abstracts an H atom from organic molecules, forming water as a product and leaving an organic radical:

$$HO\cdot + RCH_3 = RCH_2\cdot + H_2O \tag{2.30}$$

The O–H bond formed in this reaction is stronger than the C–H bonds of organic molecules. Reaction (2.30) initiates the oxygen radical cycle described in the preceding section. The sum of reactions (2.26)–(2.30) is

$$RCH_3 + 2O_2 + 2X = RCHO + 2XO + H_2O \tag{2.31}$$

The aldehyde product is subject to additional H-atom abstraction reactions with $HO\cdot$, followed by further O_2 reactions. C–C bonds are cleaved, via rearrangement reactions, and further oxidation products ensue, leading eventually to CO_2.

Hydroxyl radicals are continually produced in the atmosphere via the action of sunlight on ozone (O_3):

$$O_3 + h\nu\,(\lambda < 325 \text{ nm}) = {}^*O_2 + {}^*O \tag{2.32}$$

Ozone is a meta-stable, high-energy form of oxygen. It exists mainly in the stratosphere where it provides a screen for UV light (see section on Stratospheric Ozone, starting on p. 141). When O_3 absorbs a UV photon, it dissociates into O_2 and O atoms, a few of which (2%) are produced in electronically excited states, ${}^*O_2$ and ${}^*O$. The ${}^*O$ atoms have enough energy to react with water, producing hydroxyl radicals:

$$^*O + H_2O = 2HO\cdot \tag{2.33}$$

There are not many O_3 molecules in the lower atmosphere, and they do not often encounter UV photons, most of which are filtered out by the stratospheric ozone; thus the number of hydroxyl radicals is limited. Nevertheless, hydroxyl radicals are so reactive that reactions (2.32) and (2.33) produce enough of them to ensure that radical chemistry consumes organic molecules in short order, even though the organic molecules in the presence of O_2 by itself would be stable indefinitely.

Hydroxyl radicals carry out other important atmospheric reactions. They convert NO_2 to nitric acid:

$$HO\cdot + NO_2 = HNO_3 \tag{2.34}$$

and thereby provide the mechanism for washing nitrogen oxides out of the atmosphere. Likewise, they convert SO_2 to H_2SO_4 via the reactions

$$HO\cdot + SO_2 = HSO_3\cdot \tag{2.35}$$

$$HSO_3\cdot + O_2 + H_2O = H_2SO_4 + HO_2\cdot \tag{2.36}$$

and the resulting hydroperoxyl radicals produce additional hydroxyl radicals, via reaction (2.29).

Finally, hydroxyl radicals convert carbon monoxide to carbon dioxide:

$$HO\cdot + CO = CO_2 + H\cdot \tag{2.37}$$

The great stability of CO_2 induces the rupture of the O–H bond, and H· atoms are left as a co-product. The H· atoms immediately attack O_2 molecules,

$$H\cdot + O_2 = HO_2\cdot \tag{2.38}$$

producing hydroperoxyl radicals, which produce still more hydroxyl radicals. Carbon monoxide is a natural product of plant decay and an important component of combustion emissions. It is present in the atmosphere at levels exceeding those of organic compounds, and is therefore an important participant in atmospheric oxygen radical chemistry.

The reactions (2.26)–(2.38) constitute the main pathways for the oxidation and removal of most oxidizable molecules in the atmosphere. The hydroxyl radical is nature's vacuum cleaner. If anthropogenic emissions to the atmosphere continue to increase, there is some concern that the vacuum cleaner may become clogged; the oxidizable molecules might use up the available HO·. If the HO·-consuming reaction rates exceed the HO·-producing reaction rates, then the steady-state HO· concentration would diminish, thereby increasing the lifetime, and the concentrations, of pollutants. However, the factors controlling HO· production are poorly understood. According to some evidence, O_3 levels and, therefore, the rates of HO· production, have been increasing in the lower atmosphere, thanks, ironically, to increased pollution. Whether more pollution changes the balance between the rates of HO· consumption and production is an unanswered question.

5. Transition Metal Activation of O_2

We complete our discussion of oxygen reactivity by considering the interactions of oxygen with transition metal ions in aqueous solution. Transition metal activation is important for atmospheric chemistry because a good deal of it actually occurs in raindrops. For example, raindrops are the site of a substantial fraction of atmospheric H_2SO_4 production. Transition metals activate oxygen for further reduction by donating single electrons and forming oxygen-containing free radicals. Because water is a neutral medium for radicals, they can persist in raindrops, where they are brought into close proximity to other more reactive molecules.

Transition metal ions enter raindrops in trace concentrations because dust particles are often included in condensation nuclei. Manganese and iron are particularly important, since they occur commonly in minerals. They are very reactive with O_2 in their divalent state, M^{2+},* because they have unpaired d electrons (5 for Mn^{2+} and 4 for Fe^{2+}), which are readily donated to O_2. The first step in the reaction involves formation of a dioxygen adduct

$$M^{2+} + O_2 = (MO_2)^{2+} \tag{2.39}$$

Biology exploits this adduct formation by using the iron-containing protein hemoglobin to transport O_2 from the lungs to the tissue. The protein protects the bound O_2 from reacting further. However, when O_2 binds to free Fe^{2+} or Mn^{2+} ions in solution, it reacts immediately. If an additional M^{2+} ion is encountered, then a *binuclear* complex is formed

$$(MO_2)^{2+} + M^{2+} = (MO_2M)^{4+} \tag{2.40}$$

This complex readily breaks apart by cleaving the O–O bond to form a pair of *ferryl* or *manganyl* ions

$$(MOOM)^{4+} = 2(MO)^{2+} \tag{2.41}$$

If the oxygen is considered to have its normal 2– valence in these ions, then the metal valence is 4+. But other resonance forms contribute to the electronic structure:

$$(M^{4+}O^{2-}) <-> (M^{3+}O^-) <-> (M^{2+}O^0)$$

The metal and oxygen atom compete on an even footing for the valence electrons. When a proton is added, a hydroxyl radical is generated via the second resonance form

$$(M^{3+}O^-) + H^+ = M^{3+} + HO\cdot \tag{2.42}$$

Alternatively, the original O_2 adduct can dissociate to produce the *superoxide* ion, O_2^-

$$(MO_2)^{2+} = M^{3+} + O_2^- \tag{2.43}$$

The M^{3+} ion is stabilized by strong attraction to the solvent water molecules. At low pH, O_2^- is protonated and becomes the hydroperoxyl radical, $HO_2\cdot$, which, as we have seen, is a good O-atom donor. In addition, superoxide rapidly disproportionates to dioxygen and hydrogen peroxide

$$2O_2^- + 2H^+ = O_2 + H_2O_2 \tag{2.44a}$$

This reaction occurs spontaneously because it takes more energy to add the first electron to O_2 and form a superoxide radical than to add the second electron. The net result is the transfer of two electrons from M^{2+} to O_2, forming the highly reactive oxidant hydrogen peroxide.

---

*M^{n+} represents a metal cation of given charge, n+.

Additional H_2O_2 reaches raindrops from the atmosphere as a result of the analogous dispro-portionation of hydroperoxyl radical (the protonated form of superoxide) in the gas phase:

$$2HO_2\cdot = O_2 + H_2O_2 \tag{2.44b}$$

When hydrogen peroxide encounters an M^{2+} ion, an additional electron is transferred, and the O–O bond breaks, leaving water and a hydroxyl radical

$$M^{2+} + H_2O_2 + H^+ = M^{3+} + H_2O + HO\cdot \tag{2.45}$$

Thus, the M^{2+} ions react with O_2 to produce hydroxyl radicals through either the binuclear or the mononuclear pathway. (The relative importance of the two pathways depends on the M^{2+} concentration, since binuclear complex formation, reaction (2.40), increases with in-creasing M^{2+} concentration.) The hydroxyl and hydroperoxyl radicals carry out rapid oxidations in solution, just as they do in the gas phase.

Since raindrops contain only traces of Fe^{2+} and Mn^{2+}, it might be thought that radical production would diminish when these traces are converted to M^{3+} ions and used up. However, the M^{3+} ions can be reduced back to M^{2+} by the action of sunlight. Under ambi-ent pH conditions, the M^{3+} ions form hydroxide complexes, $M^{3+}OH^-$; when these complexes absorb a UV photon, the hydroxide transfers an electron to the metal ion, leaving hydroxyl radical and M^{2+}, which is ready to react with O_2 to produce more radicals:

$$M^{3+}OH^- + h\nu = M^{2+} + HO\cdot \tag{2.46}$$

Thus a *photocatalytic* cycle is set up, in which trace concentrations of metal ions can induce sustained rates of radical production under the influence of sunlight and ambient O_2.*

C. STRATOSPHERIC OZONE

High in the atmosphere, a layer of ozone, O_3, acts as a solar filter, screening out the ultra-violet rays. These photons are energetic enough to induce photochemical damage in biological molecules. They are harmful to plants and animals alike, so much so that it is unlikely that complex life forms would have evolved without this ultraviolet shield.

The ozone layer developed as Earth's atmosphere grew rich in oxygen, through the ad-vent of green plants. O_3 is a high-energy form of oxygen. The conversion of O_3 to O_2 releases energy; likewise the conversion of O_2 to O_3 requires the input of energy. This energy is pro-vided by ultraviolet photons that encounter O_2 high in the atmosphere. Thus, the ultraviolet shield itself is created by ultraviolet photons.

Although O_3 is inherently unstable, its decomposition is slow, as are other gas-phase molecular reactions in the stratosphere. It is subject, however, to *catalytic* destruction by chemicals that are avid for oxygen atoms. Scientists have long been aware of this possibility and have investigated potential threats to the ozone shield from gases emitted by human

---

*Y. Zuo and J. Hoigné (1993). Evidence for photochemical formation of H_2O_2 and oxidation of SO_2 in authentic fog water. *Science* 260: 71–73; and B.C. Faust et al. (1993). Aqueous-phase photochemical formation of peroxides in authentic cloud and fog waters. *Science* 260: 73–75.

activity. In the early 1970s, attention focused on the potentially destructive effects of supersonic aircraft, which inject one of the ozone-destroying agents, NO, directly into the stratosphere.*

In a research article in 1974,[†] Mario Molina and F. Sherwood Rowland (who, along with Paul Crutzen, won the 1995 Nobel Prize in chemistry for their work on stratospheric ozone) called attention to what has turned out to be a far greater danger. They pointed out that chlorine atoms, Cl·, are very efficient ozone-destroyers, and that CFCs deliver Cl· directly to the stratosphere. These CFC molecules undergo no chemistry until they drift into the stratosphere, where the UV photons can break the C–Cl bonds. It is a great irony that CFCs came into widespread production precisely because their lack of reactivity in the troposphere makes them safe to use in many applications. It is the same lack of reactivity that preserves them until they migrate to the stratosphere, where they become potent ozone-destroyers. Moreover, because it takes about a century for the CFCs to complete their journey back to the troposphere in degraded form, the effect is felt long after they are produced.

Four years after Molina and Rowland's warning, the U.S. government banned the use of CFCs as propellants in aerosol cans, but it took another decade to produce comprehensive international action. In 1987, the Montreal protocol committed the governments of the world to cutting back CFC production to 50% of 1986 levels by 1999. The London amendments of 1990 strengthened the terms, including a complete phase-out of the main CFCs by 2000. In 1992, the Copenhagen agreements accelerated the timetable for phase-out to 1996 (with a ten-year grace-period for developing countries). In view of the technological importance of CFCs and their commercial value, the delay between warning and action was hardly surprising. A major impetus to action was the discovery of a dramatic "hole" in the ozone layer over the south pole. Figure 2.27 illustrates the sharp drop in the springtime ozone level over Antarctica, from the mid-1970s onward. This effect had *not* been predicted by computer modeling at the time; the discrepancy revealed a role for an additional class of chemical reactions that had not been anticipated. These reactions, discussed on pp. 154–158, are now well understood, but the episode is a cautionary example of potential surprises that we might face, due to our incomplete knowledge of environmental chemistry.

1. Atmospheric Structure

Our gaseous envelope extends many miles from the surface of Earth. The atmosphere is quite uniform throughout its extent with respect to its major chemical constituents (Table 2.1), except for water vapor, which is concentrated in the lower region. The air is far from uniform in other respects, however. It grows thinner with increasing altitude; the density falls off, roughly logarithmically, with increasing distance from the surface. Even

---

*This concern was a consideration in the cancellation at that time of the U.S. effort to develop a civilian supersonic transport plane. This issue has resurfaced in the 1990s as the airline industry reconsiders supersonic jets to meet the expected increase in worldwide demand for air transport. New findings suggest that flights in the lower stratospere may cause less damage to the ozone layer than initially expected (see p. 154), although more research is needed.

[†]M. J. Molina and F. S. Rowland (1974). Stratospheric sink for chlorofluoromethanes: chlorine-catalyzed destruction of ozone. *Nature* 249: 810–812.

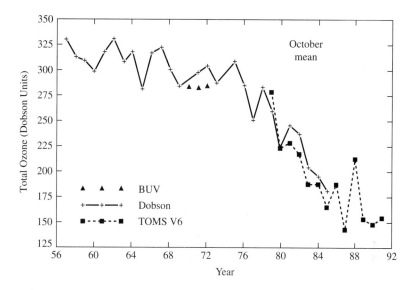

Figure 2.27 October mean ozone values in the Antarctic (symbols used: pluses, Dobson ground observations from Halley Bay; squares and triangles are from satellite observations). *Source:* WHO Global Ozone Research and Monitoring Project (1992). *Scientific Assessment of Stratospheric Ozone: 1991* (Report No. 25, NASA/NOAA/UKDOE/UNEP/WMO) (Geneva, Switzerland: World Meteorological Organization).

less uniform than the density is the temperature, which depends on altitude, as shown in Figure 2.28. The slope of the curve of temperature versus altitude is called the *lapse rate*. The lapse rate of the atmosphere reverses several times. Up to about 10 km, temperature decreases uniformly with increasing altitude. This reflects the fact that the atmosphere in this lower region is heated from below by convection and radiation from Earth's surface. Above 10 km, however, the temperature increases again with increasing altitude, reaching a maximum near 50 km; beyond this altitude, the temperature decreases once again. The maximum in the temperature profile reflects a heating process high in the atmosphere, which is due to the absorption of ultraviolet solar photons by the ozone layer. Beyond about 90 km, the temperature rises once more due to the absorption of solar rays in the far-ultraviolet region by atmospheric gases, principally oxygen. These far-ultraviolet rays are sufficiently energetic to ionize molecules and break them into their constituent atoms. Due to the thinness of the atmosphere at such high altitudes, fragments recombine only rarely; an appreciable fraction of the gases in this region, called the *thermosphere,* exist as atoms or ions.

The structure of the temperature profile reflects a physical layering of the atmosphere as well. A negative lapse rate leads to convection of the air; warm air rises from below and cool air sinks. But a positive lapse rate provides a region of stability with respect to convection, since warm air overlies cool air. The change from negative to positive lapse rate is called a *temperature inversion,* the point at which the lapse-rate changes marks a stable boundary between two physically distinct layers of air. Local temperature inversions in the troposphere are quite common over many cities, particularly those nestled in surrounding mountains or,

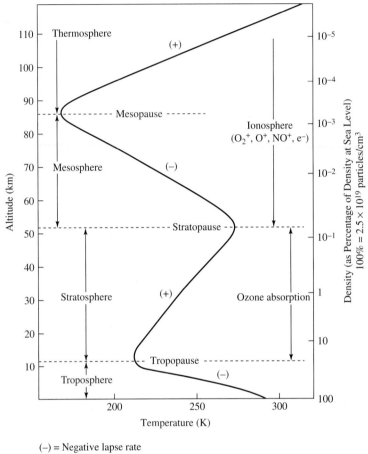

Figure 2.28 Layer structure of the atmosphere.

as in the case of Los Angeles, where prevailing winds from the ocean blow cool air into a region ringed on three sides by mountains. Warm air flowing over the mountaintops traps the cool air below, allowing pollutants to build up for considerable periods. Anyone flying into Los Angeles on a sunny day is likely to see the local temperature inversion which produces a distinct boundary between the clear air above and the brown smoggy air below.

The temperature inversion shown in Figure 2.28 at an altitude of about 10 km is called the *tropopause*. It marks a global layering of the atmosphere into the troposphere below and the stratosphere above. The troposphere constitutes about 10% of the atmosphere's height but contains 80% of its mass. Due to its negative lapse rate, the air it contains is mixed rapidly by convection. It is also a region of much turbulence, due to the global energy flow that results from the imbalances of heating and cooling rates between the equator and the poles.

The stratosphere, on the other hand, is a quiescent layer; due to its positive lapse rate, it mixes slowly. The tropopause itself is a stable boundary, and the flux of air across it is low. As we saw in the case of sulfate aerosols from volcanic explosions, residence times of molecules or particles in the stratosphere are measured on a scale of years or more. The air in the stratosphere is quite thin, so that pollutants in the stratosphere have a relatively greater global impact than they would in the much denser troposphere.

2. Ultraviolet Protection by Ozone

Concern about pollution of the stratosphere centers on possible threats to the ozone layer, which serves two essential functions: it protects living matter on Earth from the harmful effects of the sun's ultraviolet rays, and it provides the heat source for layering the atmosphere into a quiescent stratosphere and a turbulent troposphere. The ozone layer extends over tens of kilometers in altitude but is very thin; if compressed to 1 atm pressure at 0°C its thickness would be only 3 mm on average. Ozone thicknesses are measured in *Dobson units* (DU), named after G.M.B. Dobson, who conducted pioneering measurements of the stratosphere in the 1920s and 1930s. One Dobson unit equals one-hundredth of a millimeter thickness of the ozone layer at standard temperature and pressure. Therefore, ozone columns typically have thicknesses of 300 DU.

Absorption of ultraviolet radiation by stratospheric ozone is strong enough to eliminate much of the ultraviolet tail of the solar radiation spectrum at the surface of Earth (Figure 2.29). Ozone absorbs light with wavelengths between 200 and 300 nm, in the UV range of the solar spectrum. These ultraviolet rays are harmful to life. They carry enough energy to break the bonds of organic molecules and produce reactive fragments. Ultraviolet radiation that passes through the ozone layer produces sunburn and also skin cancer, particularly in people with light pigmentation. The damage done to living tissue depends upon the solar wavelength, as shown in Figure 2.30. The solid curve in the figure shows the spectrum of the skin's sunburn sensitivity at constant light intensity. Also shown is the spectrum of the solar rays at ground level. The product of these two curves gives the action spectrum for sunburn, that is, the response of the skin to solar radiation. The overlap of ground-level radiation with the sunburn sensitivity curve would be much greater without the filtering effects of the ozone layer.

Exposure to solar ultraviolet radiation seems clearly linked to the incidence of skin cancer (Figure 2.31). Both the incidence of skin cancer and the solar ultraviolet flux decrease with increasing distance from the equator. The UV flux at ground level depends upon latitude, because the sun's rays fall more directly at lower latitudes, and because the ozone layer is thinner over the equator than toward the poles. Between 30 and 46 degrees latitude north, the annual ultraviolet flux is reduced by a factor of three, and the incidence of malignant skin cancer falls roughly by a factor of two.

Ultraviolet rays also damage green plants. Their light-harvesting photosynthetic apparatus is tuned to visible radiation and can be destroyed by sufficiently intense ultraviolet radiation. Particularly susceptible are the light-harvesting phytoplankton that float at the surface of the ocean and provide the starting material for marine food chains. Antarctic

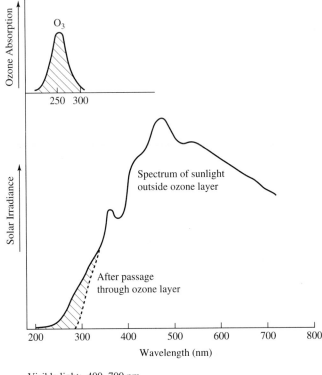

Visible light: 400–700 nm
Ultraviolet light: < 400 nm

Figure 2.29   Absorption of sun's ultraviolet light by ozone.

phytoplankton populations have been shown to decrease in concert with the ozone depletion, although effects on the rest of the ecosystem have not yet been detected.

How much does a given change in the ozone layer affect the amount of UV light reaching Earth's surface? We can work out the relationship in the following way. For an absorbing constituent of the atmosphere, the total number of molecules through which the light passes can be expressed as an equivalent thickness, l, of a global layer of the gas, brought to standard temperature (0°C) and pressure (1 atm). The fraction of light transmitted through the layer decreases exponentially with increasing thickness (the Beer-Lambert law):

$$T = I/I_0 = e^{-\varepsilon l} \tag{2.47}$$

where ε is the *absorptivity* (also called the *extinction coefficient*) expressed in units of inverse length. If the product εl is 1.0, then 37% of the incident light is transmitted, whereas if it is 10.0, only 0.005% is transmitted.

On average, the equivalent thickness of ozone at 0°C and 1 atm is 0.34 cm. The value of ε depends on the wavelength because the absorption of light by ozone varies according to the wavelength (Figure 2.29). The product εl goes from 1 to 10 between 310 and 290 nm, the critical region for sunburn and perhaps for skin cancer.

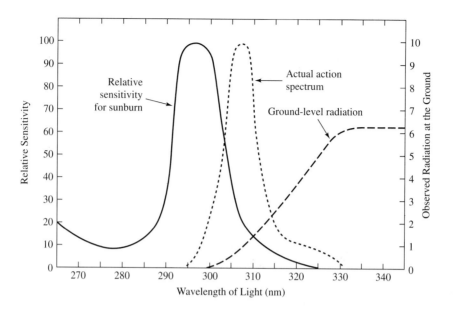

Figure 2.30   Action spectrum of ultraviolet radiation damage to living tissue.

The effect of a small decrease in the ozone thickness can be estimated by differentiating equation (2.47):

$$dT = -\varepsilon e^{-\varepsilon l}\, dl \tag{2.48}$$

Dividing through by equation (2.47) gives the relative response:

$$dT/T = -\varepsilon\, dl = -\varepsilon l\, dl/l \tag{2.49}$$

Thus, a given fractional decrease in l produces a fractional increase in T, which is amplified by the product εl. A 1% decrease in the ozone layer gives a 1% increase in ultraviolet transmittance at 310 nm, a 3% increase at 300 nm, and 10% at 290 nm. Thus, ozone changes affect the UV flux most sensitively at the shortest wavelengths where damage to biological molecules is greatest.

These calculations help to explain the concern over ozone depletion. During the 1980s, globally averaged total ozone was declining at a rate of 0.4% per year. Using the above calculations, the ozone loss per decade (4%) would cause an increase of 12% and 40% in the transmittance of UV light at 300 and 290 nm, respectively. These perturbations would shift the action spectrum shown in Figure 2.30 to the left and enhance the damage of sunlight to living tissue.

3. Ozone Chemistry

The reactions responsible for the steady-state concentrations of ozone in the stratosphere are shown in Figure 2.32. Ozone is formed in the stratosphere when O_2 molecules absorb solar radiation. The far-ultraviolet photons of the sun have enough energy to split oxygen molecules into oxygen atoms high in the atmosphere (reaction a). The oxygen atoms combine

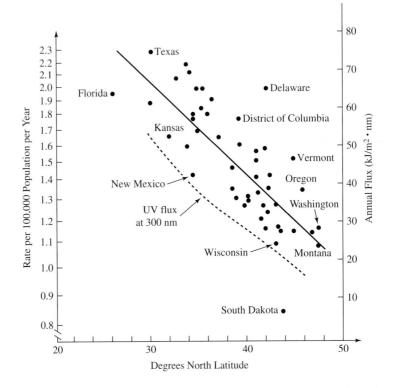

Figure 2.31   Variation with latitude of UV flux at 300 nm (dashed line) and death from skin cancer among white males in the United States (solid line) (excluding Alaska and Hawaii). *Source:* Adapted from F. S. Rowland (1982). Chlorofluorocarbons and stratospheric ozone. In *Light, Chemical Change, and Life.* J. D. Coyle et al., eds. (Washington, DC: Open University Press). Copyright © 1982 by Open University Press. Reprinted with permission.

with other oxygen molecules to form ozone (reaction *b*). Because one molecule is created out of two, a third molecule labeled *M* must be present at the encounter in order to carry away some of the energy released in the reaction; otherwise the ozone molecules would fly apart as fast as they were formed. *M* can be any molecule that happens to be present. In the atmosphere, it is most likely to be nitrogen or another molecule of oxygen. (Likewise, many of the reactions we have previously considered involve a single product formed from two reactants. They all require a "third body" to carry off the excess energy. Usually, we don't bother to include *M* in the chemical equation because it appears on both sides and does not itself enter into the reaction. It does, however, influence the rate of the reaction, and is therefore included here.)

Ozone is also destroyed by solar radiation. When O_3 absorbs a solar UV photon, it dissociates into O_2 and O (reaction *c*). (We have already encountered this process in reaction (2.32), noting that a small percent of the O atoms are electronically excited.) And when O_3 encounters an O atom, they can combine to form two O_2 molecules (reaction *d*). These reactions limit the extent to which O_3 can build up.

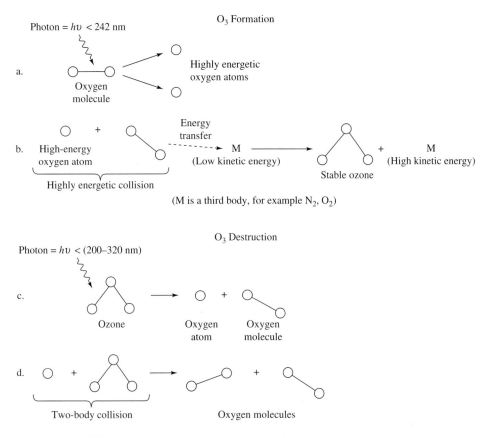

Figure 2.32   Reactions of ozone formation and destruction in the stratosphere.

We can calculate a *steady-state* concentration of ozone by setting the formation rate equal to the destruction rate. Reaction rates are given by the products of the reactant concentrations times the rate constants, k. For the four reactions in Figure 2.32:

$$\text{rate } a = k_a[O_2]$$

$$\text{rate } b = k_b[O][O_2][M]$$

$$\text{rate } c = k_c[O_3]$$

$$\text{rate } d = k_d[O][O_3]$$

The rate of ozone production is rate b, while the loss rate is rate c + rate d. The ozone steady-state condition is then

$$\text{rate } b = \text{rate } c + \text{rate } d$$

or

$$k_b[O][O_2][M] = k_c[O_3] + k_d[O][O_3] \qquad (2.50)$$

In this expression, $[O_2]$ and $[M]$ (the concentration of air molecules) are known, and the rate constants are known from experiments. There are two unknowns, $[O_3]$ and $[O]$. In order to eliminate one of them, we need an additional equation. This can be provided by assuming that the concentration of O atoms is at steady-state; the production and loss rates of O atoms must also balance. The rate of O production is twice the rate of reaction a (two oxygen atoms produced per O_2 molecule split) plus the rate of reaction c, while the loss rate is rate b + rate d: At steady state,

$$2(\text{rate } a) + \text{rate } c = \text{rate } b + \text{rate } d$$

or

$$2k_a[O_2] + k_c[O_3] = k_b[O][O_2][M] + k_d[O][O_3] \tag{2.51}$$

Equations (2.50) and (2.51) can now be solved for the steady-state concentrations of O and O_3. For example, subtraction of equation (2.50) from equation (2.51) gives:

$$2k_b[O][O_2][M] = 2k_a[O_2] + 2k_c[O_3] \tag{2.52}$$

which can be solved for $[O]$. The expression is simplified if we assume that $k_a[O_2] \ll k_c[O_3]$, that is, that photolysis of O_2, reaction a, produces far fewer O atoms than does photolysis of O_3, reaction c. This assumption turns out to be valid for the lower reaches of the stratosphere, where O_3 is important because the flux of solar photons is much smaller in the far-ultraviolet region (O_2-splitting) than in the near ultraviolet (O_3-splitting). Dropping $k_a[O_2]$ from equation (2.52), we obtain

$$[O] = k_c[O_3]/k_b[M][O_2] \tag{2.53}$$

Addition of equations (2.50) and (2.51) gives

$$2k_a[O_2] = 2k_d[O][O_3] \tag{2.54}$$

After substituting equation (2.53), we can solve for the $[O_3]/[O_2]$ ratio:

$$[O_3]/[O_2] = [k_a k_b[M]/k_c k_d]^{1/2} \tag{2.55}$$

The value of this quantity depends on the altitude. As the altitude increases, several variables change: the concentration of air molecules $[M]$ decreases; both k_a and k_c increase because the photon flux increases; and k_b and k_d increase slightly because the temperature increases (in the stratosphere). At an altitude of 30 km, the concentration of air molecules is $10^{17.7}$ molecules/cm^3, while the approximate values of the rate constants, expressed with this concentration unit (and averaged over Earth's surface), are $k_a = 10^{-11}$, $k_b = 10^{-32.7}$, $k_c = 10^{-3}$, and $k_d = 10^{-15}$. Inserting these numbers into equation 2.55 gives $[O_3]/[O_2] = 10^{-4}$.

Thus, even in the ozone layer, the number of O_3 molecules is much smaller than the number of O_2 molecules. But the O_3 population is much larger than it would be at equilibrium. The equilibrium ratio of $[O_3]/[O_2]$ can be obtained from K_{eq}, the equilibrium constant for the reaction

$$3O_2 = 2O_3 \tag{2.56}$$

From the standard free-energy change at 298 K and 1 atm pressure, 2×163.4 kJ (Table 2.5),

K_{eq} is estimated to be $10^{-57.0}$. From the equilibrium expression

$$[O_3]^2/[O_2]^3 = K_{eq} \qquad (2.57)$$

we obtain

$$[O_3]/[O_2] = [K_{eq}[O_2]]^{1/2} \qquad (2.58)$$

Since at sea level the concentration of oxygen $[O_2]$ is 0.21 atm,

$$[K_{eq}[O_2]]^{1/2} = (10^{-57.0} \times 0.21)^{1/2} = 10^{-28.8} \qquad (2.59)$$

Because the $[O_2]$ concentration and temperature are much lower in the stratosphere, this ratio in the stratosphere is much smaller. Thus, the steady-state ratio of $[O_3]/[O_2]$ exceeds the equilibrium value by many orders of magnitude.

At any given place in the atmosphere, the ozone concentration is subject to large variations because the solar flux varies throughout the day as well as seasonally. These variations are fairly regular, however, and the time-averaged ultraviolet flux depends only on the altitude, becoming smaller at lower altitudes as most of the rays are absorbed. Consequently, the steady-state calculation should give a reasonable estimate of the average ozone concentration at different altitudes. The calculation can be repeated at other altitudes, with appropriate modification of the constants, and the result is the dotted curve in Figure 2.33. The $[O_3]/[O_2]$ ratio is predicted to peak at about 30 km. At higher altitudes, ozone photolysis (reaction c in Figure 2.32) is increasingly rapid, whereas at lower altitudes, net ozone formation is limited by the decreasing supply of O atoms from reaction a.

The measured ozone profile (solid line in Figure 2.33), has the same shape as the calculated curve, but the calculated values are about a factor or two too high. Since the rate of ozone production depends only on the UV flux, this discrepancy reflects the existence of mechanisms for ozone destruction other than the one considered so far.

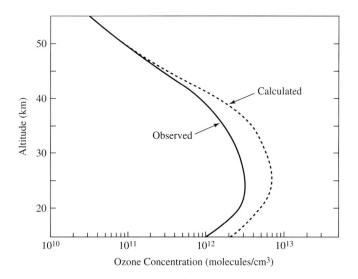

Figure 2.33 Plot of observed and calculated atmospheric ozone concentrations (not including reactions with OH, C1, Br, and NO).

4. Catalytic Destruction of Ozone

Because it is a meta-stable molecule, ozone can be destroyed by catalytic chain reactions. In such reactions, the ozone is converted to dioxygen by a chain carrier, X, which is itself restored in the process. The general reactions are

$$X + O_3 = XO + O_2 \qquad\qquad (e)$$

$$XO + O = X + O_2 \qquad\qquad (f)$$

Adding these two reactions together produces reaction d in Figure 2.32

$$O_3 + O = 2O_2 \qquad\qquad (d)$$

Reactions e and f provide an additional route to the reaction of ozone with oxygen atoms, thereby speeding up the destruction process. Since formation is unaltered, accelerating the destruction reaction lowers the steady-state concentration of ozone. Because X is regenerated, it is a catalyst; a single X molecule can destroy many ozone molecules. The chain reaction continues until X is removed by some side reaction that inactivates it.

There are many possible candidates for the chain carrier X, but four species have been identified as important for destruction of stratospheric ozone: hydroxyl radical, chlorine and bromine atoms, and nitric oxide. They are considered in turn as follows.

a. Hydroxyl radical. It was assumed formerly that most of the destruction of ozone in the lower stratosphere (16 to 20 km) was due to nitric oxide. However, according to recent research,* hydroxyl radical may be more important, accounting in this region of the atmosphere for nearly one-half of total ozone destruction. We have already seen how hydroxyl radical is formed in the troposphere, where it is produced by the reaction (2.33) of water with excited O generated from ozone photolysis (reaction 2.32). Stratospheric HO· is formed in similar fashion: excited O atoms react with a hydrogen source, supplied principally by H_2O and CH_4. Hydroxyl radical is capable of accepting an O atom from O_3 to produce HO_2· (reaction e), which in turn can react with O atoms, regenerating HO· (reaction f). The combined reactions lead to the ozone-destruction reaction (reaction d). Since water and methane are components of the natural atmosphere, the HO· cycle of reactions constitutes a natural mechanism for ozone loss. There is some concern that this loss mechanism could be accelerated due to the rising atmospheric concentration of methane which, as we saw earlier, is increasing at a rate of about 1% a year.

b. Chlorine and bromine. Chlorine and bromine atoms are highly efficient chain carriers in reactions e and f. However, Cl or Br in the stratosphere comes from few natural sources, because potential sources in the troposphere are unable to reach the stratosphere in significant quantities. For example, the oceans are abundantly endowed with chloride and bromide ions, but sea spray is efficiently cleared from the troposphere. Similarly, there are many

---

*P. O. Wennberg et al. (1994). Removal of stratospheric O_3 by radicals. *In situ* measurements of OH, HO_2, NO, NO_2, ClO, and BrO. *Science* 266: 398–404.

organochlorine and organobromine natural products, and the more volatile molecules are emitted into the troposphere where they are destroyed by hydroxyl radicals via reaction (2.30). However, the most abundant species, methylchloride and methylbromide, are long-lived enough to contribute a significant background level of halogen-induced ozone destruction.

Much larger sources of stratospheric chlorine and bromine have been inadvertently created by human manufacturing of organic compounds that contain one or two carbon atoms attached to only fluorine, chlorine, and/or bromine. These are the chlorofluorocarbons, CFCs, and the bromine-containing halons. CFCs have been widely used as refrigerants, as blowing agents for plastic foams, as propellants for aerosol sprays, and as solvents for cleaning microelectronic components. The halons have been used as fire extinguishers; the heavy bromine-containing molecules provide a blanket of gas that effectively smothers flames. The CFCs and halons have been enormously useful in these applications because they are nontoxic and nonflammable. These desirable properties are directly related to the low chemical reactivity of these molecules in the troposphere. Because they lack H atoms and thus contain no C–H bonds, CFCs and halons are not subject to oxidation either in a flame or biochemically. Even hydroxyl radicals are unable to attack these molecules, which therefore escape the tropospheric fate of most organic species. Only in the stratosphere are CFCs and halons destroyed, by the action of UV photons. The result of absorbing UV photons is to break the weakest bond in the molecule, either C–Br or C–Cl

$$RX + h\nu = R\cdot + X\cdot \tag{2.60}$$

Once released, chlorine and bromine atoms destroy ozone via the reactions $(e) + (f)$.

Satellite measurements indicate that high levels of CFCs do reach the stratosphere; above about 20 km, their concentration decreases, while concentrations increase for HCl and HF, the ultimate products of CFC destruction. These molecules are formed by a variety of mechanisms, principally the slow attack of Cl· or F· on methane molecules, which drift up from the troposphere

$$Cl\cdot (F\cdot) + CH_4 = HCl\ (HF) + CH_3\cdot \tag{2.61}$$

Since HF has no natural source, the data establish CFCs as the source of most of the stratospheric chlorine. It is for this reason that CFC production is being curtailed and phased out. Even with the ban on CFCs, CFC-catalyzed destruction of the ozone layer will persist for decades. It takes many years for these molecules to be transported up to the altitude of peak photoefficiency. The time-lag between production, consumption, and eventual effects on the ozone layer is considerable.

c. Nitric oxide. Although NO is produced abundantly in the lower atmosphere by combustion and lightning, almost all of it is oxidized to NO_2 and converted to nitric acid [reaction (2.34)] in the troposphere, after which it is rained out before reaching the stratosphere. On the other hand, nitrous oxide, N_2O, although much less abundant, is also much less reactive; it does eventually reach the stratosphere. Above 30 km, most of the N_2O is photolyzed by UV photons to produce dinitrogen and excited oxygen atoms:

$$N_2O + h\nu(UV) = N_2 + {}^*O \tag{2.62}$$

A small percentage, 10% or less, of the N_2O molecules react with excited oxygen atoms formed via reaction (2.32), as well as (2.62), to produce NO:

$$N_2O + *O = 2NO \tag{2.63}$$

This is the main source of NO in the stratosphere.

NO can act as X in ozone-destroying reactions e and f, cycling through NO_2 in the process. However, NO_2 also reacts with other carriers in O_3 destruction chains, HO· and ClO·. The reaction with HO· produces nitric acid [reaction (2.34)], as we have already seen, while the reaction with ClO· produces an analogous molecule, chlorine nitrate

$$ClO· + NO_2 = ClONO_2 \tag{2.64}$$

Nitric acid and chlorine nitrate do not participate in O_3 destruction directly. Rather, they are *reservoir* molecules; they sequester the chain-carrying species HO· and ClO· in less reactive forms, releasing them in response to UV light:

$$HONO_2 + hv = HO· + NO_2 \tag{2.65}$$

$$ClONO_2 + hv = ClO· + NO_2 \tag{2.66}$$

Although HO· and ClO· thereby remain available for O_3 destruction, binding to NO_2 reduces their concentration significantly and thereby reduces ozone destruction. Furthermore, the reservoir molecules can also be rained out when stratospheric and wet tropospheric air mix in the upper troposphere.

Thus, NO has a two-sided effect, on the one hand providing another catalytic chain mechanism for O_3 destruction, but on the other, inhibiting two other major mechanisms for O_3 destruction. Which of these effects dominates depends upon altitude.* Above about 25 km, the net effect of nitrogen oxides is to lower O_3 concentrations via reactions e and f. In fact, in the middle and upper stratosphere, nitrogen oxides account for over 50% of total ozone destruction. However, in the lower stratosphere, the overall effect of nitrogen oxides is to protect O_3 from destruction via reactions (2.34) and (2.64). These findings have given new impetus to the feasibility of supersonic jet transport in the lower stratosphere. Uncertainties persist, however, because jet emissions would not necessarily remain at low altitudes.

5. Polar Ozone Destruction

Although the reactions described so far account for the observed average levels of stratospheric ozone, they are unable to account for the ozone hole over Antarctica or for the substantial ozone reduction that has also been detected in the Arctic region. Each spring the ozone layer thins markedly over the poles. The phenomenon is limited both regionally and seasonally. Although several explanations have been considered, the cause was firmly established as catalytic destruction by Cl· when the O_3 loss was observed to coincide with a sharp rise in ClO· (Figure 2.34). But the effect is too great and too sudden to be explained by the chain reactions e and f, with X = Cl·.

---

*See Wennberg et al., op. cit., p. 152.

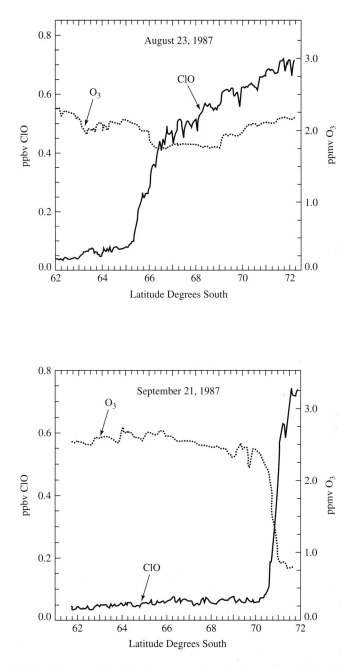

Figure 2.34 Chlorine oxide and ozone concentrations over Antarctica at 18-km altitude, August 23 and September 21, 1987.   *Source:* F. S. Rowland (1991). Stratospheric ozone in the 21st century: the chlorofluorocarbon problem. *Environmental Science and Technology* 25: 622–628. Copyright © 1991 by the American Chemical Society. Reprinted with permission.

Additional chemistry is involved, including an important role for reactions at the surface of cloud particles (Figure 2.35). In the frigid winter temperatures over the poles, the stratospheric air becomes trapped in a vortex in which clouds form, even though the air is very dry. These polar stratospheric clouds (PSCs) form first at 193 K, when particles of nitric acid trihydrate ($HNO_3 \cdot 3H_2O$) condense, and again at 187 K, when water ice particles condense. Under these conditions, the stage is set for the subsequent emergence of the ozone hole. First, the cloud particles efficiently absorb HNO_3 and $ClONO_2$, as well as HCl [the latter via reaction (2.61)]. Reactions on the surface of the particles then convert HCl and $ClONO_2$ to the more reactive Cl_2 and HOCl.

$$HCl + ClONO_2 = Cl_2 + HNO_3 \qquad (2.67)$$

$$H_2O + ClONO_2 = HOCl + HNO_3 \qquad (2.68)$$

These reactions are too slow in the gas phase to be of any importance, but they are greatly accelerated at the cloud-particle surfaces because 1) the reactant molecules are concentrated there, and 2) the formation of HNO_3 is assisted by hydrogen bonding to the water molecules in the particles. These reactions are further pushed to the right as the chlorine-containing gas

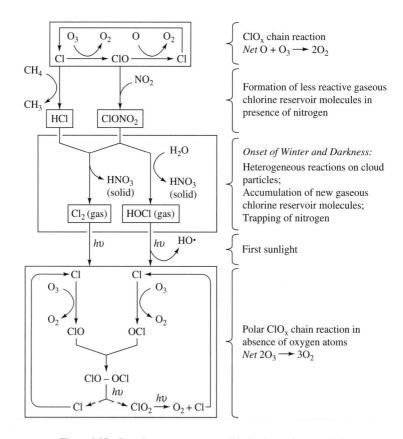

Figure 2.35 Reaction sequence responsible for Antarctic ozone hole.

products escape while the HNO_3 remains adsorbed on the ice particles. As the cloud particles grow over the winter, they sink to lower altitudes, thus physically separating the nitrogen from the active chlorine. Because nitrogen is unavailable, it cannot sequester the chlorine in the relatively unreactive molecule $ClONO_2$; the result is that, over the dark of winter, the vortex accumulates Cl_2 and $HOCl$.

As daylight returns in the springtime, Cl_2 and $HOCl$ molecules are converted to $Cl\cdot$ by UV light:

$$Cl_2 + hv = 2Cl\cdot \tag{2.69}$$

$$HOCl + hv = HO\cdot + Cl\cdot \tag{2.70}$$

We can now understand why the ozone hole appears seasonally. During the dark polar winter, Cl_2 and $HOCl$ accumulate; in spring, when the air is bathed in sunlight, the photolysis reactions produce a burst of $Cl\cdot$ atoms that react with the O_3 and produce $ClO\cdot$ (Figure 2.34).

One more reaction is needed to complete the ozone-hole mechanism. In order to continue the catalytic cycle, the $Cl\cdot$ needs to be regenerated from $ClO\cdot$. Normally this occurs via attack of O atoms in reaction *f*. But in the lower stratosphere, where the polar ozone levels decline most steeply, there are not enough O atoms to keep the reaction going at the required rate. Instead, the $ClO\cdot$ builds up to a concentration (ppb range) sufficient to form dimers:

$$2ClO\cdot = ClO–OCl \tag{2.71}$$

Radicals (like $ClO\cdot$) form dimers readily; the two odd electrons pair up in making a new bond. Normally, dimerization terminates a radical chain reaction because the dimer is unreactive. Once again, however, photolysis intercedes, this time to split the O–Cl bond, regenerating the $Cl\cdot$ atoms:

$$ClOOCl + hv = ClOO\cdot + Cl\cdot \tag{2.72}$$

$$ClOO\cdot + hv = O_2 + Cl\cdot \tag{2.73}$$

This step completes the chain reaction, leading to the rapid destruction of a large quantity of O_3.

The chain reaction is broken when sunlight evaporates PSCs, leading to the conversion of HNO_3 to NO_2. The latter reacts with available ClO, converting it to $ClONO_2$, thus interrupting the chain sequence by eliminating reaction (2.71).

The extent of O_3 reduction depends on the extent of Cl_2 and $HOCl$ formation during the dark of winter. This production rate in turn depends not only on the amount of chlorine in the stratosphere, but also on the number of cloud particles available to promote the reactions on the ice surface. Following the eruption of Mount Pinatubo in 1991, the polar ozone reduction became even more severe over the next two years because the sulfate aerosol from the volcano's SO_2 emission increased the surface available for reactions (2.67) and (2.68).

Although ozone depletion has been observed in both polar regions, the effect is more pronounced in Antarctica than in the Arctic region. This difference appears to be because temperatures in the Arctic do not fall low enough, or stay low long enough, to cause the removal of HNO_3 via the precipitation of large cloud particles.* Although ClO concentra-

*The temperature is higher over the North Pole than the South Pole because there is more air movement into the stratosphere in the northern hemisphere.

tions in the Arctic region are elevated, just as they are in Antarctica, HNO_3 provides a source of NO_2, which sequesters the reactive ClO. These findings suggest, however, that further cooling of the Arctic stratosphere could, in principle, cause an ozone hole comparable to that in Antarctica. This cooling could result from climate change, because Earth's surface temperature increases at the expense of decreasing temperature in the stratosphere. Figure 2.36 shows just such an effect in the temperature profile of the atmosphere projected for a two-fold increase in CO_2. Thus, the complex chemistry of polar clouds may link the dangers of climate change and those of ozone destruction.

6. Ozone Projections

Now that the chemistry of ozone is understood in detail, atmospheric scientists are able to project changes in the ozone layer, with some confidence, using global atmospheric models. In Figure 2.37 are projected trends in chlorine levels in the atmosphere up to the year 2080, assuming implementation of the Montreal protocol and the London (1990) and Copenhagen (1992) revisions. In Figures 2.38a, b, and c are the calculated percent ozone changes up to the year 2030 for three scenarios: 1) assuming constant CFC emissions at 1974 levels; 2) assuming CFC emissions as specified under the Montreal protocol; and 3) assuming CFC emissions based on the London amendments of 1990. These data suggest that, although the Montreal protocol and its subsequent revisions were significant steps in slowing down stratospheric ozone losses, ozone losses will continue well into the twenty-first century. This continued decline in ozone concentrations, despite reductions in the production and use of CFCs, reflects the long lag-times between the emissions of CFC in the troposphere and their eventual migration to the stratosphere.

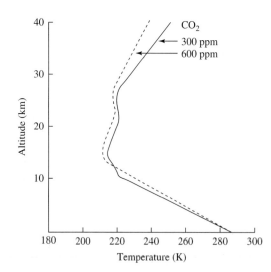

Figure 2.36   Effect of doubling the carbon dioxide concentration on the temperature profile of the atmosphere.   *Source:* S. Manabe and R. T. Wetherald (1967). Thermal equilibrium of the atmosphere with a given distribution of relative humidity. *Journal of the Atmospheric Sciences* 24: 241–259. Copyright © 1967, American Meteorological Society. Reprinted with permission.

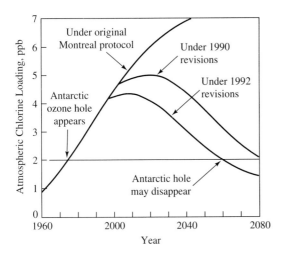

Figure 2.37   Measured and projected abundances of chlorine in the atmosphere under the terms of the Montreal protocol and its subsequent revisions. *Source:* Projections by the World Meteorological Organization, Geneva, Switzerland.

7. CFC Substitutes

A great deal of effort has been applied to finding substitutes for CFCs. The main strategy has been to explore the suitability of hydrochlorofluorocarbons, or HCFCs; these differ from CFCs because they have not only chlorine and fluorine substituents on the carbon, but also hydrogen. The presence of C–H bonds allows the HCFCs to be attacked by hydroxyl radicals and thereby be destroyed in the troposphere. At the same time, the Cl and/or F substituents lend HCFCs some of the desirable properties of CFCs, such as low reactivity and fire suppression, good insulating and solvent characteristics, and boiling points suitable for use in refrigerator cycles. As shown in Table 2.6, CH_2FCF_3 (HFC-134a) is being substituted for CCl_2F_2 (CFC-12) in compressors of home refrigerators and automobile air-conditioners; their boiling points are similar, –26 and –29°C, respectively. Also, the polyurethane foam insulation in refrigerator walls is being blown with CH_3CCl_2F (HCFC-141b), instead of with CCl_3F (CFC-11), as is the foam in upholstery. HCFC-141b is a good insulator and has low flammability.

However, while HCFCs have significantly shorter atmospheric lifetimes than CFCs, some of the molecules survive and drift into the stratosphere, where they contribute to ozone depletion. For example, the potential for ozone destruction of HCFC-141b is reduced to only one-tenth of that of CFC-11. If enough of HCHC-14b were produced, it could still deplete the ozone layer significantly. For this reason, HCFCs are viewed as transitional CFC substitutes; the amended Montreal protocol calls for the eventual elimination of HCFCs and for their permanent replacement by substances without chlorine.

Hydrofluorocarbons (HFCs) such as CH_2FCF_3 satisfy the criterion of no chlorine, but are potent greenhouse gases. The fluorine atoms stabilize the C–H bonds so that they are less readily attacked by hydroxyl radicals than are those of unsubstituted hydrocarbons. The HFC lifetimes are correspondingly longer. The greenhouse contribution of a CH_2FCF_3 (HFC-134a) molecule, for example, is calculated to be 30% as high as that of a CO_2 molecule. For

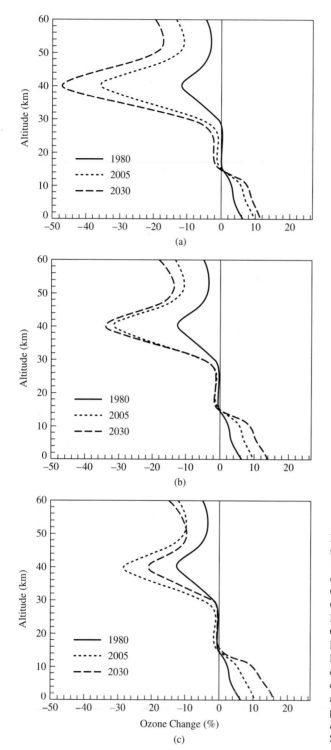

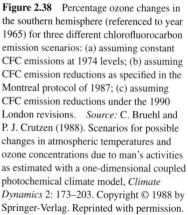

Figure 2.38 Percentage ozone changes in the southern hemisphere (referenced to year 1965) for three different chlorofluorocarbon emission scenarios: (a) assuming constant CFC emissions at 1974 levels; (b) assuming CFC emission reductions as specified in the Montreal protocol of 1987; (c) assuming CFC emission reductions under the 1990 London revisions. *Source:* C. Bruehl and P. J. Crutzen (1988). Scenarios for possible changes in atmospheric temperatures and ozone concentrations due to man's activities as estimated with a one-dimensional coupled photochemical climate model, *Climate Dynamics* 2: 173–203. Copyright © 1988 by Springer-Verlag. Reprinted with permission.

TABLE 2.6 POTENTIAL CFC SUBSTITUTES

Market	Current CFC	Alternative
Refrigerants	CFC-12 (CF_2Cl_2)	HFC-134a (CF_3CFH_2) HCFC-22 (CHF_2Cl) Blends or azeotropes
Blowing agents	CFC-11 ($CFCl_3$)	HCFC-141b (CH_3CFCl_2) HCFC-123 (CF_3CFCl_2) HCFC-22 (CHF_2Cl) Blends or azeotropes
Cleaning agents	CFC-113 ($CF_2ClCFCl_2$)	Blends or azeotropes New compounds

Source: L. E. Manzer (1990). The CFC-ozone issue: progress on the development of alternatives to CFCs. *Science* 249: 31–35.

this reason environmental groups have called for elimination of HFCs, even though HFCs are not covered by the Montreal protocol, which is limited to ozone-destroying substances.

New technologies that rely on neither HCFCs or HFCs to replace CFCs are being developed. Aerosol propellants, for example, can be isobutane or dimethyl ether (mixed with water to suppress flammability). Similarly, CFCs have been replaced by hydrocarbons as blowing agents in styrofoam production. The foam insulation in refrigerator walls, first blown with CFC-11 and now with HCFC-141b, may soon be replaced with vacuum panels that are filled with a solid filler material and sealed under vacuum in a gas-tight envelope. The electronics industry, which relies heavily on CFC solvents for cleaning circuit boards, is rapidly switching to aqueous detergent cleaners and to new imprinting methods that reduce the amount of cleaning needed.

The working fluids of refrigerators and air-conditioners are the hardest to replace. Here, too, alternatives are coming into view. There has been much interest in older materials such as ammonia and hydrocarbons, but deterrents include the toxicity and corrosiveness of ammonia and the flammability of hydrocarbons. Flammability, at least, can be managed with proper engineering (we are exposed to flammability hazards all the time, for example, in handling automobiles and gas furnaces); on the market now is a German-made domestic refrigerator (called "Greenfreeze") that uses a propane-butane mixture. Moreover, there are air-conditioners in development that do not have compressors but that rely, instead, on evaporative cooling combined with desiccant drying of the cooled air. In the longer term, there is also interest in using sound waves for cooling.

There are currently no adequate substitutes for the halons, which are used to flood enclosed spaces such as offices, airplanes, and military tanks, for example, in case of fire. Since 1994 when production ceased, halons are being carefully banked, pending the development of substitutes. The halons exhibit a combination of low reactivity and effective fire suppression that is hard to match. The most promising candidate appears to be CF_3I, which, like CF_3Br (halon-1301) is heavy enough to blanket and smother fires. The C–I bond is rapidly photolyzed by solar photons, even at ground level, so the molecule's atmospheric lifetime is short. However, toxicity and corrosiveness have not been fully evaluated.

Overall, the shift away from massive reliance on CFCs is happening more quickly than anyone thought possible a few years ago. As usual, necessity is the mother of invention.

D. AIR POLLUTION

While the greenhouse gases and the CFCs may be considered global pollutants (because they potentially harm the climate system and the stratospheric ozone layer worldwide), the term "air pollution" generally refers to substances that on local and regional scales directly harm animals, plants, and people and their artifacts. The phenomenon is hardly new. There have been complaints about air quality for centuries, especially in cities. But the steady expansion of population and industrial civilization has changed the nature of air pollution. The pervasive effects of emissions are increasingly manifest, and the need to control them is influencing to a greater degree the development of technology, particularly in the energy and transportation sectors.

1. Pollutants and Their Effects

A wide range of chemicals can pollute the air, but the ones generally viewed as needing control measures are carbon monoxide, sulfur dioxide, toxic organics, particulates, nitrogen oxides, and volatile organic compounds. The first four directly harm human welfare, whereas the last two are ingredients of photochemical smog, whose harmful effects are due to the production of ozone and other "oxidant" molecules.

a. Carbon monoxide. As shown in Figure 2.39, carbon monoxide (CO) emissions in the United States peaked around 1970 and have since been declining gradually. By far the major source of CO has been transportation. From 1940 to 1970, the increase in CO was directly proportional to the increase in motor vehicle travel. In the period 1970 to 1989, emissions declined, despite the continued increase in vehicle travel, due to increasingly stringent emission-control standards and improvements in energy efficiency.

Although carbon monoxide occurs naturally in the environment, it is an asphyxiating poison because it can displace the O_2 bound to hemoglobin (Figure 2.40). The Fe binding sites in hemoglobin bind CO 320 times more tightly than O_2. Such high affinity means that in human blood, CO occupies about 1% of hemoglobin binding sites; in smokers, this percentage doubles, on average, due to the CO in the inhaled smoke. When the ambient concentration of CO reaches 100 ppm, the percentage occupancy of the hemoglobin binding sites rises to 16%. This concentration of CO can be encountered in dense traffic in enclosed spaces (tunnels, parking garages), and may result in headaches and shortness of breath. The severity of the effects depends on the duration of the exposure and level of exertion (Figure 2.41) because it takes some time for the inhaled CO to equilibrate with the circulating blood. At concentrations higher than 750 ppm (0.1% of the air molecules), loss of consciousness and death occur quickly. At lower levels, the effects are reversed by breathing uncontaminated air, which allows O_2 to replace the CO bound to the hemoglobin molecules. However, individuals with heart problems are sensitive to even temporary oxygen insufficiency, and hospital admissions for congestive heart failure have been found to be influenced by CO levels in urban air.

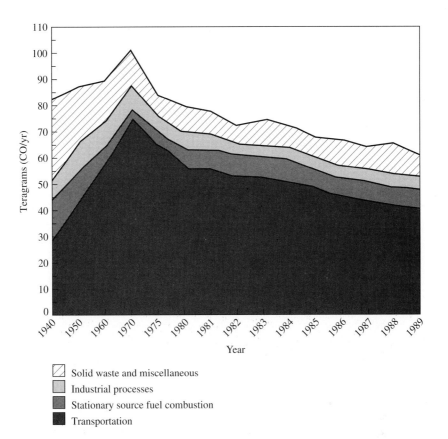

Figure 2.39   Trends in emissions of carbon monoxide in the United States, 1940–1989. *Source:* U.S. Environmental Protection Agency (1991). *National Air Pollutant Emission Estimates: 1940–1989.* (Report No. EPA-450/4-91-004) (Research Triangle Park, North Carolina: Office of Air Quality).

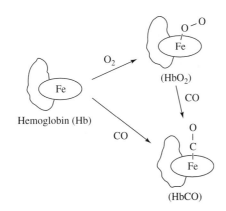

Figure 2.40   Replacement of oxygen by carbon monoxide in hemoglobin.

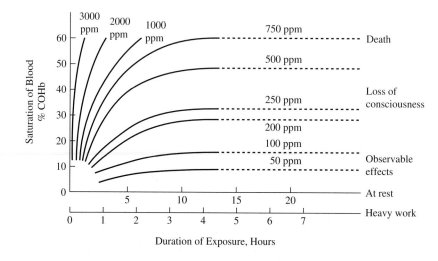

Figure 2.41   Dose-response curves of HbCO uptake in blood.

While the main source of anthropogenic CO is automotive transport, individuals may be at greater risk of CO poisoning from malfunctioning stoves and space heaters in their homes. CO is produced whenever combustion is incomplete. The problem is particularly acute in poor countries where unvented and inefficient stoves are common.

b. Sulfur dioxide. The main source of anthropogenic sulfur dioxide air emissions has been stationary source combustion, particularly from coal-burning power plants (Figure 2.42). The sulfur content of refined petroleum is generally quite low, but the sulfur content of coal is quite high. The steep rise in U.S. sulfur oxide emissions between 1940 and 1970 reflects the large increases in coal-fired electricity production that occurred without significant efforts to reduce sulfur emissions. Since 1970, emissions have declined and stabilized. Some of this decline reflects improvements in energy efficiency, but some is due to restricting sulfur emissions, both by using coal with lower content of sulfur and by implementing emission controls such as flue-gas desulfurization.

The sulfur in coal is converted to sulfur dioxide at the high temperatures of combustion. Sulfur dioxide itself is a lung irritant; it is known to be harmful to people suffering from respiratory disease. However, the most damaging health effects in urban atmospheres are caused not by sulfur dioxide but by sulfuric acid aerosol formed from its oxidation. Sulfuric acid irritates the fine vessels of the pulmonary region, causing them to swell and block the vessel passages. Breathing may be severely impaired. The effect appears to be cumulative, with older people suffering the most severe respiratory problems.

In addition, the sulfuric acid aerosol corrodes human artifacts. It steadily dissolves limestone ($CaCO_3$),

$$CaCO_3 + 2H^+ = Ca^{2+} + CO_2 + H_2O \tag{2.74}$$

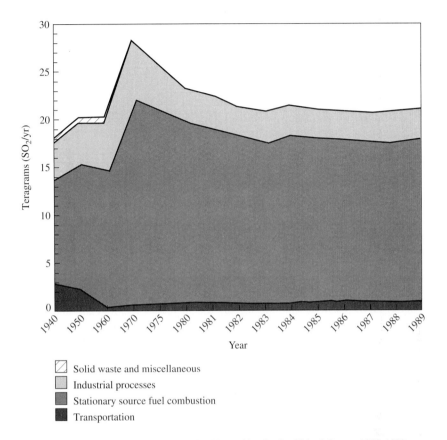

Figure 2.42   Trends in emissions of sulfur oxides in the United States, 1940–1989. *Source:* United States Environmental Protection Agency (1991). *National Air Pollutant Emission Estimates: 1940–1989.* (Report No. EPA-450/4-91-004) (Research Triangle Park, North Carolina: Office of Air Quality).

damaging outdoor monuments in cities around the world. Likewise, ancient stained glass windows are being eroded as the acid leaches out mineral constituents of the glass. Also, acid accelerates the corrosion of iron,

$$2Fe + O_2 + 4H^+ = 2Fe^{2+} + 2H_2O \qquad (2.75)$$

Protection of iron and steel structures with corrosion-inhibiting paint costs billions of dollars annually. In heavily industrialized areas, zinc coatings on galvanized steel may last as little as 5–10 years.

c. Toxic organics.   Many organic compounds are toxic, but only a limited number are of concern as air pollutants. Several environmental toxins that are dispersed and transported through the air are not inhaled in significant quantities; rather they are deposited and delivered through the food chain. Examples of deposited toxins (discussed in Part IV) are dioxins, lead, and mercury. The toxic organic compounds that act as direct air pollutants are the small aldehydes, benzene, and polycyclic aromatic hydrocarbons (PAHs).

1) Formaldehyde and Acetaldehyde. Formaldehyde, CH_2O, is a reactive molecule that irritates the eyes and lungs at quite low concentrations, just over 0.1 ppm. In laboratory animals, it causes respiratory impairment; at substantially higher levels, it causes nasal cancer. In the United States, industrial releases of formaldehyde to air in 1993 totaled about 5,170 tons (0.0052 Tg), down slightly from 5,700 tons (0.0057 Tg) in 1989, with the major industrial sources being textiles, chemicals, lumber and furniture, and stone and clay. In 1993 it was the fourth largest emitter to air among suspected carcinogens. Formaldehyde is also a source of indoor air pollution, for it is released from formaldehyde resins used in construction materials such as plywood, particle board, glass fiber insulation, and such. Levels of formaldehyde can be quite high in mobile homes (up to 10 ppm); exposed individuals develop symptoms that include drowsiness, nausea, headaches, and respiratory ailments. Because formaldehyde is potentially carcinogenic, chronic exposure, even at low doses, poses a public-health problem.

While outdoor levels of formaldehyde are low, there is concern that they may rise. Appreciable quantities of formaldehyde are produced by partial oxidation of methanol (CH_3OH). Likewise, the burning of ethanol releases significant amounts of acetaldehyde, which has similar toxic qualities. Hence, outdoor levels of formaldehyde and acetaldehyde may increase substantially if methanol and ethanol become future automotive fuels. In some parts of the United States, ethanol (C_2H_5OH) is already blended with gasoline, and its use as an additive is expected to grow significantly under current U.S. EPA regulations. In Brazil, a large fraction of the automotive fleet runs on ethanol from sugar cane.

2) Benzene. Benzene is a carcinogen in animals, and has been implicated as a causative agent in human leukemia. It is produced mainly from crude oil, and is widely used in the petroleum, chemical, and manufacturing industries. Over one-half the annual supply of benzene goes into gasoline production processes. The content of benzene in gasoline has averaged less than 1% but can range as high as 4%. Moreover, benzene is used as an intermediate in the manufacture of other chemicals and products. It has also been used as a solvent in products such as rubber cement and paint remover. In 1993, air emissions of benzene from U.S. industrial sources alone totaled about 4,910 tons (0.0049 Tg), down substantially from 1989, when air emissions were 11,220 tons (0.0112 Tg); among carcinogens, it ranked sixth-highest in total volume released to the air. Most emissions came from the primary metals industry, followed by the chemical industry, and then the petroleum industry.

Because benzene is volatile, it is released easily when benzene-containing products are used. Benzene use is strictly controlled in the workplace, and levels in gasoline are being reduced. In many applications, it is being replaced by alkylated benzenes such as toluene (methylbenzene), which are much less toxic than benzene; the alkyl groups are readily oxidized by enzymes in the liver, producing benzoic acid or related acids, which are excreted.

3) Polycyclic Aromatic Hydrocarbons (PAHs). Compounds with four or more benzene rings fused together, such as benzo[a]pyrene (Figure 2.43), are potent carcinogens. Interestingly, their carcinogenicity depends upon activation by the same class of liver en-

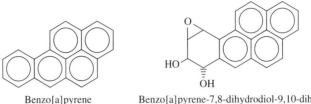

Benzo[a]pyrene   Benzo[a]pyrene-7,8-dihydrodiol-9,10-dihydroepoxide

Figure 2.43   Structures of benzol(a)pyrene and an oxygenated metabolite.

zymes, cytochromes P450 (see pp. 293–295), that metabolize toluene and other *xenobiotics* (molecules foreign to the organism). When these enzymes add oxygen to the PAHs, they produce epoxide adducts (Figure 2.43) that react strongly with the heterocyclic bases of DNA, altering genes.

PAHs are formed as side-products of carbon fuel combustion. Although they are present at low levels in automobile exhaust, the levels are much higher when large quantities of soot particles are produced, as in diesel exhaust or smoke from coal or wood fires. (The soot itself contains sheets of benzene rings, like graphite.) As long ago as 1775, soot exposure was linked to scrotal cancer in chimney sweeps in London. Coke oven workers have also been documented to have increased levels of lung and kidney cancer.

d. Particles.   Epidemiological evidence associates particles more directly with disease and mortality than any gaseous pollutant. Trends in U.S. emissions of total suspended particulates from 1940 to 1989 are shown in Figure 2.44a. Currently, industrial processes contribute the most emissions (about 37%), with the balance derived more or less equally from the three other sources indicated in the figure. Around the middle of this century, most of the emissions came from industrial processes and stationary sources; these emissions were significantly reduced by substituting new, cleaner technologies and by adopting emission-control equipment such as electrostatic precipitators to trap the larger particles before escaping from the smokestack. Currently, most of the particulate pollution consists of the smaller particles less than 10 microns in diameter (PM_{10}); as shown in Figure 2.44b, from 1985 to 1989, emissions of PM_{10} were 6 Tg out of a total emissions of around 7.5 Tg. These smaller particles are more difficult to control, and they are the most hazardous to human health.

Atmospheric particles are of concern for many reasons: they diminish visibility; they settle out on plants and interfere with photosynthesis; and they penetrate the lungs, blocking and irritating air passages. Moreover, the particles themselves can exert toxic effects. Coal miners' black-lung disease, asbestos workers' pulmonary fibrosis, and the city dwellers' emphysema are all associated with the accumulation of particles in the lung. Small particles have the greatest health impact because they penetrate most deeply into the lung. Particles larger than several microns are trapped in the nose and throat, from which they are eliminated more readily.

Asbestos fibers are especially dangerous because they can cause *mesothelioma,* a cancer of the pleural cavity, even at quite low exposures. The most hazardous form of asbestos

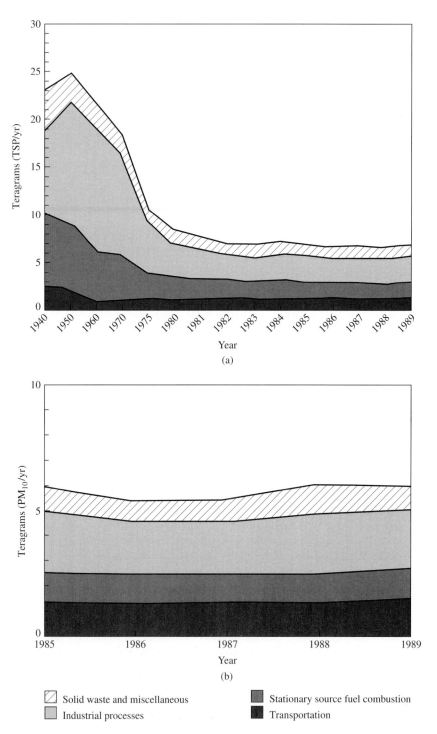

Figure 2.44 Trends in emissions in the United States of (a) total suspended particulates (TSP), 1940–1989, and (b) particulates less than 10 microns in diameter (PM$_{10}$), 1985–1989. *Source:* United States Environmental Protection Agency (1991). *National Air Pollutant Emission Estimates: 1940–1989.* (Report No. EPA-450/4-91-004). (Research Triangle Park, North Carolina: Office of Air Quality).

is *crocidolite,* in which the tiny fibers are rod-like and can penetrate deep into the lungs. The more common form of asbestos, *chrysotile,* has serpentine fibers that clump into bundles, most of which are intercepted in the upper airways, where they are less hazardous. Due to the risk of cancer, however, the use of all forms of asbestos has been sharply curtailed. Nevertheless, asbestos is still present as insulation and fireproofing material in many buildings. What to do about it is a contentious issue because, unless extraordinary precautions are taken, attempts to remove it can release large quantities of fibers into the air of a building. The released fibers settle on surfaces from which they can be re-suspended quite readily, thereby remaining much more hazardous than if they had remained locked and undisturbed in the building material.

Soot particles pose particular problems because they can adsorb significant amounts of toxic chemicals on their irregular surfaces. Soot particles are very prevalent in diesel exhaust and wood smoke. Coal fires release soot as well as SO_2; in foggy conditions, the resulting sulfate aerosol can combine with the soot to produce a toxic smog, with serious health consequences, especially for those with respiratory ailments. In London in December 1952 a heavy smog of this sort killed about 4,000 people within a few days. As a result of switching from coal to oil and gas for household heating, the London-type smog has largely disappeared in developed countries. But coal is still widely burned in developing countries, whose cities often have unhealthy air. In addition to soot, urban air is frequently polluted by the dust of smelters and other industries (which can be controlled by filters and electrostatic precipitators), as well as by dust blown from the surrounding countryside.

e. NO$_x$ and volatile organics. Nitrogen oxides (NO_x) and volatile organic compounds (VOCs) are not direct air pollutants, in that they rarely affect health directly. Rather, they are the main ingredients in the formation of *photochemical smog,* the brown haze that blankets many cities worldwide. Although most damage from smog results from the action of ozone and other oxidants, these oxidants cannot build up without the combined action of NO_x and VOCs. Controlling smog formation requires reducing emissions of NO_x and VOCs.

Trends in U.S. air emissions of NO_x compounds from 1940 to 1989 are shown in Figure 2.45. Almost all NO_x emissions are due to transportation and stationary-source fuel combustion. As with all air pollutants generated from fossil fuel consumption, the increase prior to 1970 was due to increased fuel consumption with little attention paid to controlling air pollution. Over the last two decades, emissions have stabilized through the implementation of emission controls and the increased conservation of energy, particularly in transportation.

Emissions of volatile organic compounds (VOCs) are shown in Figure 2.46. The largest sources of VOCs are transportation and industrial processes. The trends reflect the same evolution of newer, cleaner technologies and improvements in energy efficiency that stabilized NO_x emissions.

f. Ozone and other oxidants. While anthropogenic emissions are destroying ozone in the stratosphere, they are helping to create ozone in the troposphere via the phenomenon of photochemical smog (see next section). And while ozone in the stratosphere protects us from the harmful effects of UV rays, ozone at ground level is quite harmful, producing cracks in rubber, destroying plants, and causing respiratory distress and eye irritation

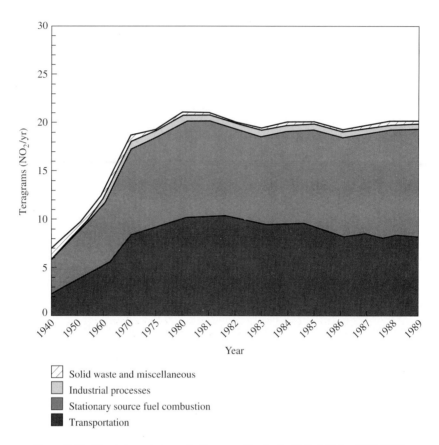

Figure 2.45 Trends in emissions of nitrogen oxides in the United States, 1940–1989. *Source:* United States Environmental Protection Agency (1991). *National Air Pollutant Emission Estimates: 1940–1989.* (Report No. EPA-450/4-91-0040) (Research Triangle Park, North Carolina: Office of Air Quality).

in humans. These effects set in at quite low concentrations, around 100 ppb. In several U.S. cities, the air is frequently in violation of the current standard of 120 ppb, measured over an hour. (A proposed tightening of the standard to 70–90 ppb, averaged over eight hours, is under review.) Moreover, the ozone can drift over entire regions, possibly affecting plant growth in widespread areas.

These effects of ozone result from its being a strong oxidant and O-atom donor. Ozone reacts especially well with molecules containing C=C double bonds, forming epoxides. Such molecules are abundant in rubber, in the photosynthetic apparatus of green plants, and in the membranes lining the lung's air passages.

Other oxidant molecules are also formed in photochemical smog and produce similar damage. An example is peroxyacetyl nitrate (PAN), $CH_3C(O)OONO_2$, a potent eye irritant.

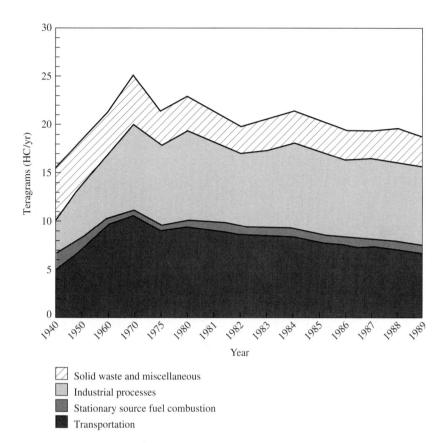

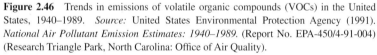

Figure 2.46   Trends in emissions of volatile organic compounds (VOCs) in the United States, 1940–1989. *Source:* United States Environmental Protection Agency (1991). *National Air Pollutant Emission Estimates: 1940–1989.* (Report No. EPA-450/4-91-004) (Research Triangle Park, North Carolina: Office of Air Quality).

2. Photochemical Smog

Nitrogen oxides and volatile hydrocarbons are key ingredients in the formation of photochemical smog, a condition that afflicts an increasing number of cities and their surroundings. Photochemical smog can form whenever large quantities of automobile and industrial exhausts are trapped by an inversion layer over a locality that is, at the same time, exposed to sunshine. The classic location for smog is Los Angeles, with its dependence on the automobile, abundant sunshine, and frequent inversion layers, but automobile traffic has introduced the problem to many other cities. It is characterized by an accumulation of brown, hazy fumes, containing ozone and other oxidants, with the harmful effects described above.

Figure 2.47 shows the time-course of the key atmospheric ingredients for a classic, smoggy day in Los Angeles. The concentration of hydrocarbons peaks during the

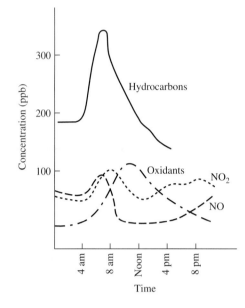

Figure 2.47   Example of a concentration-time profile of smog-forming chemicals in Los Angeles air.

early-morning rush hour. The concentration of nitric oxide reaches a peak at the same time and then begins to decrease as the nitrogen dioxide concentration increases. Subsequently, the concentration of oxidants rises, and the hydrocarbon concentration falls.

Recall that the formation of ozone from oxygen requires energy, which is provided in the stratosphere by the absorption of UV light. What drives ozone formation near the ground level, where little UV arrives? The critical ingredient is NO_2, the only common atmospheric molecule capable of absorbing visible light. Its spectrum (Figure 2.48) has maximum absorption at about 400 nm, in the blue region. It is this absorption that gives smoggy air its brown tint. The photoexcited NO_2 is unstable, and dissociates to NO and O atoms:

$$NO_2 + h\nu(<400 \text{ nm}) = NO + O \tag{2.76}$$

The O atoms react immediately with the surrounding O_2 molecules to produce ozone, just as in the stratosphere (reaction b in Figure 2.32, p. 149).

But since each molecule of ozone requires an oxygen atom from NO_2 dissociation, this mechanism cannot build up the ozone concentrations to levels greater than the NO_2 itself. Moreover, the NO produced in reaction (2.76) can react with the ozone to restore the NO_2 by the same reaction that destroys stratospheric ozone (reaction e, page 152 with X = NO). Thus, nitrogen oxides and oxygen alone cannot lead to net ozone formation in sunlit air.

Hydrocarbons are needed as well. Hydrocarbons produce peroxyl radicals, which can react with the NO before the ozone does and regenerate NO_2 [reaction (2.27), p. 138]. If this happens, NO_2 can catalyze the formation of ozone as long as the supply of peroxyl radicals lasts, thereby allowing ozone to accumulate. The peroxyl radicals are formed when O_2 reacts with organic radicals [reaction (2.26)], which in turn are produced by the action of hydroxyl radicals on hydrocarbons [reaction (2.30)]. The hydroxyl radicals themselves are pro-

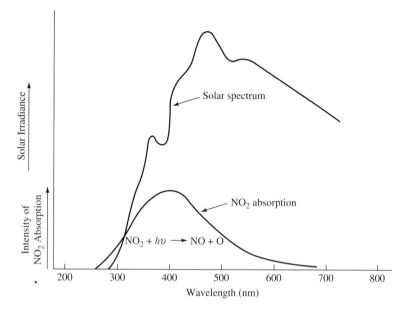

Figure 2.48   Absorption of solar light by nitrogen dioxide to form oxygen atoms. The reaction shown under the absorption curve occurs when nitrogen dioxide absorbs light at wavelengths of less than 400 nm. For wavelengths greater than 400 nm, nitrogen dioxide is excited but does not decompose.

duced from ozone [reactions (2.32) and (2.33)], thereby completing the photochemical cycle. Since a single organic radical can produce many peroxyl radicals by successive rounds of O_2 combination and fragmentation, the ozone concentration can build up rapidly to levels exceeding the concentration of nitrogen oxides and far exceeding the concentration of hydroxyl radical. The interlocking NO_x, O_3, and hydrocarbon radical cycles are diagrammed in Figure 2.49.

The combination of reactive species in these cycles produce other oxidants. For example, the reaction of peroxyl radicals with NO_2 produces peroxyalkyl and peroxyacyl nitrates:

$$ROO\cdot + NO_2 = ROONO_2 \qquad \text{(R signifies alkyl or acyl)} \qquad (2.77)$$

Similarly, peroxyacetyl nitrate (PAN) is formed from the peroxyacetyl radical, a relatively common smog constituent.

Because the initiation of the cycle depends on the formation of organic radicals, the extent of smog formation depends on the reactivity of hydrocarbons with the hydroxyl radical. Some hydrocarbons produce few radicals, others many more. As mentioned earlier, the H-atom abstraction proceeds spontaneously because the O–H bonds of the product water molecule are stronger than the C–H bond being attacked by the hydroxyl radical. But the rate of the reaction depends in some detail on the nature of the C–H bond being attacked. For

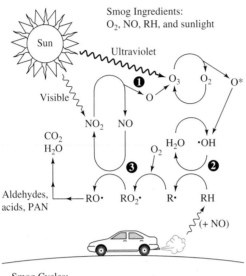

Smog Cycles:
❶ O_3 formed by O atoms from NO_2 photolysis;
❷ HC radicals made by ·OH, from O_3 UV photolysis
❸ NO oxidation to NO_2 catalyzed by HC radicals;

Figure 2.49 Smog formation from O_2, NO, hydrocarbon, and sunlight.

example, the rate is relatively slow for methane because the C–H bonds in CH_4 are particularly strong. Although hydroxyl radical attack is the mechanism for methane destruction in the atmosphere, the rate is too slow to make CH_4 a significant contributor to smog formation. Thus, switching automotive fuel to methane would help reduce smog.

The higher hydrocarbons have weaker C–H bonds on average, and react faster with hydroxyl radical. Table 2.7 lists hydroxyl radical reaction rates for a number of gasoline components. The alkanes and cyclohexanes show a limited range of rates, 3–8 in units of 10^{12} cc/molecule-sec, depending on the structure. The alcohols and ethers are also in this range, although methanol is somewhat slower. Among the aromatic compounds, benzene has a lower rate, 0.9 in the same units, reflecting the higher strength of aromatic C–H bonds. Toluene, however, has a rate comparable to the alkanes, 6, while the rate for m-xylene is much higher, 23. In these cases it is the methyl C–H bonds that are attacked; the resulting radical is stabilized by resonance with the aromatic pi electron system, aided, in the case of m-xylene, by the electron-donating propensity of the other methyl substituents.

Higher values still, 24–35, are seen for the alkenes. For these molecules the mechanism is not H-atom abstraction, but rather attack by hydroxyl radical on the C=C bond to form a radical adduct:

$$R_2C=CR_2 + OH\cdot = R_2(OH)C - CR_2\cdot \tag{2.78}$$

The pi bond electrons are loosely held, and offer a favorable site for interaction with radicals. The steps in the propagation of the smog reaction cycle subsequent to reaction (2.78) are much the same as for the alkanes depicted in Figure 2.49. Due to their high reactivity,

TABLE 2.7   PROPERTIES OF SOME COMPONENTS OF GASOLINE

Component	Research octane number (RON)	Motor octane number (MON)	Vapor pressure (psi @ 100°F)	PA*
butane			51	3.23
n-pentane	62	67	15.5	4.80
n-hexane	19	22	5.0	5.90
methyl propane			82	2.83
2-methylbutane	99	104	20	
2-methylpentane	83	79	6.6	5.82
2-methylhexane	41	42	2.2	6.85
iso-octane	100	110	1.65	3.15
1-butene	144	126	50	24.4
1-methylpropene	170	139	62	24.4
1-pentene	118	109	19	35.0
cyclohexane	110	97	3.3	8.50
methylcyclohexane	104	84	1.6	7.87
benzene	99	91	3.3	0.88
toluene	124	112	1.04	5.98
meta-xylene	145	124	0.33	22.8
ethanol	115[†]		17	3.3
methanol	123	93	60	1.0
Methyl *tert*-butyl ether (MTBE)	123	97	8	2.6
Ethyl *tert*-butyl ether (ETBE)	111[†]		4	8.1

*Photochemical activity measured as rate of reaction with OH radicals, units, cc/(molecule sec) $\times 10^{12}$.

[†]Average (RON + MON).

Source: D. Seddon (1992). Reformulated gasoline, opportunities for new catalyst technology. *Catalysis Today* 15: 1–21.

however, alkenes are the most important hydrocarbon molecules in the dynamics of smog formation.

3. Emission Control

Limiting atmospheric pollution depends upon two strategies: removing the pollutants before they are dispersed, and changing conditions to reduce the amount of pollutants produced in the first place. Both these strategies have been tried for most air pollutants, with mixed success.

   a. Sulfur dioxide.   In order to reduce the level of sulfuric acid aerosols in urban air, power plants are often built with tall smokestacks to disperse the plume over a wide area. This may alleviate the local problem, but at the expense of producing acid rain (see discussion, pp. 116–117) in areas that are downwind.

Actual abatement requires reducing the sulfur dioxide emissions or, alternatively, limiting the sulfur content of fuels. In coal-fired power plants, sulfur dioxide is currently removed from the stack gases by installing chemical scrubbers, in which the stack gas is passed through a slurry of limestone, converting it to calcium sulfite:

$$CaCO_3 + SO_2 = CaSO_3 + CO_2 \qquad (2.79)$$

Although limestone is relatively cheap, a great deal of it has to be used, and the resulting calcium sulfite sludge presents a significant waste-disposal problem. A more attractive technology currently being tested uses a regenerable amine salt as the scrubbing agent. Heating the resulting SO_2 adduct recovers the amine salt and drives off the SO_2, which can be converted to commercial-grade sulfuric acid.

Another possibility is to remove the sulfur from the coal before or during combustion. The coal can be cleansed of the major sulfide mineral, FeS_2 (iron pyrite), by grinding the coal and floating the mineral particles away with a water/oil/surfactant emulsion. However, the coal still contains the organically bound sulfur. This sulfur can be removed by pulverizing the coal and mixing it with limestone in a fluidized bed combustor, a device in which air is passed through a screen from underneath, keeping the particles suspended until they burn. The limestone captures the SO_2 before it passes into the stack gas. However, the resulting calcium sulfite remains a disposal problem.

b. Nitrogen oxides, carbon monoxide, and hydrocarbons. Combustion in air produces nitrogen oxides as inevitable byproducts. Their emission levels depend on the temperatures reached in the combustion process; the hotter the flame, the greater the NO production rate. Although all kinds of combustion contribute to NO_x emissions, the main contributors, at least in the developed world, are automotive transport and fuel combustion in stationary sources such as home furnaces, power plants, and industrial facilities. In the United States, transportation produces about 40% of the NO_x, while stationary sources account for most of the rest (Figure 2.45).

Combustion also accounts for much of the atmospheric CO and hydrocarbons, at least in urban areas, because automotive exhaust contains substantial quantities of unburned gases. In addition, some of the volatile automotive fuel escapes before combustion, adding to the hydrocarbon levels. In the United States, transportation accounts for about one-third of volatile organic compound emissions. Industrial sources are significant, largely because of solvent evaporation (Figure 2.46). In addition, substantial quantities are emitted by vegetation. Plants release numerous hydrocarbons, especially *terpenes,* which have C=C double bonds and therefore react rapidly with hydroxyl radicals. In some areas the vegetation is a major contributor to the reactive hydrocarbons responsible for photochemical smog.*

NO_x emissions are difficult to control because efficient energy conversion requires high combustion temperatures, whether in cars or power plants. Moreover, there is a

---

*W. L. Chameides et al. (1988). The role of biogenic hydrocarbons in urban photochemical smog: Atlanta as a case study. *Science* 241: 1473–1475.

trade-off between NO_x and unburned gases as the ratio of air to fuel in the combustion chamber is varied (illustrated for cars in Figure 2.50). The NO production rate is maximum near the *stoichiometric* ratio (just enough O_2 to completely oxidize the fuel), where the highest temperature is reached. If less air is admitted to the combustion zone ("fuel-rich"), the NO production rate falls along with the temperature, but the emission of CO and unburned hydrocarbons (HCs) increases.

It is possible to lower both NO and HC by carrying out the combustion in two stages, the first of which is rich in fuel and the second of which is rich in air. In this way the fuel is burned completely, but the temperature is never as high as it would be for a stoichiometric mixture. This two-stage approach is being incorporated into new power plants; it has been tried in cars via the "stratified-charge" engine, but with less success.

The other approach to reducing emissions is to remove the pollutant from the exhaust gases. In newer automobiles, this is accomplished with a three-way *catalytic converter,* so named because it reduces emissions of hydrocarbons (HCs), carbon monoxide (CO), and nitric oxide (NO). In order to deal with both NO and unburned gases, the converter has two chambers in succession (Figure 2.51). In the reduction chamber, NO is reduced to N_2 by hydrogen, which is generated at the surface of a rhodium catalyst by the action of water on unburned fuel molecules (analogous to steam reforming):

$$\text{hydrocarbons} + H_2O = H_2 + CO \tag{2.80}$$

$$2NO + 2H_2 = N_2 + 2H_2O \tag{2.81}$$

In the oxidation chamber, air is added, and the CO and unburned hydrocarbons are oxidized to CO_2 and H_2O at the surface of a platinum/palladium catalyst:

$$2CO + O_2 = 2CO_2 \tag{2.82}$$

$$\text{hydrocarbons} + 2O_2 = CO_2 + 2H_2O \tag{2.83}$$

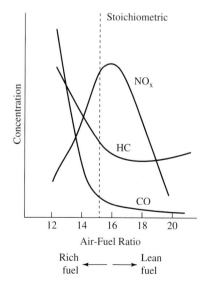

Figure 2.50 Auto exhaust composition.

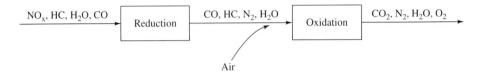

Figure 2.51   The three-way catalytic converter for removal of hydrocarbons (HCs), carbon monoxide (CO), and nitric oxide (NO) from automobile exhaust.

The catalytic converter is quite effective in reducing automotive emissions. It is credited with significant reductions in ozone levels in some urban areas. In Los Angeles, peak ozone levels were cut nearly in half between 1970 and 1990, despite a 60% increase in the number of vehicle miles driven. However, further improvements are difficult. Efforts are under way to address the problem of cold starts, since the catalytic converter only works on hot exhaust gases. Another problem is that, with the introduction of the catalytic converter, some 50% of the remaining hydrocarbon and CO emissions are produced by 10% of the cars, which have inadequate or inoperative emission controls. Reducing this percentage further has proven difficult.

In any event, it is increasingly recognized that the key to smog reduction is control of NO emissions; hydrocarbons are generally too abundant to be brought low enough to be the limiting factor. Even if the automotive contribution is reduced further, the contributions from other sources can often support substantial levels of smog production. But smog requires NO, whose only significant sources are automotive transport and power and industrial plants. These sources are therefore under increased scrutiny with respect to NO control.

In stationary power plants, it is possible to convert NO back to N_2 using catalytic converters, similar in concept to those developed for cars. Since unburned gases are generally a smaller fraction of the exhaust stream in power plants, some additional reductant is generally needed to effect substantial NO control. The reduction can be accomplished by injecting ammonia into the catalytic chamber.

$$6NO + 4NH_3 = 5N_2 + 6H_2O \qquad (2.84)$$

Careful control of conditions is required, however, to prevent oxidation of the ammonia directly to NO and NO_2 by residual O_2 in the exhaust stream. In another approach, the compound urea, $CO(NH_2)_2$, has been sprayed directly into the combustion flame in order to reduce the NO. The mechanism is complex, but the overall reaction is

$$2CO(NH_2)_2 + 6NO = 5N_2 + 2CO_2 + 4H_2O \qquad (2.85)$$

In the long run, the most effective NO_x abatement strategy will be to introduce fuel cells for electric power generation and transport (see Fuel Cells, pp. 81–84), thereby eliminating the high temperatures that produce NO in the first place.

4. Gasoline and Its Formulations

Because much of urban air pollution is produced by transportation, the liquid fuels used in transport have been scrutinized intensively, and pollution regulations have substantially changed the composition of gasoline. The challenge has been to design fuel formulations that perform well in terms of fuel efficiency, engine performance, and pollution reduction.

a. Knocking and octane. Combustion is a radical chain process, similar to the oxygen-radical chemistry discussed in the preceding sections, that allows the oxygen to combine very rapidly with the fuel molecules. The spark-ignition engine of automobiles works by igniting a mixture of gasoline and air with a spark. The air/fuel mixture is first compressed by a piston in a cylinder, then ignited. The spark fragments the fuel molecules in its path, generating enough radicals to set off the chain reaction. The force of the explosion on the piston delivers power to the drivetrain. The greater the degree of compression, the greater the power. However, the compression stroke itself heats the fuel; at sufficiently high temperatures, the fuel molecules can fragment spontaneously, setting off the explosion prematurely. Pre-ignition lowers the power generated and produces extra wear on the engine. This is the *knocking* phenomenon, recognizable from the characteristic noise produced by the engine under acceleration.

The temperature required for radical generation depends on the structure of the fuel molecule. Branched-chain hydrocarbons are more resistant to radical formation than straight chain hydrocarbons because the branching increases the fraction of the hydrogen atoms that are on methyl groups, CH_3. The C–H bonds are stronger for methyl groups (423 kJ/mol) than for methylene groups, CH_2 (410 kJ/mol). Consequently, methylene groups are more susceptible to attack by thermally activated oxygen molecules, which pull off a hydrogen atom, leaving a hydrocarbon radical.* Since straight chain hydrocarbons have more methylene groups than do branched chain hydrocarbons, they are subject to pre-ignition at lower temperature.

The performance of gasoline is largely a function of its octane rating. Gasoline is a mixture of low-boiling hydrocarbons, most of them containing seven or eight carbon atoms. Among these, 2,2,4-trimethylpentane ("iso-octane") is particularly resistant to pre-ignition due to its highly branched structure. It is assigned an "octane" rating of 100. The zero of the octane scale is set by n-heptane, a straight-chain hydrocarbon with a strong knocking tendency. The octane rating of any gasoline is then set by comparison with these two hydrocarbons in standard engine tests. Table 2.7 lists octane ratings for a number of the compounds in gasoline. In some cases the rating exceeds 100, meaning that the fuel is even less prone to preignition than iso-octane.

b. Diesels and cetane. A diesel engine works very differently from a spark-ignition engine. In diesels, air in the piston is preheated by compression, and fuel is then

---

*Methyl C–H cleavage produces a *primary* radical $RCH_2\cdot$, while methylene cleavage produces a *secondary* radical $R(R')CH\cdot$. The unpaired electron is stabilized somewhat by delocalization over carbon substituents, whereas hydrogen atoms are not effective in this role. Since the secondary radical has two carbon substituents, it is more stable than the primary radical, which has only one; this is why it takes less energy to break methylene C–H bonds than methyl C–H bonds. Still more stable is a *tertiary* radical $R(R')(R'')C\cdot$, and a tertiary C–H bond is correspondingly weaker (402 kJ/mol). Branched chain hydrocarbons do have tertiary C–H bonds, but their reactivity is compensated by the far more numerous primary C–H bonds.

sprayed into the hot chamber, burning on contact. Since pre-ignition is not an issue, the degree of compression can be very high, allowing for greater efficiency. The engine is built ruggedly to accommodate the higher compression forces, and since there are fewer moving parts than in a spark engine (no valves are needed), a diesel engine wears out more slowly. This is why diesel is generally the engine of choice for large trucks and buses.

Whereas easy fragmentation of fuel molecules is undesirable for spark engines, it is desirable for diesel engines, because it enhances the combustion of the injected fuel. Thus, straight-chain hydrocarbons are abundant in diesel fuel. The quality of diesel fuel is judged by its "cetane" number, which increases with fragmentation tendency, opposite to the octane number. Cetane (n-hexadecane, $C_{16}H_{34}$) is given a value of 100, while a highly branched isomer, heptamethylnonane is given a value of 15. Also, diesel engines work best with higher molecular weight molecules, in the C_{11}–C_{16} range, whereas the optimum range for gasoline is C_6–C_{10}. Diesel fuel therefore takes a different cut of oil refining production than gasoline (see p. 21).

Diesel emissions contain many more particles than spark engine emissions because of the combustion characteristics of the injected fuel. Molecules at the air-fuel interface burn completely, but molecules at the center of the injected plume heat up before they have access to oxygen molecules and therefore tend to decompose to solid carbon. In the spark engine, however, the fuel molecules are intimately mixed with air before combustion, and soot production is therefore much lower.

Because of the deleterious health effects of particles, there is pressure to abandon diesel engines, particularly in urban buses, whose slow progress along crowded streets is responsible for elevated particle concentrations in the street-level air. Many cities are trying natural gas-fueled bus fleets as an alternative. New diesel technology is also available, which greatly reduces particle emissions through the use of particle traps with catalysts, similar to the catalysts used in converters on cars, to oxidize the carbon particles in the exhaust.

   c. Lead in gasoline. It was discovered in the 1920s that knocking could be diminished if organo-lead compounds, particularly tetraethyl- or tetramethyl-lead were added to the gasoline. In the succeeding decades, lead was added to virtually all gasolines in order to improve performance. The lead additives suppress the radical chain reactions in the pre-ignition phase. As the air/fuel mixture becomes compressed and heated, the weak alkyl-lead bonds break, releasing Pb atoms, which combine rapidly with oxygen to form particles of PbO and PbO_2. These particles provide attachment sites for hydrocarbon radicals, which recombine with one another, thereby terminating the chain reaction. To avoid building up lead deposits on the inside engine surfaces, gasoline containing lead usually also contains ethylene dichloride or ethylene dibromide. These organo-halogens act as scavengers for the lead, producing PbX_2 compounds (X = Cl or Br). Because these compounds are volatile at the high temperature of the exhaust gases, they remove the lead from inside the engine and release it into the atmosphere.

   Beginning in the mid-1970s, unleaded gasoline was offered for sale in the U.S. and gradually displaced leaded gasoline. The reason was that lead compounds in the exhaust react with the rhodium and platinum catalysts in catalytic converters, "poisoning" their surfaces and rendering them inactive. As the fraction of the car fleet with catalysts increased, so did the lead-free fraction of the gasoline supply. This displacement had a salutary effect on lead pollution, as well as on air pollution in general, since human exposure to lead decreased dramatically when the lead was removed from gasoline, as discussed on pp. 324–325.

In Canada and some European countries, another organometallic compound has been introduced as an anti-knock additive: methylcyclopentadienyl manganese tricarbonyl (MNT). This compound releases Mn atoms, which are converted to Mn_3O_4 particles in the combustion chamber which, like PbO, captures hydrocarbon radicals and inhibits knocking. Unlike lead, the toxicity of manganese is low once released into the environment (indeed, it is a biologically essential element). Moreover, MNT does not significantly increase the human intake of Mn from natural sources. Nevertheless, MNT has not been approved as a gasoline additive by the U.S. EPA because the additive itself is toxic if ingested or inhaled. In addition, fouling of spark plugs and of onboard sensors has been reported. However, a federal appeals court has ruled (October, 1995) against the EPA's ban on MNT.

d. Reformulated gasoline. An alternative to adding free radical scavengers to reduce knocking is to alter the gasoline composition by reducing the fraction of low-octane components and increasing the fraction of high-octane components. A modern oil refinery has considerable latitude to alter the hydrocarbon molecules in oil through cracking, alkylation, and reforming reactions (see the discussion of petroleum composition and refining in Part I, pp. 20–23). In the United States, the removal of alkyl-lead additives was initially compensated for by increasing the content of aromatic compounds, principally benzene, toluene, and xylenes (sometimes called the BTX component). As seen in Table 2.7, benzene's octane rating is nearly as high as iso-octane's, and toluene's is substantially higher. The benzene ring resists fragmentation at high temperature, and methyl substituents stabilize it further. At the same time, the methyl-benzene bonds are more stable than the C–C bonds in alkanes; the former have more s orbital character because of the sp^2 hybridization of the benzene carbon atoms. Consequently, increasing the aromatic fraction of gasoline boosts its octane rating.

Despite the high-octane ratings of BTX gasoline, the BTX fraction is being decreased in the United States. As discussed above, the xylenes react rapidly with hydroxyl radicals and therefore have greater potential to form smog than do the alkanes. Benzene, although low in photochemical activity, is a carcinogen. Both of these are reasons why the BTX fraction is an undesirable way to boost engine performance.

The aromatics are expected to be replaced by "oxygenates," fuel molecules that contain one or more oxygen atoms. The four oxygenates currently in use are methanol, ethanol, and the methyl and ethyl ethers of *tertiary*-butyl alcohol, MTBE and ETBE. Unlike the aromatics, they have hydroxyl-radical reaction rates no higher than the alkanes (Table 2.7), as well as octane ratings substantially higher than 100. (All four have low fragmentation tendencies; the C–C bonds in the *t*-butyl group are all tertiary, and the ethyl C–C bond is stabilized by the O-atom substituents in ethanol and ETBE.)

There is another reason why oxygenates are added to gasoline and are, indeed, mandated by the 1990 Clean Air Act. Because the fuel molecules in oxygenate-enriched gasoline already contain some oxygen atoms, their conversion to CO_2 during combustion is more complete, even if the combustion mixture is somewhat fuel-rich. Consequently, oxygenates are expected to reduce emissions of unburned hydrocarbons and CO. The oxygenate mandate was imposed in order to bring U.S. cities into compliance with standards for CO emissions, especially in the winter months when cold starts and fuel-rich combustion produce elevated emissions.

However, the oxygenate additives, principally MTBE, have stirred enormous controversy. Their effectiveness in emission reduction has been questioned. They do work as expected in older cars, which are often tuned to fuel-rich combustion mixtures, but the improvement is minimal in newer cars, which are tuned closer to the stoichiometric fuel/air ratio. (Tuning the car is superior to adding oxygenates in terms of energy efficiency; the energy density of oxygenates is lower than that of hydrocarbons, because the former are partially oxidized to begin with.) Furthermore, there have been numerous health complaints by people exposed to gasoline that contains MTBE, although health effects have not been substantiated clinically.

It appears likely that oxygenates will comprise about one-third of the gasoline consumed in the United States in the next few years. Ethanol and MTBE are the most likely fuels to meet this demand, although their relative usage is currently a subject of intense debate. (EPA favors the use of biomass fuels, principally ethanol, while the petroleum industry favors MTBE.) Ethanol is obtained by fermentation of plant matter, primarily corn in the United States and sugar cane in Brazil, whereas methanol and MTBE are produced from petroleum fractions. (ETBE is a hybrid, being synthesized from corn-based ethanol and petroleum-based *t*-butyl alcohol; at present, there is little experience with its use.) Ethanol has been advocated as a renewable energy source and as a contribution to national energy self-sufficiency. However, using ethanol still consumes fossil fuels, which are used to grow and process corn and to distill the ethanol from the fermentation broth; the energy used to prepare ethanol is comparable to the energy content of the ethanol produced. This energy balance is more favorable for Brazilian ethanol, due to the continuous growing season in the tropics and the higher photosynthetic efficiency of sugar cane. "Gasohol," a blend of 15% ethanol in gasoline, is economically viable in the United States only because of agricultural subsidies to corn farmers. On the other hand, ethanol has low toxicity, and there have been no health complaints about gasohol.

A final complication is that emissions control is affected by the volatility of the fuel. Not all of the volatile organic compounds from gasoline emerge as unburned fuel from the tailpipe. A substantial fraction is released before the fuel is burned, by evaporation from the carburetor, crankcase, and fuel tank, and also at the filling station. Some of the evaporative losses can be recovered by devices that capture the vapors before they are released. But an additional objective in reformulating gasoline is lowering the vapor pressure. While most of the gasoline molecules are seven- and eight-carbon alkanes, a substantial fraction have six, five and even four carbon atoms. Since the vapor pressure increases rapidly with decreasing hydrocarbon size the amounts of the lightest hydrocarbons are being reduced in reformulated gasolines. But there are limits to how far this reduction can be carried, partly because of technical and economic factors at the refinery, and partly because some light molecules are needed to get car engines started, especially on cold days. Consequently, the vapor pressure of gasoline additives is also a consideration. As seen in Table 2.7, the volatility of MTBE and ETBE is comparable to that of the six-carbon alkanes, while the volatility of ethanol-gasoline mixtures, and especially methanol mixtures, is substantially higher.*

E. SUMMARY

Looking back on our survey of atmospheric issues, we see that atmospheric balances can be upset both on a global scale through augmentation of the greenhouse effect and stratospheric ozone destruction, and on a local and regional scale through the buildup of fossil-fuel exhaust gases and their oxidation products. These problems are interrelated in complex and sometimes paradoxical ways. For example, hydroxyl radicals and nitrogen oxides catalyze ozone destruction in the stratosphere, but they are responsible (together with hydrocarbons) for ozone formation in polluted urban air. The chlorofluoromethane gases are harmless locally but contribute globally to both the greenhouse effect and ozone destruction. Carbon monoxide and hydrocarbons are local pollutants whose noxious qualities can be eliminated by oxidizing them to carbon dioxide, but an increase in the global carbon-dioxide concentration adds to the greenhouse effect. The consequences of increasing the concentrations of greenhouse gases are uncertain because long-term trends in climate are currently unpredictable, but they could well be dire. While scientists have learned a great deal about the atmosphere, partly in response to recent environmental concerns, there is an urgent need to learn much more, in order to assess the human impact on it.

Despite the uncertainties, it is clear that overriding influences on air quality are the quantity of energy consumed, the kinds of fuels used, and the energy efficiencies of the applied technologies. At the end of Part I we discussed the benefits of increased energy efficiency and alternative sources of energy (Figures 1.57 and 1.58). In addition to conserving energy resources, these alternatives can provide a cleaner atmosphere, as suggested by the global emissions forecast in Figure 2.52a, b, and c.

---

*Pure ethanol is actually less volatile than gasoline because, although the molecular weight is low, ethanol is an associated liquid, with intermolecular hydrogen bonds. However, when ethanol is dispersed in gasoline, its hydrogen bonds are eliminated and its volatility increases greatly.

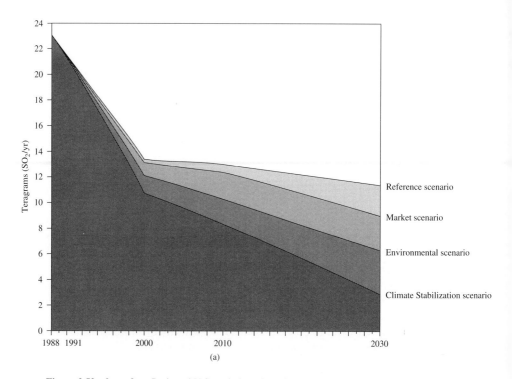

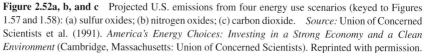

(a)

Figure 2.52a, b, and c   Projected U.S. emissions from four energy use scenarios (keyed to Figures 1.57 and 1.58): (a) sulfur oxides; (b) nitrogen oxides; (c) carbon dioxide.   *Source:* Union of Concerned Scientists et al. (1991). *America's Energy Choices: Investing in a Strong Economy and a Clean Environment* (Cambridge, Massachusetts: Union of Concerned Scientists). Reprinted with permission.

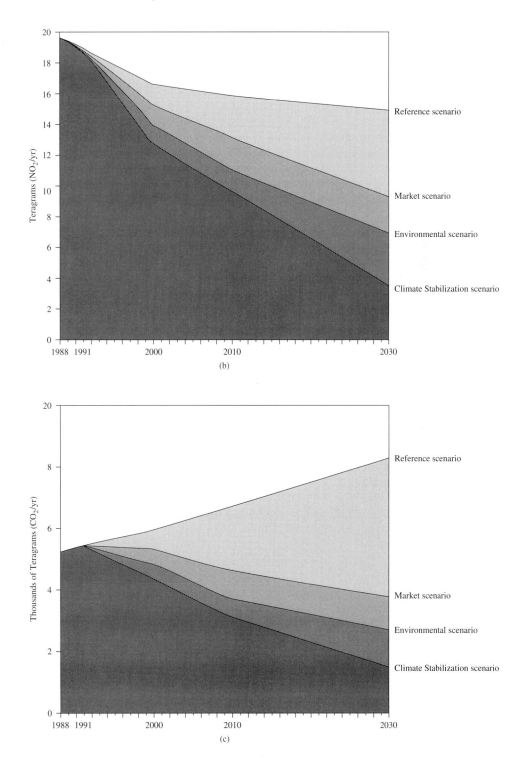

(b)

(c)

PROBLEM SET II: ATMOSPHERE

1. Sketch the heat balance of Earth in watts per square meter (Wm^{-2}), showing the following components: (1) short-wave solar radiation incident at the top of Earth's atmosphere; (2) short-wave radiation reflected by the atmosphere and Earth's surface; (3) short-wave radiation absorbed by the atmosphere and Earth's surface; (4) long-wave radiation emitted from Earth's surface; (5) the share of the long-wave radiation emitted directly to space (through the atmospheric "window") and the share absorbed by the atmosphere; (6) the long-wave radiation emitted downward from the atmosphere to Earth's surface, and outward from the atmosphere into space; (7) the sensible and latent heat transfer from Earth's surface to the atmosphere.

   The following information (in units of Wm^{-2}) is sufficient to conduct this exercise:

   - $S_0 = 1368 \ Wm^{-2}$;
   - the total albedo is 30%, 86% of which is provided by the atmosphere, and the remainder by Earth's surface;
   - 24% of the incident short-wave radiation is absorbed by the atmosphere, and 46% by Earth's surface;
   - Earth's surface temperature is 288 K;
   - about 5% of the long-wave radiation emitted from Earth's surface radiates directly to space through the atmospheric window;
   - the radiative cooling of the atmosphere, and a corresponding radiative heating of Earth's surface is about 106 Wm^{-2};
   - the Stefan-Boltzmann constant equals $0.567 \times 10^{-7} \ Wm^{-2}K^{-4}$.

2. Outgoing long-wave radiation from the top of Earth's atmosphere has been measured by satellites to be 237 Wm^{-2}. From this information, calculate the temperature at the top of the atmosphere. If S_0, the solar irradiance, is 1368 Wm^{-2}, calculate Earth's albedo.

3. Calculate the long-wave emission by Earth's surface, given that its mean global temperature is 288 K.

4. The polar ice caps have an albedo of about 0.80, while the polar seas have a maximum albedo of about 0.20. How could this difference in albedo cause the spontaneous melting of much of the ice cap if a small rise in ambient temperature were to occur? How could large-scale soot deposition on the ice cap cause a similar melting?

5. Calculate the critical radius for water droplet formation in the atmosphere at 20°C and 101% relative humidity. (See equation 2.10 and accompanying discussion for information required to do the calculation.)

6. Calculate the temperature of Earth if there were no greenhouse effect. Assume S_0, the solar irradiance, equals 1368 Wm^{-2}, and the total planetary albedo is 30%. (The Stefan-Boltzmann constant equals $0.567 \times 10^{-7} Wm^{-2} K^{-4}$.)

7. (a) What characteristic molecular properties of H_2O and CO_2 cause their absorption of infrared radiation?

   (b) Why are chlorofluorocarbons (CFCs) such as CF_2Cl_2 and $CFCl_3$ such effective greenhouse gases?

   (c) In addition to H_2O, CO_2, and CFCs, name two other greenhouse gases and identify for each of these latter two, one human activity leading to atmospheric emissions.

8. Which of the molecules in the following table have the potential to cause a greenhouse effect similar to CFCs (see Figure 2.13)? For polyatomic molecules, identify the specific vibrations involved.

S-C-O (linear)	H-Cl	F-F	H$_2$S (nonlinear)	Cl$_2$O (nonlinear)
$\lambda_1 = 11,641$ nm	$\lambda = 3,465$ nm	$\lambda = 11,211$ nm	$\lambda_1 = 3,830$ nm	$\lambda_1 = 14,706$ nm
$\lambda_2 = 18,975$ nm			$\lambda_2 = 7,752$ nm	$\lambda_2 = 30,303$ nm
$\lambda_3 = 4,810$ nm			$\lambda_3 = 3,726$ nm	$\lambda_3 = 10,277$ nm

9. **(a)** Describe the two opposing effects on climate resulting from combustion of coal with high sulfur content.

   (b) Cite two processes by which sulfate particles in the troposphere affect climate.

10. **(a)** The total mass of carbon contained in fossil fuels that was burned in the world from 1860 to 1992 is estimated to be 2.25×10^{14} kg C. The amount of carbon released as CO_2 from agricultural expansion and deforestation over this period is estimated at 1.18×10^{14} kg C. The concentration of CO_2 in the atmosphere in 1992 was 355 ppm, corresponding to a total mass of 2.78×10^{15} kg of CO_2. If the concentration of CO_2 in 1860 was 295 ppm, calculate the percentage of CO_2 from these two sources that remained in the atmosphere over the period 1860 to 1992.

   (b) Plant studies indicate that the net primary production (NPP) of organic carbon by photosynthesis can increase with increasing CO_2 concentration in the atmosphere. Assume that the increase of NNP in the biosphere is 0.27 of the percentage increase of atmospheric CO_2. Given that the global NPP of the biosphere is currently estimated to be 7.32×10^{13} kg C/yr, estimate how much more carbon is being absorbed per year in NPP compared to the amount that would be absorbed if the atmospheric CO_2 concentration were 295 ppm.

   (c) Currently, about 6.0×10^{12} kg C/yr of fossil fuel carbon and 1.7×10^{12} kg C/yr of carbon from destruction of land vegetation are released to the atmosphere. Assuming the value obtained in part (b), what percentage of the emitted carbon might be taken away by increased absorption in the biosphere? The effect of deforestation may be reducing NPP. How does such a reduction affect the potential of biomass to serve as a sink for CO_2?

11. Name two strategies for reducing greenhouse gas emissions related to energy consumption.

12. The amount of CH_4 emitted annually is estimated to be 25 to 50 times greater than the amount of N_2O emitted annually, yet the atmospheric concentration of CH_4 is estimated to be only 6 times higher. Can you offer an explanation?

13. **(a)** From Figure 2.25, calculate the bond order of O_2 (defined as one-half the sum of electrons in the bonding orbitals, minus one-half the sum of electrons in the anti-bonding orbitals).

   (b) What property of the bonding in O_2 prevents rapid oxidation reactions with most molecules at ambient temperatures in the atmosphere and the biosphere?

   (c) Name the two classes of molecules (or atoms) that can bond effectively with O_2, even at ambient temperatures, and explain why they can?

14. When an electron donor such as an organic radical R· or a heavy metal M· reacts with O_2 to form RO$_2$· or MO$_2$·, the O–O bond is weakened considerably. Use the molecular orbital diagram shown in Figure 2.25 to explain why this weakening occurs. Explain, with at least two examples from atmospheric or biological chemistry, how these electron donors can serve as catalysts for oxidation at ambient temperatures.

15. Biological compounds such as carbohydrates, fats, and proteins are thermodynamically unstable in the presence of O_2, since, upon combustion, they will undergo exothermic reactions with oxygen to form CO_2, H_2O, and other simple gases and compounds. On the other hand, aerobic life forms require O_2 for generating energy, via oxidation reactions, needed for a myriad of metabolic

activities. Explain how the special electronic structure of O_2 allowed life on the planet to evolve in the presence of an oxidizing atmosphere, and how these life forms are able to harvest energy from oxidation reactions under controlled conditions within cells.

16. **(a)** Why is hydroxyl radical such an important component of the atmosphere?

(b) How is it generated?

(c) Describe two of its reactions with air pollutants.

17. **(a)** Write the equations for ozone formation and ozone destruction in the absence of catalytic chains, including the net reactions.

(b) Write the equations of ozone destruction involving NO, OH, and Cl. For each of these, describe explicitly how they lead to ozone destruction, and describe the sources of these reactants.

(c) Describe the two reactions by which NO_2 serves to protect the ozone layer.

(d) How might stratospheric-flying aircraft perturb the ozone layer?

18. **(a)** How does stratospheric ozone shield Earth's surface from harmful UV radiation?

(b) Calculate the fractional increase in transmission (dT/T) for a 1% decrease in the thickness (l) of the ozone layer for the following three wavelengths [in nanometers (nm)]: 310 nm, 295 nm, and 285 nm. (Assume $l = 0.34$ cm; $\varepsilon = 3, 18,$ and 56 cm^{-1} for 310 nm, 295 nm, and 285 nm, respectively.)

(c) As a result of the 1% decrease, describe what happens to the position of the curve of the *actual action spectrum* relative to the curve of *relative sensitivity to sunburn,* as shown in Figure 2.30.

(d) Will the calculated increase in transmission at 285 nm cause more damage to biological tissue than the calculated increase in transmission at 295 nm? Explain your reasoning.

19. **(a)** In the ozone hole phenomenon over Antarctica, how is nitrogen removed from the atmosphere, and which molecules serve as "chlorine reservoirs" in the darkness of the Antarctic winter?

(b) How does the sunlight at the beginning of spring initiate the polar ClO_x chain reaction?

(c) Write the four equations involved in this chain reaction, and describe the equations that finally stop the chain reaction.

20. Give two reasons why the stratosphere is more susceptible to chemical pollution than the troposphere.

21. Assume that the concentration of suspended particulates in a polluted atmosphere is 170 $\mu g/m^3$. The particulates contain adsorbed sulfate and hydrocarbons comprising 14% and 9% of the weight, respectively. An average person respires 8,500 liters of air daily and retains 50% of the particles smaller than 1 μm in diameter in the lungs. How much sulfate and hydrocarbon accumulate in the lungs in one year if 75% of the particulate mass is contained in particles smaller than 1 μm?

22. Choose one of the following four air pollutants, SO_2, NO_x, CO, VOC, and provide the following information: (1) environmental effects; (2) human health effects; (3) major sources; (4) atmospheric reactions; (5) atmospheric lifetime.

23. Describe the contribution of transportation to air pollution, including the four pollutants cited in problem 22, as well as particulates. Do the same for stationary-source fuel combustion.

24. Why is NO_2, unlike the higher oxides of carbon and sulfur (CO_2 and SO_3, respectively), unstable at 25°C in the presence of sunlight? Describe briefly how the reaction of NO_2 with sunlight plays a key role in smog formation and in the regeneration of NO_2 itself.

25. For the reaction of carbon monoxide to carbon dioxide:

$$CO + \tfrac{1}{2}O_2 \rightarrow CO_2$$

the equilibrium constant K_{eq} (at 25°C) = 3×10^{45}. Given this enormous value, why doesn't carbon monoxide convert spontaneously to carbon dioxide in air? How does the use of platinum in the catalytic converter of automobiles facilitate the conversion to carbon dioxide? (Hint: platinum contains unpaired d electrons in its outer orbital.)

26. Why was lead added as a component of gasoline before the 1970s? Why was it removed with the advent of the catalytic converter? Why did the new, unleaded gasolines contain higher concentrations of aromatics and liquid olefins? (Refer to Table 2.7; the aromatics are the fifth cluster of compounds, and the liquid olefins are the third cluster.) Among the properties listed in Table 2.7, in which are aromatics clearly superior to liquid olefins (discounting considerations of toxicity, particularly for benzene, which is a human carcinogen)? For reasons of minimizing vapor pressure, photochemical activity, and toxicity, the newly formulated gasolines will have reduced levels of aromatics and liquid olefins. To compensate for these reductions, oxygenated compounds have been added. The two most common oxygenates are ethanol and MTBE. In addition to reducing CO, what other favorable advantages do ethanol and MTBE provide with respect to the properties listed in Table 2.7? Why do the producers of MTBE claim that their compound is superior to ethanol? (Hint: conventional gasoline has a vapor pressure of about 9 psi.)

SUGGESTED READINGS II: ATMOSPHERE

J. T. HOUGHTON, G. J. JENKINS, and J. J. EPHRAUMS, eds. (1990). *Climate Change: The IPCC Scientific Assessment* (Cambridge, UK: Cambridge University Press).

T. E. GRAEDEL and P. J. CRUTZEN (1993). *Atmospheric Change: an Earth System Perspective* (New York: W. H. Freeman and Company).

S. H. SCHNEIDER (1989). The changing climate. *Scientific American* 261(3): 70–79.

R. M. WHITE (1990). The great climate debate. *Scientific American* 262(7): 36–43.

J. E. HANSEN and A. A. LACIS (1990). Sun and dust versus greenhouse gases: an assessment of their relative roles in global climate change. *Nature* 346: 713–719.

R. J. CHARLSON, S. E. SCHWARTZ, J. M. HALES, R. D. CESS, J. A. COAKLEY, JR., J. E. HANSEN, and D. J. HOFMANN (1992). Climate forcing by anthropogenic aerosols. *Science* 255: 423–430.

R. J. CHARLSON and T. M. L. WIGLEY (1994). Sulfate aerosol and climate change. *Scientific American* 270(2): 48–57.

J. E. HANSEN, A. LACIS, R. RUEDY, and M. SATO (1992). Potential climate impact of Mount Pinatubo eruption. *Geophysical Research Letters* 19: 215–218.

B. HILEMAN (1992). Role of methane in global warming continues to perplex scientists. *Chemical and Engineering News* 70(6): 26–28.

J. S. LEVINE, W. R. COFER III, D. R. CAHOON, JR. and E. L. WINSTEAD (1995). Biomass burning—a driver of global change. *Environmental Science and Technology* 29: 120A–125A.

W. M. POST, T.-H. PENG, W. R. EMANUEL, A. W. KING, V. H. DALE, and D. L. DEANGELIS (1990). The global carbon cycle. *American Scientist* 78: 310–326.

A. M. THOMPSON (1992). The oxidizing capacity of the earth's atmosphere: probable past and future changes. *Science* 256: 1157–1165.

R. A. KERR (1991). Hydroxyl, the cleanser that thrives on dirt. *Science* 253: 1210–1211.

F. S. ROWLAND (1991). Stratospheric ozone in the 21st century: the chlorofluorocarbon problem. *Environmental Science and Technology* 25: 622–628.

P. S. ZURER (1993). Ozone depletion's recurring surprises challenge atmospheric scientists. *Chemical and Engineering News* 71(21): 8–18.

P. S. ZURER (1995). NASA cultivating basic technology for supersonic passenger aircraft. *Chemical and Engineering News* 73(7): 10–16 (discusses aspects related to ozone depletion and reduction of NO_x emissions).

A. RUSSELL, J. MILFORD, M. S. BERGIN, S. McBRIDE, L. McNAIR, Y. YANG, W. R. STOCKWELL, and B. CROES (1995). Urban ozone control and atmospheric reactivity of organic gases. *Science* 269: 491–495.

F. M. BLACK (1991). Control of motor vehicle emissions—the U.S. experience. *Critical Reviews in Environmental Control* 21(5,6): 373–410.

J. M. LENTS and W. J. KELLY (1993). Clearing the air in Los Angeles. *Scientific American* 269(4): 32–39.

J. V. HALL, A. M. WINER, M. T. KLEINMAN, F. W. LURMANN, V. BRAJER, and S. D. COLOME (1992). Valuing the health benefits of clean air. *Science* 255: 812–816.

E. V. ANDERSON (1992). Ethanol's role in reformulated gasoline stirs controversy. *Chemical and Engineering News* 70(44): 7–13.

D. SEDDON (1992). Reformulated gasoline, opportunities for new catalyst technology. *Catalysis Today* 15:1–21.

H. A. MOONEY, P. M. VITOUSEK, and P. A. MATSON (1987). Exchange of materials between terrestrial ecosystems and the atmosphere. *Science* 238: 926–932.

Part III

Hydrosphere

A. WATER RESOURCES

1. Global Perspective

We live, quite literally, in a watery world. All living things depend absolutely on a supply of water. The biochemical reactions of every living cell take place in aqueous solution, and water is the transport medium for the nutrients a cell requires and for the waste products it excretes. Water is abundant on the planet's surface, but about 97% of Earth's supply of water is in the oceans, where it is too salty to be used by humans or other land creatures. Every day, however, the sun's rays distill a large quantity of water that falls back to the surface as rain. Proportionately more rain falls on land than on the oceans, providing a continual supply of fresh water. We treat water as if it were free and, in a sense, it is free—a by-product of the enormous flux of solar energy on Earth. The hydrologic cycle accounts for about half of the solar energy absorbed by Earth's surface.

The global movement and storage of water is illustrated in Figure 3.1. Every year, 111,000 km^3 of water falls on land, and 71,000 km^3 returns to the atmosphere via evaporation from wet surfaces and transpiration from plants; these two processes are called collectively *evapotranspiration*. The remainder, 40,000 km^3, is the *runoff,* which eventually reaches the oceans. Some of the runoff is carried to the ocean directly in surface waters, but much of the water falling on land percolates into permeable rock layers and is stored as groundwater in aquifers. Wells dug into these aquifers

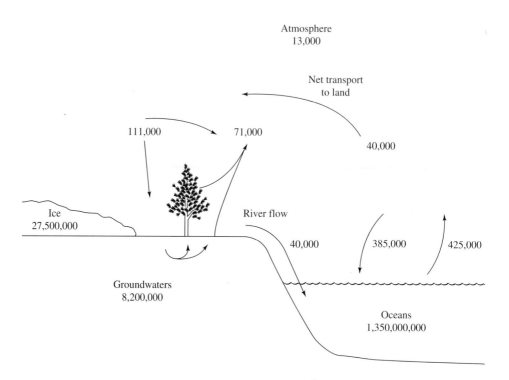

Figure 3.1 The global water cycle. The numbers are km^3 for the water reservoirs, and km^3/yr for the flows. *Source:* D. H. Speidel and A. F. Agnew (1982). *The Natural Geochemistry of Our Environment* (Boulder, Colorado: Westview Press).

supply a substantial fraction of the water used by humans. If the runoff were divided evenly, it could provide each person with 7,690 m^3 a year of fresh water (1990 population). But of course it is not divided evenly. Some continents are rainier than others, and the variation within continents is even greater (Table 3.1). The runoff per km^2 of land area in South America, for example, is over four times as much as in Africa, while within Africa, the runoff per km^2 in Zaire is 17 times as much as in Kenya.

The amount of water withdrawn for human uses is much less than the total runoff, averaging about 9% worldwide; but in some countries the fraction is considerably higher. Water is used for many purposes: drinking, cooking, washing, carrying wastes, cooling machines, and irrigating crops. As might be expected, utilization patterns depend on the level of economic development in a region (Tables 3.1 and 3.2). Per capita water utilization varies widely; the continental averages range from 1,692 m^3 per year in North America to 244 m^3 per year in Africa. In Canada, 80% of the water utilization is for industry and power generation, while 8% is for agriculture; in Mexico, however, the percentages are reversed. Interestingly, the European per capita utilization, 726 m^3 per year, is less than half of that of North America, despite similar economic conditions; however, a greater fraction of the runoff is withdrawn in Europe (16% versus 10%).

TABLE 3.1   ANNUAL WATER SUPPLY AND WITHDRAWAL FOR CONTINENTS AND
VARIOUS COUNTRIES (1990)

Continents/ Countries	Water supply			Water withdrawal		Per capita use/Supply ratio (%)	Status*
	Total (km³)	Per km² (m³)	Per capita (m³)	Total (km³)	Per capita (m³)		
WORLD	40,673	309,799	7,690	3,240	660	9	Potential problems
Africa	4,184	141,154	6,460	144	244	4	Potential problems
Kenya	15	26,330	590	1.1	48	8	Scarcity
Zaire	1,019	449,374	28,310	0.7	22	> 0	Surplus
North America†	6,945	324,882	16,260	697	1,692	10	Surplus
Mexico	357	187,039	4,030	54	901	22	Potential problems
Canada	2,901	314,609	109,370	42	1,752	2	Surplus
South America	10,377	591,982	34,960	133	476	1	Surplus
Peru	40	31,250	1,790	6.1	294	16	Stress
Brazil	5,190	613,728	34,520	35	212	1	Surplus
Asia	10,485	383,893	3,370	1,531	526	16	Potential problems
China	2,800	300,223	2,470	460	462	19	Potential problems
Indonesia	2,530	1,396,579	14,020	17	96	1	Surplus
Europe‡	2,321	490,746	4,660	359	726	16	Potential problems
Poland	49	160,946	1,290	17	472	37	Stress
Sweden	176	427,579	21,110	4.0	479	2	Surplus
Oceania	2,011	238,639	75,960	23	907	1	Surplus
Australia	343	45,025	20,480	18	1,306	6	Surplus
Papua New Guinea	801	1,768,759	199,700	0.1	25	> 0	Surplus

*• Water surplus: ≥ 10,000 m³/capita
  • Potential water management problems: ≥ 2,000 m³/capita, < 10,000 m³/capita
  • Water stress: ≥ 1,000 m³/capita, < 2,000 m³/capita
  • Water scarcity: < 1,000 m³/capita

†Includes Central America

‡Excludes the former USSR

Source: Data drawn from World Resources Institute (in collaboration with the United Nations Environment Programme and
the United Nations Development Programme) (1993). *World Resources 1992–93* (Oxford, UK: Oxford University Press).

    Worldwide, agriculture accounts for the lion's share of water use, 69%, and agricul-
tural demand is growing as population continues to increase (Figure 3.2). Large inputs of
water are needed in agriculture because growing plants transpire rapidly. In order to capture
atmospheric CO_2, their leaves are designed for efficient exchange of gases (Figure 3.3). This

TABLE 3.2 USES OF WATER FOR CONTINENTS AND COUNTRIES (1990)

Continents/Countries	Domestic (%)	Industry/Power (%)	Agriculture (%)
WORLD	8	23	69
Africa	7	5	88
Kenya	27	11	62
Zaire	58	25	17
North America*	9	42	49
Mexico	6	8	86
Canada	11	80	8
South America	18	23	59
Peru	19	9	72
Brazil	43	17	40
Asia	6	8	86
China	6	7	87
Indonesia	13	11	76
Europe[†]	13	54	33
Poland	16	60	24
Sweden	36	55	9
Oceania	64	2	34
Australia	65	2	33
Papua New Guinea	29	22	49

*Includes Central America
[†]Excludes the former USSR

Source: World Resources Institute (in collaboration with the United Nations Environment Programme and the United Nations Development Programme) (1992). *World Resources 1992–93* (Oxford, UK: Oxford University Press).

exchange takes place in tiny pores called stomata; the cells in the stomata absorb CO_2 and release O_2 during photosynthesis. While the stomata are open, water vapor is released as well. Plants can conserve water by closing the stomata (in droughts, for example), but doing so curtails the CO_2 supply. Furthermore, as the seedlings develop, the leaf area expands until it is about three times the surface area of the soil below, and the amount of water lost by transpiration exceeds that lost by evaporation. Because the exchange of gases makes the evaporation of water from the moist cells unavoidable, photosynthesis is always accompanied by copious transpiration. The amount of water required to produce a bushel of corn, for example, is 5,400 gallons (20 m³). Moreover, the amount of water agriculture requires is even greater because the efficiency of traditional irrigation systems is not high; the global

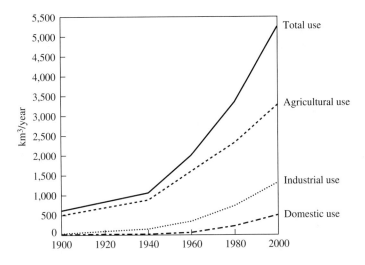

Figure 3.2   Global water use, 1900–2000.   *Source:* M. K. Tolba et al. (eds.) (1992). *The World Environment 1972–1992, Two Decades of Challenge* (London: Chapman & Hall on behalf of The United Nations Environment Programme).

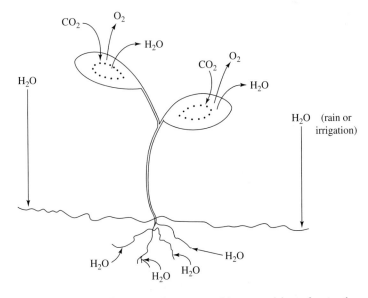

Figure 3.3   Exchange of gases on the stomata of leaves, and loss of water through transpiration.

average irrigation efficiency is estimated to be 37%. Thus, most of the water runs off or evaporates before it even reaches the plants. New irrigation technologies are much more efficient but are also more complex, and most of the world's farmers lack the needed capital and maintenance facilities.

Irrigation may dwarf domestic uses of water, and yet many people around the world are chronically short of water for personal needs. In many places, freshwater aquifers are being drained faster than they can be replenished. Local reservoirs might be insufficient, especially in times of drought. Water resources are further strained by increasing population; for example, unless the water supply in Africa and Asia increases significantly, the expected increases in population are predicted to place both continents under "water stress" (see Table 3.1 notes) by 2025. In addition, water quality is just as important as quantity. The water in many places is contaminated because of failure to separate wastewater from the water supply. Waterborne diseases continue to be scourges of humankind. In many parts of the world, the critical need for sanitary water is all too often neglected.

2. U.S. Water Resources

To gain a more detailed perspective on water resources, we examine the pattern of distribution and use in the continental United States. A schematic of the water flows is given in Figure 3.4. As in the global water cycle, about two-thirds of the precipitation is returned to the atmosphere via evapotranspiration; the balance goes to runoff, amounting to about 2,000 km^3/yr. Most of the water is in the east of the country, and flows to the Atlantic and Gulf coasts. A quarter of the annual runoff (468 km^3) is withdrawn for various uses (1990 figures), of which three-quarters (338 km^3) is subsequently returned to the streamflow and the remainder is "consumed," mostly through evapotranspiration after or during use; eventually, of course, all the water is returned to the global cycle. The pattern of water utilization is diagrammed in Figure 3.5. The 468 km^3 of fresh water used in 1990 was actually drawn partly (110 km^3) from groundwater, which is eventually recharged from the surface. Water usage is dominated by agriculture and by the cooling of electric generators, each of which requires about 40% of the supply. The remaining 20% is divided between industry and mining on the one hand (8%), and domestic and commercial uses on the other (12%). Major industrial users are the steel, chemicals, and petroleum industries; the mining uses include water extraction of minerals and fossil fuels, and milling and related activities. Most of the industrial and mining water is obtained directly from surface and groundwater, but about 20% of it comes from public supplies, which also provide most of the domestic and commercial water. Domestic uses (drinking, food preparation, bathing, washing clothes and dishes, flushing toilets, and watering lawns and gardens) amounted to about 380 liters (100 gallons) per day per person in 1990. The total was about three times higher than for commercial uses (hotels, restaurants, office buildings, and such).

While agriculture is a heavy user of water throughout the country, irrigation is especially important in the arid West, where extensive diversion of the major waterways has permitted large-scale crop production. California and Idaho are the largest users of irrigation water, together accounting for 34% of the national total.

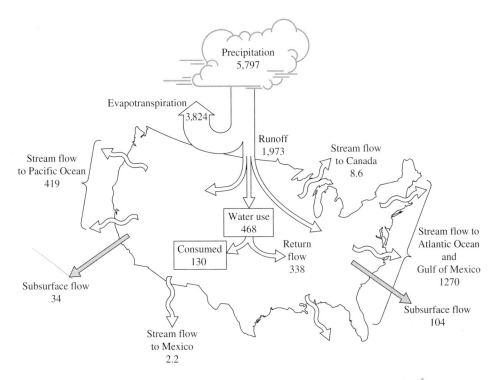

Figure 3.4   Annual water flows in the United States. The numbers are in units of km³/yr.
Source: Adapted from U.S. Water Resources Council (1978). *The Nation's Water Resources,
1975–2000: Second National Water Assessment, vol. 1: Summary* (Stock number 052-045-
00051-7) Washington, DC: Superintendent of Documents).

3. Water as Solvent and as Ecosystem

In considering issues surrounding water use and quality, it can be helpful to look at the two
distinct roles water plays in Earth's environment. On the one hand, water is a remarkable solvent; its unusual molecular properties allow it to dissolve and transport a wide range of materials. On the other hand, water houses ecosystems; a large percentage of the biosphere lives
in some form of aqueous environment. We often define water quality in terms of the ability
of the aqueous environment to support the normal range of biological species with their accompanying biochemical processes. These two perspectives on water are interrelated, of
course: the materials dissolved in water affect its ability to support life, and the biological
processes of aqueous environments often play critical roles in clearing dissolved substances
from contaminated waters.

    We shall follow these two perspectives on water by tracing the course of water through
the hydrological cycle. From atmosphere to soil, the dominant aspect of water is its role as a
solvent. Water is almost pure when it enters the atmosphere, but immediately begins to

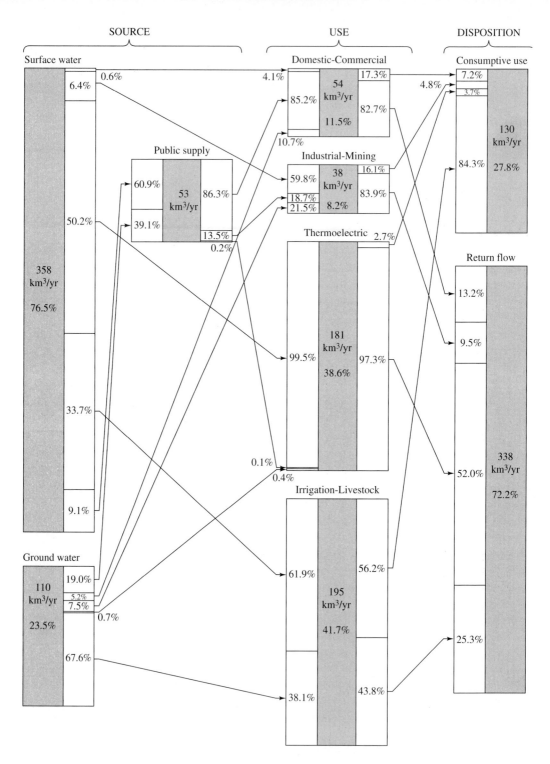

Figure 3.5 Sources, uses, and dispositions of fresh water in the United States, 1990. *Source:* W. B. Solley et al. (1993). *Estimated Use of Water in the United States in 1990,* U.S. Geological Survey, Circular 1081 (Washington, DC: U.S. Department of the Interior).

interact and form solutions with other substances. As discussed in Part II, water condenses in clouds through the nucleation of atmospheric particles, including nitrates and sulfates. Atmospheric water, whether in clouds or rain, is in equilibrium with other constituents in the atmosphere, particularly carbon dioxide.

When rain falls to Earth, the number of interactions increases as water enters the soil, where it absorbs increasing numbers of ions. Eventually, water and its load of dissolved materials accumulate in surface waters and the oceans; at this point, the role of water as a solvent is augmented by its role as ecosystem. Most critical to considerations of the aqueous biosphere is the redox potential, determined in large part by the amount of dissolved oxygen in the water relative to other nutrients necessary for life. The balance changes with time and place, leading to distinct ecological niches in freshwater lakes, rivers, wetlands, and coastal areas of oceans.

B. FROM CLOUDS TO RUNOFF: WATER AS SOLVENT

1. Unique Properties of Water

Water is the most commonplace of substances, and yet its properties are unique. Perhaps most remarkable are its high melting and boiling temperatures. Table 3.3 lists the melting and boiling points for the hydrides from carbon to fluorine, all of which are elements in the first row of the periodic table. Compared to the neighboring hydrides, water has by far the highest melting and boiling points. One characteristic that makes Earth uniquely suitable for the evolution of life is its surface temperature which, over most of the planet, lies within the liquid range of water.

The high temperature required for boiling implies that liquid water has a high cohesive energy; the molecules associate strongly with one another. One source of this cohesion is water's high dipole moment (1.85 Debye*). The O–H bonds are highly polar; negative charge builds up on the oxygen end while positive charge builds up on the hydrogen end. The dipoles tend to align with one another, increasing the cohesive energy. But if the dipole-dipole interaction were the only factor, then HF would have a higher boiling point than water, because it has a higher dipole moment (1.91 Debye). The greater cohesion of water derives,

TABLE 3.3   LIQUID TEMPERATURE RANGE FOR WATER AND NEIGHBORING FIRST ROW ELEMENT HYDRIDES

	CH_4	NH_3	H_2O	HF
Melting point (°C)	−182	−78	0	−83
Boiling point (°C)	−164	−33	100	20

*A dipole moment, μ, is defined as the charge times the distance separating the charge. 1 Debye (D) = 3.336×10^{-30} coulomb (C) × meters (m). For example, for one electron (1.60×10^{-19} C) separated from a proton by 1 Å (10^{-10} m), $\mu = 1.6 \times 10^{-29}$ C × m = 4.80 D.

in addition, from the molecule's bent structure, with hydrogen at each end. Water is able to donate two hydrogen bonds, one from each of its H atoms, while simultaneously accepting two hydrogen bonds, one to each of the electron lone pairs on the oxygen atom. Thus water forms a three-dimensional network of H-bonds, whereas HF, having only one donor site, is limited to forming linear H-bond chains (Figure 3.6).

The H-bond network of water is demonstrated strikingly by the crystal structure of ice, in which each water molecule is H-bonded to four other molecules in connected six-membered rings (Figure 3.7). These six-membered rings produce a very open structure, which accounts for another unusual property of water, its expansion upon freezing. Most substances contract upon solidification because the molecules take up more room in the chaotic liquid state than they do in the close-packed solid. The liquid continues to expand as it is heated due to the increased motion of the molecules. When ice melts, however, the open lattice structure collapses on itself and the density *increases*. H-bonded networks still exist, but they fluctuate very rapidly, allowing the individual water molecules to be mobile and closer together. As liquid water is heated from 0°C, the H-bonded network is further disrupted and the liquid water continues to contract. When the temperature reaches 4°C, the density of liquid water reaches a maximum. As the temperature rises above 4°C, the water slowly expands, reflecting the increasing thermal motion. The lower density of ice relative to water is of great ecological significance to the world's temperate-zone lakes and rivers. Because winter ice floats on top of the water, it protects the aquatic life below the surface from the inhospitable climate above. If ice were denser than water, the lakes and rivers would freeze from the bottom up, creating arctic conditions.

The ability of water molecules to hydrogen-bond to each other accounts for the existence of crystalline hydrates (or clathrates) of nonpolar molecules. Examples of these are compounds with the formula $8X \cdot 46H_2O$, where $X = Ar$, Kr, Xe, or CH_4. The crystals consist of arrays of dodecahedra (Figure 3.8) formed from 20 water molecules H-bonded together. Within the dodecahedra are the nonpolar molecules. By filling the voids, the guest molecules stabilize the clathrate H-bond network. The hydrates are stable to temperatures appreciably higher than the melting point of ice. For example, methane hydrate melts at 18°C. Clogging of natural gas pipelines by methane hydrate was once a problem for gas distribution. It was solved by removing water vapor before the gas was fed into the lines.

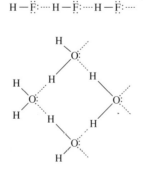

Figure 3.6   Linear H-bonding in HF versus network H-bonding in H_2O.

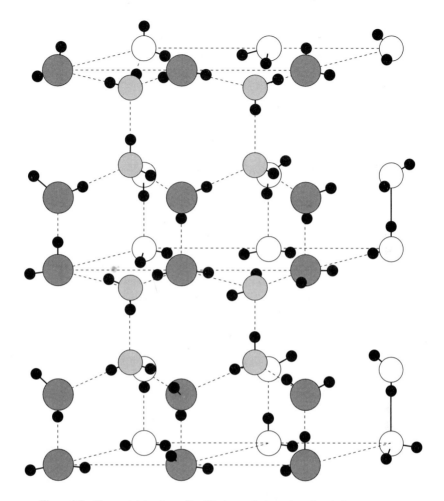

Figure 3.7   The crystal structure of ice. The larger circles and smaller circles represent oxygen atoms and hydrogen atoms, respectively. Oxygens with the same color shading indicate they lie in the same plane.

This tendency of water to form extended H-bond networks around nonpolar molecules explains why oil and water do not mix. Water molecules actually attract hydrocarbon molecules; as shown in Table 3.4, the enthalpy of dissolution in water is negative for simple hydrocarbons, meaning that molecular contact releases heat. The heat release reflects an attractive force created by the dipole/induced dipole interaction between water and the guest molecule. Nevertheless, the free energy values are positive; the reaction is disfavored by large negative entropy changes ($\Delta G = \Delta H - T\Delta S$). The negative entropy implies that mixing increases molecular order, consistent with the water molecules packing around the nonpolar solute, as they do in the clathrate structures. Thus, oil and water do not mix because the packing tendency of the water inhibits it.

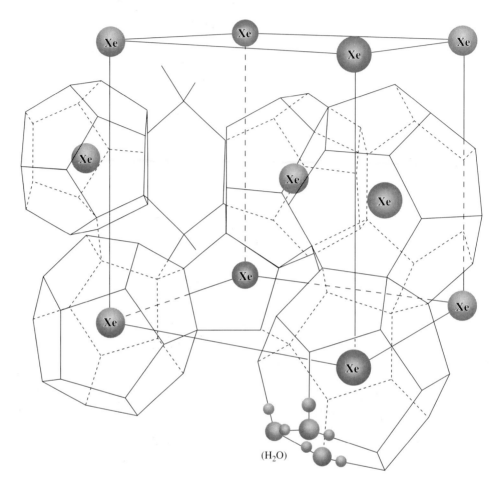

(H₂O)

Figure 3.8 The structure of a clathrate crystal, xenon hydrate; the xenon atoms occupy cavities (eight per unit cube) in a hydrogen-bonded three-dimensional network formed by water molecules (46 per unit cube).

This tendency of water and nonpolar molecules to separate is responsible for many important structures in biology. For example, proteins keep their shape in water by folding in such a way that *hydrophobic* ("water-hating") alkyl and aryl groups of the amino acids are in the interior, away from the water, while *hydrophilic* ("water-loving") polar and charged groups face the exterior, where they contact the water. Likewise, biological membranes are bilayers of lipid molecules (Figure 3.9), with the hydrocarbon chains of the lipids packed together away from the water in the interior of the bilayer.

2. Acids, Bases, and Salts

a. Ions, autoionization, and pH.   While water mixes poorly with oil, it mixes very well with ionic compounds. The ready solvation of ions stems from water's large dipole

TABLE 3.4   FREE ENERGY, ENTHALPY, AND ENTROPY OF SOLUTION IN LIQUID WATER AT 298 K

Process	ΔG (joules/mol)	ΔH (joules/mol)	ΔS (joules/K/mol)
CH_4 in benzene $\rightarrow$ CH_4 in water	+10,878	−11,715	−75
C_2H_6 in benzene $\rightarrow$ C_2H_6 in water	+15,899	−9,205	−84
C_2H_4 in benzene $\rightarrow$ C_2H_4 in water	+12,217	−6,736	−63
C_2H_2 in benzene $\rightarrow$ C_2H_2 in water	+7,824	−795	−29
Liquid propane $\rightarrow$ C_3H_8 in water	+21,129	−7,531	−96
Liquid n-butane $\rightarrow$ C_3H_{10} in water	+24,476	−4,184	−96
Liquid benzene $\rightarrow$ C_6H_6 in water	+17,029	0	−59
Liquid toluene $\rightarrow$ C_7H_8 in water	+19,456	0	−67
Liquid ethyl benzene $\rightarrow$ C_8H_{10} in water	+23,012	0	−79

Source: W. Kauzmann (1959). Some factors in the interpretation of protein denaturation. *Advances in Protein Chemistry* 14: 1–63.

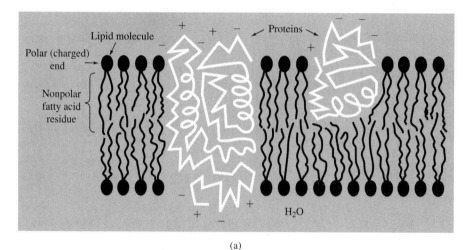

(a)

(b)

Figure 3.9   (a) The hydrophilic-hydrophobic bilayer structure of a biological membrane; (b) Molecular structure of a lipid molecule.

moment and H-bonding ability. Anions interact with the positive end of the dipole via H-bonds (Figure 3.10), while cations interact with the negative end via coordinate bonds from the oxygen lone-pair electrons. For some transition metal ions, the resulting complex ion is sufficiently long-lived to isolate, as in $Cr(H_2O)_6^{3+}$; but for most ions, the bonds are broken and reformed very rapidly, producing an averaged solvation environment for the ion. Water can also H-bond with amine, alcohol, carbonyl, and other polar groups of organic molecules, thereby enhancing their solubility.

Among the ions that are stabilized in water are the hydrogen and hydroxide ions. Strong acids such as HCl, HNO_3, and H_2SO_4 release H^+ ions quantitatively when dissolved in water, while strong bases such as NaOH or KOH release OH^- quantitatively. These ions are special because they react together to form water

$$H^+ + OH^- = H_2O \tag{3.1}$$

Reaction (3.1) is a *neutralization* reaction: The strong base neutralizes the strong acid and vice versa. The reverse of reaction (3.1) is *autoionization:* the water ionizes itself because the ions are stabilized by water molecules. This reverse reaction does not proceed to any great extent; the position of equilibrium for reaction (3.1) lies far to the right. Nevertheless, autoionization is a key factor in acid-base chemistry because it sets the range of available acid and base strengths in water.

The equilibrium expression for the autoionization reaction is $K = [H^+][OH^-]/[H_2O]$. Since the concentration of the water is constant (55.5 mol/l), it is customarily included in the equilibrium constant, which is then written

$$K_w = [H^+][OH^-] \tag{3.2}$$

Thus, the existence of autoionization requires that $[OH^-] = K_w/[H^+]$ and reciprocally that $[H^+] = K_w/[OH^-]$. Taking negative logarithms, which are symbolized by the letter "p," we have

$$pH + pOH = pK_w \tag{3.3}$$

The pH scale is set by the fact that, experimentally, $K_w = 10^{-14}\,M^2$ (M = molar, moles per liter), and pK_w is therefore 14. In pure water, the autoionization reaction is the only source of ions, and consequently $[H^+] = [OH^-]$. Then $K_w = [H^+]^2$, $[H^+] = 10^{-7}$, and the pH equals 7. A pH of 7 defines neutrality; pH values below 7 are acidic, while pH values above 7 are basic, or *alkaline*.

b. pH versus concentration: weak acids and bases.

Because water dissolves so many different substances, the only place in the environment where pure water is

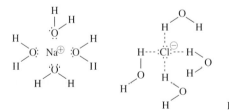

Figure 3.10   Solvation of ions in water.

found is at the start of the hydrological cycle, in water vapor. Even raindrops include other substances; as detailed in Part II, the formation of raindrops in saturated air is nucleated by atmospheric particles, particularly nitrates and sulfates. In addition, the water in cloud vapor and raindrops is in equilibrium with other constituents of the atmosphere, especially carbon dioxide. These substances tend to lower the pH of rainwater significantly below neutrality.

The relationship between the concentration of an acidic species in water and the pH of the solution depends upon the degree to which the acid donates a proton to water. Strong acids transfer a proton completely to water:

$$HA = H^+ + A^- \tag{3.4}$$

Hence in a 1 M solution of a strong acid, $[H^+] = 10^0$, and the pH = 0; since pH + pOH = 14, pOH = 14. Conversely, a 1 M solution of a strong base has pOH = 0 and pH = 14.

However, proton transfer is only partial for some acids; these are referred to as *weak acids*. The degree of transfer is determined by the equilibrium constant for the *dissociation* reaction

$$K_a = [H^+][A^-]/[HA] \tag{3.5}$$

K_a is often referred to as the *acidity constant*. When a weak acid, HA, is dissolved in water, an equal number of H^+ and A^- ions are produced, so $[H^+] = [A^-]$. Rearranging equation (3.5) to solve for $[H^+]$ yields the following equation:

$$[H^+]^2 = K_a[HA] \tag{3.6}$$

If the degree of proton transfer is small (specifically if $K_a \ll [HA]$), then only a small fraction of HA is dissociated, and [HA] is essentially the same as C_{HA}, the initial concentration of dissolved HA. For example, the acidity constant of acetic acid, HAc, is $K_a = 10^{-4.75}$ ($pK_a = 4.75$). If HAc is dissolved in water to a concentration of 0.10 M, we can use equation (3.6) to determine the pH of the solution:

$$[H^+]^2 = K_a C_{HAc} = 10^{-4.75} \times 10^{-1.0} = 10^{-5.75}$$

$$[H^+] = 10^{-2.88} \text{ M or pH} = 2.88$$

Therefore, a 0.1 M solution of acetic acid has a pH close to 3.0.*

Strong bases act by dissociating water completely to OH^-. Analogously to weak acids, weak bases partially remove a proton from water,

$$B + H_2O = BH^+ + OH^- \tag{3.7}$$

with the degree of transfer being determined by the base *hydrolysis* equilibrium constant:

$$K_b = [BH^+][OH^-]/[B] \tag{3.8}$$

---

*Since the value of $[H^+]$ is only 1% of the value of C_{HA}, the approximation that (HA) = C_{HA} is a good one in this case. When this approximation is poor, then allowance must be made for the loss of HA. Since every molecule of HA that ionizes produces one H^+ ion, $C_{HA} = [HA] + [H^+]$, or $[HA] = C_{HA} - [H^+]$. Substituting this into equation (3.6) and rearranging, we have

$$[H^+]^2 + K_a[H^+] - K_a C_{HA} = 0$$

which may be solved with the quadratic formula.

(the constant K_b again includes the concentration of water). When a weak base is dissolved in water, it produces an equal number of HB^+ and OH^- ions. Consequently, as long as the extent of the reaction is slight ($K_b \ll C_B$),

$$[OH^-]^2 = K_b[B] = K_bC_B.$$

For example, ammonia, NH_3, has a pK_b of 4.75. For a 0.10 M solution of NH_3:

$$[OH^-]^2 = 10^{-4.75} \times 10^{-1.0} = 10^{-5.75}$$

$$[OH^-] = 10^{-2.88} \text{ or pOH} = 2.88$$

Since pH = 14 – pOH, the pH of 0.1 M ammonia is 14 – 2.88 = 11.12.

The strength of a base, B, is equivalently expressed in terms of the pK_a of its *conjugate* acid, BH^+, since the base hydrolysis reaction (3.7) is just the difference between the autoionization reaction (3.1) and the BH^+ dissociation reaction

$$BH^+ = H^+ + B \tag{3.9}$$

Adding reactions (3.9) and (3.7) gives reaction (3.1). When reactions are added, their equilibrium constants multiply, that is, $K_{aBH^+} \times K_{bB} = K_w$. Therefore

$$pK_{aBH^+} + pK_{bB} = 14 \tag{3.10}$$

The sum of the pK_a and pK_b of a conjugate acid-base pair is always $pK_w = 14$.

3. Water in the Atmosphere: Acid Rain

Although the pH of pure water is 7.0 or neutral, rainwater is naturally acidic because it is in equilibrium with carbon dioxide. When dissolved in water, carbon dioxide forms carbonic acid, a weak acid:

$$CO_2 + H_2O = H_2CO_3 \tag{3.11}$$

We can gauge the amount of carbonic acid from the equilibrium constant for reaction (3.11).

$$K_s = [H_2CO_3]/P_{CO_2} = 10^{-1.5} \text{ M/atm} \tag{3.12}$$

where P_{CO_2} is the partial pressure of CO_2, in atmospheres.*

Since the atmospheric concentration of CO_2 is currently 350 ppm, $P_{CO_2} = 350 \times 10^{-6}$ atm $= 10^{-3.5}$ (at sea level), and therefore $[H_2CO_3] = K_sP_{CO_2} = 10^{-1.5} \times 10^{-3.5} = 10^{-5.0}$ M.

Rainwater is thus a dilute solution of carbonic acid. The acid dissociates partially to hydrogen and bicarbonate ions:

$$H_2CO_3 = H^+ + HCO_3^- \qquad K_a = 10^{-6.4} \tag{3.13}$$

---

*Actually, dissolved CO_2 is mostly unhydrated; only a small fraction exists as H_2CO_3 molecules:

$$CO_2 \text{ (s)} + H_2O = H_2CO_3 \text{ (s)} \quad K_h = [H_2CO_3]/[CO_2] = 10^{-2.81}$$

The total concentration of dissolved CO_2 is:

$$[CO_2]_T = [CO_2] + [H_2CO_3] = [H_2CO_3][1 + K_h^{-1}] = 10^{2.81}[H_2CO_3]$$

Most equilibrium measurements are made with respect to total dissolved CO_2, and $[H_2CO_3]$ is to be understood as $[CO_2]_T$.

Given K_a, we can calculate $[H^+]$ and pH using the equilibrium expression,

$$[H^+]^2 = K_a[H_2CO_3]$$

$$[H^+]^2 = 10^{-6.4} \times 10^{-5.0} = 10^{-11.4}$$

$$[H^+] = 10^{-5.7} \text{ and pH} = 5.7.$$

Thus, atmospheric CO_2 depresses the pH of rainwater by 1.3 units from neutrality. Even in the absence of anthropogenic emissions, rainwater is naturally acidic.*

To the acidifying effect of carbon dioxide must be added the contributions from other acidic constituents of the atmosphere, particularly HNO_3 and H_2SO_4. These acids can both be formed naturally: HNO_3 derives from NO produced in lightning and forest fires, while H_2SO_4 derives from volcanoes and from biogenic sulfur compounds. At the natural background concentrations, these acids rarely influence rainwater pH significantly; they do not reach equilibrium with raindrops beyond the area in which they are generated.

In polluted areas, however, the concentrations of these acids can be much higher and can reduce the pH of rainwater substantially over extended regions, producing what is known as *acid rain*. It is not uncommon in polluted areas to find the pH of rainwater in the pH 5–3.5 range. Some fogs, which can be in contact with pollutants over many hours, have had pH readings as low as 2.0, a degree of acidity equivalent to a 0.01 M solution of a strong acid. Moreover, the acid rain can fall quite far from the sources of pollution, due to long-range atmospheric transport. In particular, acid rain is a pressing problem for areas downwind of coal-fired power plants, whose tall smokestacks ameliorate local pollution by lofting SO_2 and NO emissions high into the air. Thus, power plants in western and central Europe affect the rain falling on Scandinavia, and power plants in the U.S. midwestern industrial belt similarly affect rain falling on northeastern regions of the United States and on southeastern Canada (Figure 3.11). While acid rain is a phenomenon in the hydrosphere, it depends upon atmospheric conditions—the extent of acidic emissions and the prevailing weather patterns.

4. Water in Soil: Neutralization and Watershed Buffering

Rainfall is normally acidic; whether it remains acidic or is neutralized depends upon events in the next stage of the hydrologic cycle, when rainwater interacts with the soil. Once the rain has fallen, it percolates through the topsoil (Figure 3.12), where its pH may drop further, to about 4.7 or less. The increased acidity of topsoil derives from the large amount of carbon dioxide produced by bacterial action, up to 100 times the concentration of carbon dioxide in the atmosphere; furthermore, plants exude a variety of organic acids, and the decay of plant matter produces other acids *en route* to the eventual conversion of the organic molecules to CO_2. This acidification process is counteracted by neutralization reactions with basic minerals, such as aluminosilicate clays. Protons in solution displace the *base cations,* Na^+, K^+,

---

*Rainwater can occasionally be neutral or alkaline as a result of contact with alkaline minerals in windblown dust from soils or from industrial emissions.

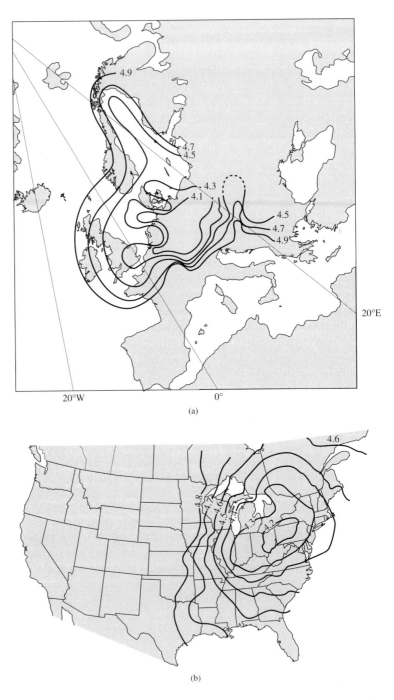

(a)

(b)

Figure 3.11 Contour lines of long-term, volume-weighted averages of pH in precipitation in (a) Europe, 1978–1982, and (b) Eastern North America, 1980–1984. *Sources:* EMEP/CCC (1984). *Summary Report,* Report 2/84 (Lillestrøm, Norway: Norwegian Institute for Air Research); National Acid Precipitation Assessment Program (NAPAP) (1987). *NAPAP Interim Assessment* (Washington, DC: U.S. Government Printing Office).

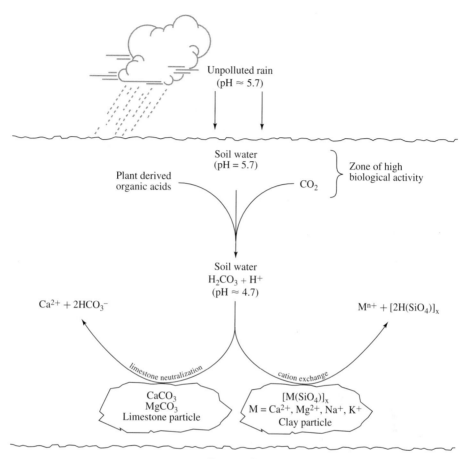

Figure 3.12 Percolation of rainwater through soil and neutralization by limestone and clays.

Mg^{2+}, and Ca^{2+}, from *ion exchange* sites on the clay surfaces (see discussion beginning on p. 212). This process serves to neutralize the soil solution, and also to break up mineral aggregates, thereby helping to form new soil. In addition, many soils contain limestone, made of calcium and magnesium carbonates, which neutralize the acids in the soil solution even more effectively. As water passes through the soil to aquifers and surface waters, these two soil constituents, calcium carbonate or aluminosilicate clays, neutralize most of the naturally occurring acidity in rainwater.

a. Neutralization by calcium carbonate.   Large parts of Earth's crust contain calcium carbonate, $CaCO_3$, which is derived from marine organisms. The chemistry of calcium carbonate neutralization of water is dominated by three considerations: 1) $CaCO_3$ is only sparingly soluble in water; 2) while the carbonate anion is itself a moderate strong base, the bicarbonate anion is only a weak base; and 3) dissolved carbonate is in equilibrium with carbon dioxide in the soil gases. The net result of these conditions is that following exposure to calcium carbonate, the pH of soil water increases, as does the concentration of calcium ions in solution.

1) Solubility Product. Although water is generally an excellent solvent for ions, many ionic compounds are only sparingly soluble because the ion-water forces are outweighed by the forces that hold the ions together, particularly when the ions can arrange themselves in an energetically favorable way in a crystalline lattice. The energies stabilizing a lattice are generally maximal when the positive and negative ions have equal size and/or charge. For example, lithium fluoride is less soluble than lithium iodide because fluoride is closer in size to the small lithium ion than is iodide; in contrast, cesium iodide is less soluble than cesium fluoride because the cesium ion is closer in size to iodide. Similarly, both sodium carbonate and calcium nitrate are soluble because the positive and negative ions differ in charge, but calcium carbonate is sparingly soluble because both the cation and anion are doubly charged and therefore have a large lattice energy.

The dissolution of a sparingly soluble salt is governed by the equilibrium constant for the reaction, called the solubility product, K_{sp}. For example, barium sulfate dissolves with a solubility product K_{sp} of 10^{-10}:

$$BaSO_4 = Ba^{2+} + SO_4^{2-}$$

$$K_{sp} = [Ba^{2+}][SO_4^{2-}] = 10^{-10} \ M^2 \ [\text{or } pK_{sp} = 10] \tag{3.14}$$

Equation (3.14) is valid as long as there is solid barium sulfate in equilibrium with the solution. If the concentration product of Ba^{2+} and SO_4^{2-} exceeds the value of K_{sp}, barium sulfate will precipitate. When pure water is equilibrated with barium sulfate, the concentration of barium matches the concentration of sulfate in solution:

$$[Ba^{2+}] = [SO_4^{2-}] = K_{sp}^{1/2} = 10^{-5} \ M$$

When either Ba^{2+} or SO_4^{2-} is present in excess, the concentration of its partner decreases in inverse proportion, via equation (3.14).

2) Solubility and Basicity. For calcium carbonate $pK_{sp} = 8.34$; if the only reaction were

$$CaCO_3 = Ca^{2+} + CO_3^{2-} \tag{3.15}$$

the solubility would be $10^{-4.17}$. However, the solubility equilibrium is shifted if the anion has an appreciable affinity for protons. This is the case for the carbonate ion, which is a moderately strong base:*

$$CO_3^{2-} + H_2O = HCO_3^- + OH^- \quad pK_b = 3.67 \tag{3.16}$$

---

*Equivalently $pK_a = 10.33$ for HCO_3^-, via equation (3.10), corresponding to the bicarbonate dissociation equilibrium

$$HCO_3^- = H^+ + CO_3^{2-}$$

A large fraction of the carbonate is immediately converted to bicarbonate via the carbonate hydrolysis reaction (3.16) increasing the solubility of the calcium carbonate. Adding reactions (3.15) and (3.16) gives the net reaction

$$CaCO_3 + H_2O = Ca^{2+} + HCO_3^- + OH^- \qquad (3.17)$$

Its equilibrium constant is the product of the constants for reactions (3.16) and (3.15):

$$K_{(3.17)} = K_{(3.16)} \times K_{(3.15)} = 10^{-12.01} = [Ca^{2+}][HCO_3^-][OH^-] \qquad (3.18)$$

If this were the only reaction occurring, then the concentrations of all three ions would be the same; $[Ca^{2+}] = K_{(3.17)}{}^{1/3} = 10^{-4}$. This value is about 50% higher than the value obtained by considering reaction (3.15) only. Taking the contributions of both reactions into account would produce a more accurate calculation and would give a solubility that is higher again by 15%. The hydroxide concentration remains close to $10^{-4.0}$, which yields pH = 10.0 for pure water saturated with calcium carbonate.

Natural waters, however, are not pure; they are in equilibrium with the atmosphere. We consequently must take account of the effect of dissolved CO_2, which reacts with the CO_3^{2-} to form HCO_3^-. This is a neutralization reaction between carbonic acid and carbonate to yield bicarbonate; it reduces the alkalinity of the pure calcium carbonate solution, while increasing the solubility of the calcium. The overall reaction is obtained by adding reaction (3.13) to reaction (3.17) and subtracting reaction (3.16) to yield

$$CaCO_3 + H_2CO_3 = Ca^{2+} + 2HCO_3^- \qquad (3.19)$$

The equilibrium constant of reaction (3.19) is therefore

$$K_{(3.19)} = K_{(3.13)}K_{(3.17)}/K_{(3.16)} = 10^{-4.41} = [Ca^{2+}][HCO_3^-]^2/[H_2CO_3] \qquad (3.20)$$

We can eliminate one of the variables by recognizing that in reaction (3.19), two HCO_3^- ions are produced for every Ca^{2+} ion, and consequently $[HCO_3^-] = 2[Ca^{2+}]$. Substituting this equivalence into equation (3.20) yields

$$4[Ca^{2+}]^3 = 10^{-4.4}[H_2CO_3] \text{ or } [Ca^{2+}] = 10^{-1.67}[H_2CO_3]^{1/3} \qquad (3.21)$$

If the solution is open to the atmosphere, then $[H_2CO_3] = 10^{-5}$ M (see p. 206), and $[Ca^{2+}] = 10^{-3.33}$ M. Note that this concentration is a factor of five higher than the concentration of calcium ions in the absence of CO_2. We can calculate the pH as well. The bicarbonate concentration is twice the calcium concentration, or $10^{-3.03}$ M; this value, along with the carbonic acid concentration, allows the pH to be calculated from the equilibrium expression for reaction (3.13):

$$pH = pK_a - \log[H_2CO_3]/[HCO_3^-] = 6.40 + 1.97 = 8.37 \qquad (3.22)$$

Thus, when a solution of calcium carbonate is in equilibrium with air-saturated water, its pH decreases from 10.00 to 8.37, while the calcium concentration increases from $10^{-4.41}$ M to $10^{-3.33}$ M.

The concentration of calcium ions increases even more in the presence of more acid. When the solution is in contact with biologically productive soils, the concentration of H_2CO_3 rises by a factor of 100; then $[Ca^{2+}]$ rises to $10^{-2.67}$ [equation (3.21)], and $[HCO_3^-]$

rises to $10^{-2.37}$, while the pH falls to 8.03 [equation (3.22)]. If stronger acids are added, the solubility of calcium carbonate increases further as HCO_3^- is converted to H_2CO_3. We can calculate the solubility as a function of pH by recognizing that the calcium ion concentration must remain equal to the total concentration of carbonate in all its protonation states:

$$[Ca^{2+}] = [CO_3^{2-}] + [HCO_3^-] + [H_2CO_3] \qquad (3.23)$$

Substituting the equilibrium expressions $K_{a1} = [H^+][HCO_3^-]/[H_2CO_3]$ and $K_{a2} = [H^+][CO_3^{2-}]/[HCO_3^-]$, we obtain

$$[Ca^{2+}] = [CO_3^{2-}]\{1 + [H^+]K_{a2}^{-1} + [H^+]^2 K_{a1}^{-1} K_{a2}^{-1}\} \qquad (3.24)$$

Since $[CO_3^{2-}] = K_{sp}/[Ca^{2+}]$, we calculate $[Ca^{2+}]$ in terms of $[H^+]$ and constants:

$$[Ca^{2+}] = \{K_{sp}(1 + [H^+]K_{a2}^{-1} + [H^+]^2 K_{a1}^{-1} K_{a2}^{-1})\}^{1/2} \qquad (3.25)$$

With the equilibrium constants given above, this equation gives $[Ca^{2+}] = 10^{-0.82}$ (0.15 M) at pH 5.

As these calculations indicate, the solubility of calcium carbonate increases rapidly as the pH decreases; magnesium carbonate behaves similarly. This behavior reveals some of the consequences of acid rain: because calcium carbonate is solubilized by acidic solutions, acid rain has pronounced effects on buildings and statues composed of limestone and marble. Conversely, as rainwater becomes more acidic, it can still be neutralized by exposure to *calcareous* soils, but only with a concomitant significant increase in the concentration of calcium and magnesium ions in solution.

b. Ion exchange: clays.

In noncalcareous soils, neutralization occurs by a different mechanism, ion exchange, in which the protons are exchanged for other positively charged cations in soil. As in carbonate neutralization, the increase in pH comes at the expense of increasing base cations concentrations.

Many solids have ions that are loosely held at fixed charge sites. These can be exchanged with ions that are free in solution. The exchanging ions can be either positively charged (cations) or negatively charged (anions):

$$R^-M^+ + M'^+ = R^-M'^+ + M^+ \quad \text{cation exchange} \qquad (3.26)$$

$$R^+X^- + X'^- = R^+X'^- + X^- \quad \text{anion exchange} \qquad (3.27)$$

In these reactions R represents a fixed charge site. It attracts ions of the opposite charge, the strength of attraction being proportional to the charge/radius ratio. (Multiply-charged ions can occupy more than one ion exchange site.) This is the underlying mechanism of ion exchange resins, which are organic polymers with charged groups covalently attached. Cation exchangers commonly have sulfonate groups, $-OSO_3^-$, whereas anion exchangers commonly have quaternary ammonium groups, $-N(CH_3)_3^+$. Some anionic groups, such as carboxylate, $-COO^-$, have especially high affinities for protons (weak acid cation exchangers), while some cationic groups, such as ammonium, $-NH_3^+$, have especially high affinity for hydroxide (weak base anion exchangers). Cation and anion exchange resins are widely used in tandem to deionize water:

$$R^-H^+ + M^+ + X^- = R^-M^+ + H^+ + X^- \tag{3.28}$$

$$R^+OH^- + H^+ + X^- = R^+X^- + H_2O \tag{3.29}$$

The spent resins can be regenerated by washing them with strong acids and bases.

The availability of exchangeable ions depends upon the structure and composition of the soil. The major components of soils (63% on average) are silicates, loose networks of polymerized silicon dioxide. Pure silicon dioxide, or silica, is a polymeric solid with a three-dimensional network of silicon atoms bound tetrahedrally to four oxygen atoms, which in turn are bound to two silicon atoms each (Figure 3.13). This three-dimensional network is incomplete in many silicate minerals that contain, instead, sheets of polymerized silicate tetrahedra arranged in layers (Figure 3.14). Three of the oxygen atoms around each silicon atom are linked to neighboring silicon atoms in the sheet, while the fourth oxygen atom extends upward out of the sheet, bound to a second parallel layer.

It is this fourth oxygen atom that provides a site for binding cations. In the common clay minerals kaolinite and pyrophyllite, the fourth oxygen atom is bound to an aluminum ion, Al^{3+} (Figure 3.15). Aluminum prefers octahedral coordination and is surrounded by six oxygen atoms. In kaolinite, two of these oxygen atoms are provided by neighboring silicate groups, while the remaining oxygens come from hydroxyl groups; the aluminum hydroxyl groups form hydrogen bonds with adjacent silicate oxygen atoms to hold the layers together. In pyrophyllite, the aluminum octahedra are sandwiched between two silicate sheets (Figure 3.16); the next triple layer is held only weakly to the first one, since the facing silicate oxygen atoms lack protons with which to form hydrogen bonds. The interlayer space can be filled with water molecules, and pyrophyllite swells considerably in water. In many clays, some of the Al^{3+} ions are replaced by Fe^{3+} ions.

Other aluminosilicates have the kaolinite and pyrophyllite structures, but with some of the aluminum or silicon ions substituted by metal ions of lower charge. Thus, the common clay mineral montmorillonite has the pyrophyllite structure, but about one-sixth of the Al^{3+} ions are replaced with Mg^{2+}. Likewise, the illite clays share this structure, but Al^{3+} ions replace some of the Si^{4+} ions in the silicate sheet. The substitution of cations with lower charge produces excess negative charge; this electrical imbalance is neutralized by the adsorption of other cations, commonly Na^+, K^+, Mg^{2+}, or Ca^{2+}, in the medium between the aluminosilicate layers.

It is the presence of these other cations, adsorbed between the aluminosilicate layers, that makes clay particles good ion exchangers, since the adsorbed cations are readily exchanged for other cations in solution. The order of exchange depends upon the affinity of the

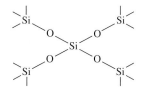

Figure 3.13 Polymeric structure of silicon dioxide.

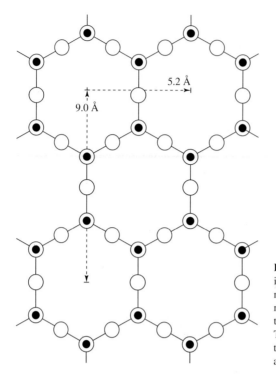

Figure 3.14 Layer structure of polymerized silicate tetrahedra. (The black circles represent silicon atoms and the open circles represent oxygen atoms. Each silicon atom is tetrahedrally bound to four oxygen atoms. The oxygen atoms shown superimposed on the silicons are directed upward and bound to a second parallel layer.)

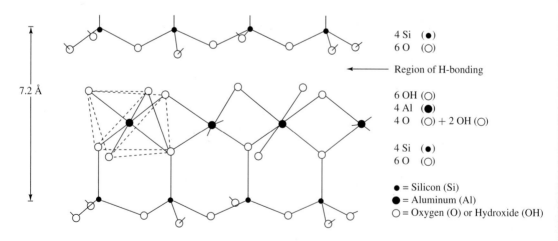

Figure 3.15   Structure of kaolinite, $Al_4Si_4O_{10}(OH)_8$. The plates contain phyllosilicate sheets bound to octahedral aluminum layers. (Dashed lines show sixfold coordinate positions in octahedral layer.) The distance between two successive plates is 7.2 Å.

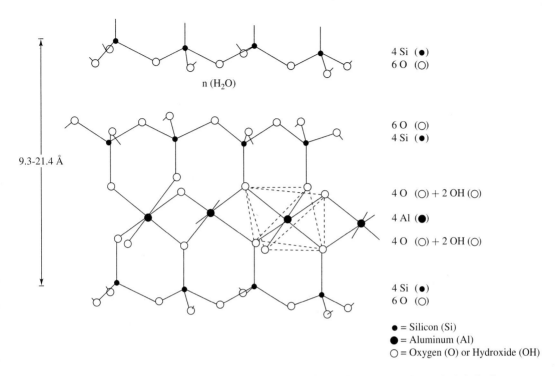

n (H$_2$O)

4 Si (●)
6 O (○)

6 O (○)
4 Si (●)

4 O (○) + 2 OH (○)

4 Al (●)

4 O (○) + 2 OH (○)

4 Si (●)
6 O (○)

9.3-21.4 Å

● = Silicon (Si)
⬤ = Aluminum (Al)
○ = Oxygen (O) or Hydroxide (OH)

Figure 3.16   Structure of pyrophyllite, Al$_2$Si$_4$O$_{10}$(OH)$_2$. The plates contain octahedral aluminum layers sandwiched between two phyllosilicate sheets. The distance between the plates can vary up to 21 Å, depending on the amount of water present between plates.

cations for anionic sites on the clay compared to their attraction for water molecules. Generally, aluminum cations are the most difficult to exchange and sodium cations the least difficult, with other ions between, in the order Al^{3+} > H$^+$ > Ca^{2+} > Mg^{2+} > K$^+$ > NH$_4^+$ > Na$^+$. Since protons are more tightly held than most other cations, if the soil solution is acidic, the protons will exchange with the adsorbed cations. The exchange of protons for other ions increases both the solution's pH and its base cation concentration. Thus, like limestone, clays neutralize the acids in soil water while increasing the concentration of base cations.

c. Hardness and detergents.   The neutralization reactions that raise the pH of natural water as it percolates through the soil (Figure 3.12) also bring into solution appreciable quantities of calcium and magnesium ions. Water with relatively high concentrations of Ca^{2+} and Mg^{2+} is considered "hard," while water with low concentrations is "soft." Soft water also has lower pH because the low Ca^{2+} and Mg^{2+} concentrations reflect poor availability of limestone or clays for neutralization.

The hard and soft appellations reflect the fact that the doubly charged Ca^{2+} and Mg^{2+} ions can precipitate detergents. Detergents are molecules with long hydrocarbon chains and ionic or polar head groups (Figure 3.17a). When added to soft water, the detergent molecules aggregate into micelles (Figure 3.17b), with the hydrocarbon tails pointing to the inside and the polar groups pointing out toward the water. The hydrophobic interior of the micelles

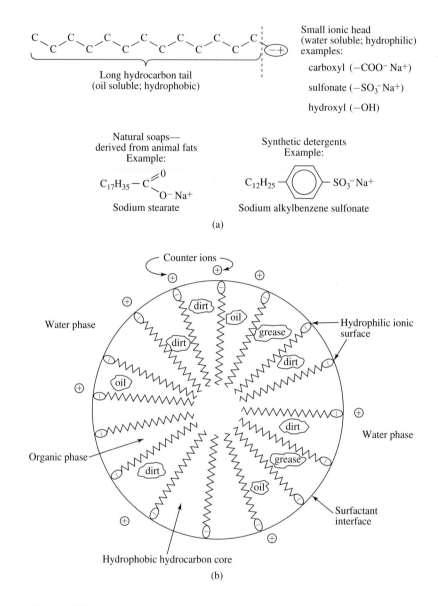

Figure 3.17   (a) Structures of detergent molecules; (b) Schematic diagram of a detergent micelle.

solubilizes grease and dirt particles and removes them from clothing, dishes, or other items being washed. In soft water, the micelles float freely and are prevented from aggregating with one another by the mutual repulsions of the polar hydrophilic head groups.

This solubilization process is prevented by hard water. The polar head groups of most detergents are negatively charged; they interact with divalent cations such as Ca^{2+} and Mg^{2+}

and precipitate. The micelles are then unavailable for solubilizing dirt; worse still, the precipitates themselves tend to be scummy and stick to the items being washed. Consequently, in areas with hard water, a clean wash requires the addition of agents—called *builders* in the laundry trade—that tie up polyvalent cations and thereby prevent them from precipitating the detergent.

Detergent products contain a variety of kinds of builders. One class comprises *chelating agents*, molecules with several donor groups that can bind the cations through multiple coordinate bonds. A particularly effective chelating agent is sodium tripolyphosphate (STP), which is shown bound to a Ca^{2+} ion in Figure 3.18a. This compound is relatively inexpensive and has the advantage that it rapidly breaks down in the environment to sodium phosphate, a naturally occurring mineral and nutrient for plants. However, STP-containing detergents release so much sodium phosphate into natural bodies of water that plant growth is overfertilized, leading to eutrophication (see discussion beginning on p. 231). In response to warnings from environmentalists, use of phosphate in detergents has been substantially reduced.

A variety of alternative chelating agents have been explored; most have not yet proven to be fully satisfactory. Sodium nitrilotriacetate (NTA—Figure 3.18b) is used in Canada and parts of Europe, but not in the United States because of concern that it might pose a health hazard in drinking water. NTA does not break down as readily as STP does and might mobilize metals other than calcium and magnesium. Sodium citrate, a naturally occurring tricarboxylate, is currently being introduced. Alternative builders are sodium carbonate or sodium silicate, which precipitate Ca^{2+} ions before they can precipitate the detergent. But the $CaCO_3$ or $CaSiO_3$ precipitates produce a scummy residue and may be detrimental to the operation of automatic washing machines. Moreover, sodium carbonate and sodium silicate are strongly alkaline and are harmful if accidentally lodged in the eyes or swallowed. The most promising development has been with zeolites, which are synthetic aluminosilicate minerals. Zeolites trap Ca^{2+} or Mg^{2+} by ion exchange, releasing Na^+ ions in their place.

d. Watersheds and buffer systems.   The accumulation of dissolved ions from soil has another effect on water: if some of the dissolved ions are conjugate acid-base pairs,

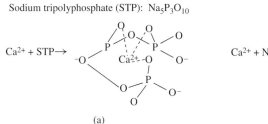

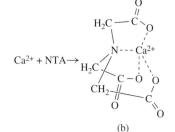

Figure 3.18   Chelating agents: chemicals that tie up positive ions in solution so they can no longer react with detergents (a) sodium tripolyphosphate (STP); (b) sodium nitrilotriacetate (NTA).

they make the solution a pH buffer. In a buffer solution, the pH is relatively insensitive to small addition of other acids or bases; the conjugate base is available to neutralize any extra acid, while the conjugate acid is available to react with any extra base. Consider for example a solution of 0.1 M acetic acid and 0.1 M sodium acetate. The pH of the solution is 4.75:

$$[H^+] = K_a[HAc]/[Ac^-] = 10^{-4.75}(0.1)/(0.1) = 10^{-4.75} \qquad (3.30)$$

If a small amount of hydrochloric acid is then added, equivalent to about 10% of the acetate present, the new acetate concentration is 0.09 M, while the new acetic acid concentration is 0.11 M. We can calculate the new pH:

$$[H^+] = 10^{-4.75}(0.11)/(0.09) = 1.22 \times 10^{-4.75} = 10^{-4.66} \qquad (3.31)$$

Adding 0.01 M of acid has decreased the pH of the buffered solution by only 0.09 units, from 4.75 to 4.66. In contrast, the same amount of hydrochloric acid added to pure, unbuffered water would give a solution that is 0.01 M in H^+. The pH is lowered by 5 units, from 7.0 to 2.0. Similar considerations apply to the addition of extra base.

The pH of a buffer depends mainly on the pK_a of the buffer acid, not its concentration. On the other hand, the concentration of the buffer acid and its conjugate base determine the *buffer capacity*, that is, how much acid or base the solution can tolerate while maintaining the pH. The buffer capacity is exhausted when added bases or acids completely neutralize either the buffer acid or its conjugate base. In the above example, the acetic acid/acetate buffer capacity is 0.1 M: the buffer solution can neutralize up to 0.1 mole per liter of either acid or base with little change in pH, but addition of more acid or base moves the pH of the solution out of the buffering range, and changes the pH as rapidly as in a sample of pure water.

In soils, the buffer capacity, as well as the buffer pH, depend on the types and amounts of minerals. The relationship is illustrated in Figure 3.19, which shows how the continuous infusion of acid over time alters the pH in a hypothetical soil having equivalent amounts of carbonates and clays. Initially the pH stays around 8, thanks to the carbonate buffering system. In calcareous soils, where the carbonate solids are distributed in fine particles and evenly dispersed, the buffering capacity is nearly inexhaustible for any reasonable level of acid deposition. The pH can be maintained for a long time.

If the carbonate rock is depleted or absent, however, the buffering processes come under the control of the cation exchange reactions in the clay minerals. The buffer pH is now slightly acidic (about 5.5) because the proton affinity is a good deal lower for the silicate anion sites than it is for carbonate. Moreover, the buffer capacity of clay soils is usually limited because it depends upon the availability of exchangeable sites occupied by the base cations Na^+, K^+, Mg^{2+}, and Ca^{2+}. The density of exchangeable cation sites in clay soils is defined as the *Cation Exchange Capacity* (*CEC*) in units of *acid equivalents* (*eq*) per square meter. This is a measure of the moles of H^+ per square meter that would be required to occupy all of the exchange sites. The fraction of the sites occupied by the base cations (rather than by H^+ or Al^{3+}) is defined as the *base saturation* (β).

The exchangeable pool of base cations on the clay surface is tiny compared to the pool trapped inside the clay particles. Natural weathering reactions such as:

cation-Al-silicate + H_2CO_3 + H_2O = HCO_3^- + cation + $Si(OH)_4$ + Al-silicate   (3.32)

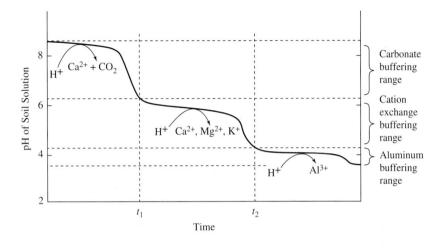

Figure 3.19 Scheme showing progression of decline in soil solution pH over time in response to atmospheric acid inputs. For a given soil, the time scale over which the soil solution passes from one buffering range to the next depends on the intensity of acid deposition and the concentration of exchangeable cations. Arrows indicate the chemistry by which soils retain H^+ from the soil solution by exchanging it for cations. The period from t_1 to t_2 is the time over which soils lose 90–95% of their original exchangeable base cations. *Source:* W. M. Stigliani and R. W. Shaw (1990). Energy use and acid deposition: the view from Europe. *Annual Review of Energy* 15: 201–216. Copyright © 1990 by Annual Reviews Inc. Reproduced with permission.

release trapped base cations, which become available for adsorption at the surface exchange sites, thus replenishing the exchangeable cation pool.

Weathering rates vary for different soil types, and often they are relatively slow compared to the rate of acidification. When silicate weathering is insufficient to replenish exchangeable base cations lost during acidification, β declines (see problem 10, end of Part III). When it is reduced to about 5%, a new buffering range (pH = 3.8 to 4.2) is encountered, which is governed by the aluminum hydroxide solubility:

$$Al(OH)_3 + 3H^+ = Al^{3+} + 3H_2O \tag{3.33}$$

At pHs above 4.2, Al^{3+} precipitates as $Al(OH)_3$, but when the pH drops below 4.2 the H^+ dissolves the $Al(OH)_3$. Since aluminum compounds are abundant in soils, buffer capacity in this range is rarely depleted. The *iron-buffer range* occurs only at an extreme stage of acidification, in soil solutions with pHs lower than 3.8. The iron buffering mechanism is similar to that of aluminum: H^+ is neutralized by dissolution of iron oxides. Soils in this extreme pH range often cannot support flora and fauna because they leach heavy metals and nutrients.

As we can see, even under preindustrial conditions, the pH of water varies at each stage of the hydrologic cycle. Rainwater is in equilibrium with carbon dioxide in the atmosphere, which lowers its pH to about 5.7; if the concentration of sulfate and nitrate in the atmosphere are high, the pH of rainwater may be lower than this. The water becomes more acidic as it comes in contact with the decomposing vegetation in the topsoil. This process is counterbalanced by the buffering reactions of calcium carbonate and aluminosilicate clays, which increase the pH. Groundwater and surface waters are protected from acidification, provided that acid inputs do not outstrip the buffering capacities of soils. But as we shall see in the next section, industrialization has increased the rate of acidification, at least in some areas, to the point where acid deposition has outrun the neutralization capacity of the surrounding environment.

5. Effects of Acidification in the Environment

a. Acid deposition and the biosphere. As described in Part II, the atmosphere receives substantial inputs of SO_2 and NO_x from both natural and anthropogenic sources. These emissions are cleared from the air within a few days by oxidation reactions and subsequent transfer to the soil, either directly by dry deposition in aerosols, or indirectly by wet deposition in rainfall. Such reactions are vital to the health of the biosphere because they cleanse the air of noxious fumes. If SO_2 and NO accumulated in the atmosphere as CO_2 does, the air would quickly become toxic.

The problem with these reactions is that the sulfur and nitrogen oxides are not destroyed, but rather transformed in the atmosphere to sulfuric and nitric acids, which are then deposited on the soils. Hence, soils can be described as *sinks* or *reservoirs* for atmospheric pollutants. Pollutants emitted to the air flow through the environment, mediated by a series of physical and chemical processes, as illustrated schematically in Figure 3.20. The first step is air transport and then deposition onto soil (1). Soils, in their capacity as *chemical filters*, may adsorb, neutralize, or otherwise retain and store the pollutant. When buffering capacities are diminished, the soil may release the pollutant to rivers and lakes (2a) or to the groundwater (2b). Eventually, pollutants are discharged to the oceans through stream (3a) and subsurface (3b) flow, and deposited in the ocean sediment (4), their ultimate repository.

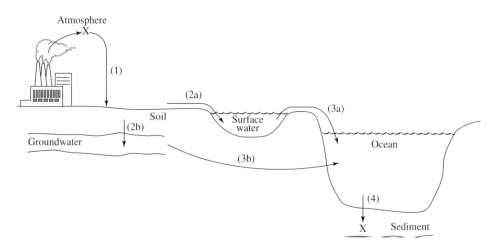

Figure 3.20   Flow of pollutant X from sources to sinks.

The effectiveness of the soil as a chemical filter of acidic inputs depends upon its buffering capacity and the rate of acid deposition. Although the buffering capacity of most soils is sufficient to neutralize naturally occurring acids, over time the capacity can be overwhelmed by high inputs of acid deposition. The decrease in a watershed's buffering capacity over time depends upon many factors, including the rate of deposition, the nature of the soil, the size of the watershed, and the flow characteristics of the lake or groundwater. These variables often make it difficult to discern acidification while it is occurring. One case where the historical record of acidification has been well documented, however, is the watershed of Big Moose Lake in the Adirondack Mountains of New York state. This area has received some of the highest inputs of acid deposition in North America because it is downwind from western Pennsylvania and the Ohio Valley, historically the industrial heartland of the United States. Pollutants carried by the westerly winds are trapped in the mountains and deposited via wet and dry deposition.

The difficulties of recognizing acidification while it is occurring are illustrated in Figure 3.21, which shows the historical trends in lake water pH (dashed line), SO_2 emissions upwind from the lake (solid line), and the extinction of different fish species. The pH of the lake remained nearly constant at around 5.6 over the entire period from 1760 to 1950. Then, within the space of 30 years, from 1950 to 1980, the pH declined more than one whole pH unit to about 4.5. The decline in pH lagged behind the rise in SO_2 emissions by some 70 years, and the peak years of sulfur emissions preceded the decline in pH by 30 years. The deposition rate is estimated to have been about 2.5 grams of sulfur $m^{-2}yr^{-1}$ during the peak period. These quantities, deposited year after year, were large enough to deplete the capacity for base cation exchange in the watershed. Thus, beginning around 1950, atmospheric acid deposition moved through the buffer-depleted soils of the watershed and percolated into the lake with diminished neutralization (see problem 10, Part III). At that point, acid-sensitive fish species such as smallmouth bass, whitefish, and longnose sucker began to disappear, followed in the late 1960s by the more acid-resistant lake trout.*

From the shape of the historical pH curve, we can see that Big Moose Lake was the subject of an inadvertent titration experiment conducted over four generations of industrial activity. The coal-driven industrialization of the Ohio Valley, upwind from the lake, supplied the acid inputs, mostly as sulfuric acid formed from SO_2 released during coal combustion. The soils of the watershed provided the supply of buffering chemicals. Thus, the watershed's natural buffering capacity delayed recognition of the deleterious effects of coal burning for about three generations. During this time, there was no direct evidence of how pollution was affecting the pH of lake water or fish mortality. As this example suggests, polluting activities may be far displaced in time from their environmental effects.

Although acidification of soils and freshwaters is itself deleterious to biota, the effect is magnified when soils are also polluted by toxic metals such as cadmium, copper, nickel, lead, and zinc. As cations, these metals compete with hydrogen, aluminum, and the base cations for cation exchange sites. At high pH, metal ions in well-buffered soils are generally

---

*It might be thought that carbonic acid itself could threaten fish, since, as noted in the preceding section, the pH of soil water can fall to 4.7 just from the CO_2 generated in biologically active soils. However, excess CO_2 is expelled back to the atmosphere from surface waters, raising the pH back toward neutrality.

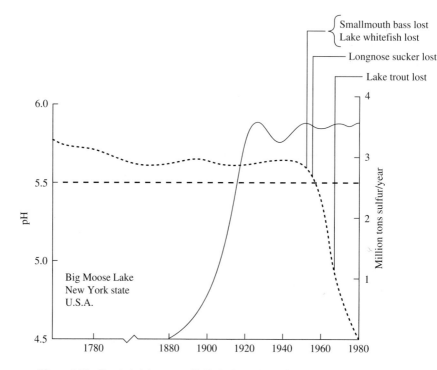

Figure 3.21   Trends in lake water pH (dashed curve), upwind emissions of SO_2 from the U.S. industrial midwest (solid curve), and fish extinctions for the period from 1760 to 1980. *Source:* W. M. Stigliani (1988). Changes in valued capacities of soils and sediments as indicators of nonlinear and time-delayed environmental effects. *Environmental Monitoring and Assessment* 10: 245–307. Copyright © 1988 by Kluwer, Academic Publishers. Reprinted with permission of Kluwer.

retained at the exchange sites; their concentrations in soil solution are low. However, as the pH declines from 7 to 4, the leaching velocity at which an ion migrates through the soil may increase by an order of magnitude as shown by the example of cadmium (Figure 3.22). Thus, at near neutral pH, the soil will accumulate heavy metals such as cadmium, only to release them as the soil acidifies. Once in the aqueous phase, cadmium ions are mobile and biologically active. They can be transported to lakes via surface or subsurface flow, transferred to groundwaters, or taken up by vegetation, with toxic effects. Al^{3+} is also toxic to plants and aquatic organisms. Some of the deleterious effects of strong acidification are probably due to the Al^{3+} leached from clay particles in the soil.

b. Acid mine drainage.   A problem related to acid rain is acid mine drainage. Coal mines, especially those that have been abandoned, are known to release substantial quantities of sulfuric acid and iron hydroxide into local streams. The first step in the process is the oxidation of pyrite (FeS_2), which is common in underground coal seams:

$$FeS_2 + 7/2O_2 + H_2O = Fe^{2+} + 2HSO_4^-$$ (3.34)

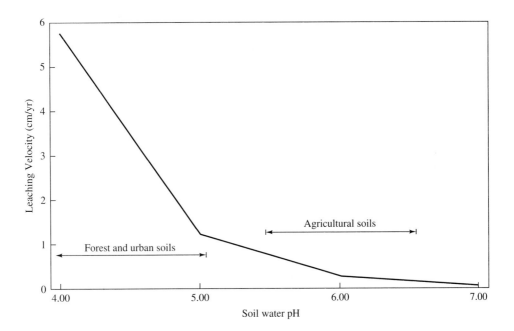

Figure 3.22   Leaching velocity of cadmium in soil as a function of soil water pH.   *Source:* W. M. Stigliani and P. R. Jaffe (1992), private communication (Laxenburg, Austria: International Institute for Applied Systems Analysis).

This reaction is analogous to the first step in the generation of acid rain, in which sulfur is oxidized during coal combustion. In acid mine drainage, the oxidation step occurs spontaneously at ambient temperatures once iron sulfide, which is stable in the absence of air, is exposed to the atmosphere. In the second step, the ferrous iron (Fe^{2+}) formed from reaction (3.34) combines with oxygen and water in the overall reactions:

$$Fe^{2+} + 1/4O_2 + 1/2\,H_2O = Fe^{3+} + OH^- \tag{3.35}$$

$$Fe^{3+} + 3H_2O = Fe(OH)_3 + 3H^+ \tag{3.36}$$

The sum of reactions (3.34), (3.35), and (3.36) yields the following reaction:

$$FeS_2 + 15/4O_2 + 7/2H_2O = Fe(OH)_3 + 2H^+ + 2HSO_4^- \tag{3.37}$$

Thus, one mole of pyrite produces two moles of sulfuric acid and one mole of ferric hydroxide, which is removed from solution as a brown precipitate. The pH of streams receiving this drainage can be as low as 3.0.

These reactions continue long after coal mining operations have ceased. The resulting pollution problem can be quite severe locally; in coal mining areas, streams are often highly polluted by sulfuric acid. The problem is difficult to prevent because sealing up mines effectively is arduous and expensive. Since coal mining is a major activity on most continents, the problem of acid mine drainage has global implications.

c. Global acidification.   Thus far we have been discussing acidification on local and regional scales. Acid mine drainage is tied to particular watersheds; acid rain depends upon industrial activity and prevailing weather patterns. But acid deposition on a global level from industrial sources is of the same order of magnitude as deposition from natural sources (Table 3.5; see problem 12, Part III). Acid deposition is not the whole story, because there are also natural and anthropogenic sources of alkaline chemicals in the atmosphere. These include ammonia (NH_3) and alkaline particles derived from ash, as well as windblown alkaline minerals. These chemicals have been crudely estimated to neutralize between 20 and 50% of the generated acidity. Thus, overall it appears that the atmosphere has probably acted as an acidic medium throughout geologic time. Nevertheless, natural sources of acidity, although of the same order of magnitude as anthropogenic sources, are spread rather evenly across the globe; polluting sources are concentrated near industrial and urban centers, with levels of acidity exceeding 50 to 100 times the natural background. It is the excessive concentration of acidity in particular regions that causes problems for the biosphere.

C. LAKES, RIVERS, WETLANDS, AND SEAS: WATER AS ECOLOGICAL MEDIUM

We turn now from rainwater and soil water to surface waters, where considerations of solvation and acid-base chemistry are supplemented by considerations of water as an ecological medium. The two are, of course, related; as we saw in the example of Big Moose Lake, the ability of a body of water to support its normal complement of biological species can be critically affected by the pH of the water or by the metal ions present in it. Acidified lakes

TABLE 3.5 ESTIMATED NATURAL AND ANTHROPOGENIC SOURCES OF GLOBAL ATMOSPHERIC ACIDITY

Source	10^{12} Moles of H^+ generated per year
NATURAL	
Unpolluted rainwater	1.0
Lightning*	1.4
Volcanoes[†]	1.3
Biogenic sulfur	4.1
TOTAL NATURAL	7.8
POLLUTION	
Coal combustion/metal smelting[†]	5.8
Combustion processes*	1.4
TOTAL FROM POLLUTION	7.2

*Refers to acidity generated from NO_x emissions.

[†]Refers to acidity generated from SO_2 emissions.

Source: Adapted from W. H. Schlesinger (1991). *Biogeochemistry, an Analysis of Global Change* (San Diego, California: Academic Press, Inc.).

can be crystal clear—and virtually sterile. Their ability to support life is limited by pH (and metal ion concentration), whereas in healthy aqueous environments, biological activity is limited by the availability of essential elements: nitrogen, phosphorus, and, most important, oxygen.

1. Biochemical Energy and Redox Potential in Aquatic Ecosystems

All of biology depends upon oxidation and reduction processes. The primary source of energy in the biosphere is the energy stored in reduced carbon compounds generated through photosynthesis. These reduced compounds, principally carbohydrates (represented as CH_2O), are the main reductants in the biosphere. When there is adequate oxygen, energy is released via oxidation of carbohydrate molecules to carbon dioxide and water:

$$(CH_2O) + O_2 = CO_2 + H_2O \tag{3.38}$$

This reaction is the reverse of photosynthesis. In reality, it is carried out in separate oxidation and reduction steps (as in photosynthesis; see Part I, pp. 68–69), which can be represented as

$$O_2 + 4e^- + 4H^+ = 2H_2O \tag{3.39}$$

and

$$(CH_2O) + H_2O = CO_2 + 4H^+ + 4e^- \tag{3.40}$$

Each of these steps is called a *half-reaction*, to which can be assigned a standard electrochemical potential, $E°[w]$ (in volts); the symbol [w] means that the standard potential is evaluated at pH = 7, a condition more representative of natural waters than the usual convention of pH = 0. Equivalently, one can specify $pE°[w] = E°(2.303RT/F)^{-1}$ as the negative logarithm of the "effective concentration" of electrons produced by the redox agents in the half-reaction. (Appendix C contains a full discussion of redox reactions, electrochemical potential, and the Nernst equation.) The $pE°[w]$ values for these two half-reactions are 13.75 and –8.20, respectively. The difference between these two values, 21.95, can be used to determine the free energy change (ΔG) according to the equation

$$\Delta G = 2.303RT\Delta pE \tag{3.41}$$

where R is the gas constant (8.314 J/K), and T = 298 K. From this equation, the free energy potentially available to the organism via aerobic respiration is calculated to be 125 kJ per mole of electrons.

O_2 is the most powerful oxidant in the biosphere. However, the effectiveness of aerobic respiration depends upon access to oxygen. This is not a problem for organisms living in contact with air, but in water, oxygen is about 150 times less concentrated than in air. The solubility of O_2 in water is only 9 mg/l (about 0.3 millimoles) at 20°C, and less at higher temperatures. About 2.7 mg of dissolved O_2 is required to oxidize 1 mg of organic carbon. Thus, at 20°C, only about 3.4 mg of CH_2O can be oxidized by the O_2 in a liter of water. When the concentration of organic compounds is high, water is easily depleted of oxygen. The rate of depletion depends on access to air. In standing water or in waterlogged soils, the diffusion of oxygen from the atmosphere is slow relative to the speed of microbial metabolism. In contrast, turbulent waters, as in rapidly flowing streams, readily support aerobic respiration.

Given the centrality of oxygen to metabolism, a parameter called *biological oxygen demand* (BOD) has been defined to measure the reducing power of wastewaters containing organic carbon. BOD is the number of milligrams of O_2 required to carry out the oxidation of organic carbon in one liter of water. Values for various industrial wastes and municipal sewage are given in Table 3.6.

When water is depleted of oxygen, organisms depending upon aerobic respiration cannot survive. In the microbial world, however, other organisms have evolved that can utilize oxidants other than O_2. Because no alternative oxidant in nature has a reduction potential as high as O_2, none of them can produce as much energy. Nevertheless, bacteria are quite capable of surviving on lower energy processes; in doing so, they can fill ecological niches that are not available to aerobic organisms. The oxidizing power of anaerobic environments in the biosphere is controlled by five molecules. In decreasing order of energy produced, they are nitrate (NO_3^-), manganese dioxide (MnO_2), ferric hydroxide ($Fe(OH)_3$), sulfate (SO_4^{2-}) and, under extreme conditions, carbon dioxide (CO_2). Microbial populations first use the oxidant that produces the most energy until it is depleted; only then does another agent become the dominant oxidant. Thus, the redox potential of a body of water tends to fall in a stepwise pattern as BOD increases.

Nitrate ion can be reduced to N_2

$$2NO_3^- + 12H^+ + 10e^- = N_2 + 6H_2O \qquad (3.42)$$

with a $pE^\circ[w]$ value, 12.65, only one unit lower than for O_2 reduction. Nitrate ion is ordinarily quite inert, but *denitrifying* bacteria have evolved specialized enzymes that catalyze this half-reaction and harness its potential for energy production. When the O_2 concentration has fallen sufficiently that aerobic bacteria no longer survive, denitrifiers can take over—provided, as is usually the case, that the water contains nitrate ions.

Once the nitrate is depleted, other organisms appear that are able to use weaker oxidants. If manganese dioxide is available, it is the next oxidant of choice. It can support a reduction half-reaction

$$MnO_2 + 4H^+ + 2e^- = Mn^{2+} + 2H_2O \qquad (3.43)$$

at $pE^\circ[w] = 6.71$. Likewise, when MnO_2 is depleted or unavailable and $Fe(OH)_3$ is present, it can be reduced to Fe^{2+}

$$Fe(OH)_3 + 3H^+ + e^- = Fe^{2+} + 3H_2O \qquad (3.44)$$

at a lower potential, $pE^\circ[w] = -3.08$.

TABLE 3.6 TYPICAL BODS FOR VARIOUS PROCESSES

Type of discharge	BOD (mg O_2/liter wastewater)
Domestic sewage	165
All manufacturing	200
Chemicals and allied products	314
Paper	372
Food	747
Metals	13

At even lower potentials, when SO_4^{2-} is available, sulfate reducing bacteria come into play,

$$SO_4^{2-} + 9H^+ + 8e^- = HS^- + 4H_2O \qquad (3.45)$$

with a $pE°[w]$ of -3.64. And at a still lower potential than that, the *methanogens* flourish, reducing CO_2 itself to methane

$$CO_2 + 8H^+ + 8e^- = CH_4 + 2H_2O \qquad (3.46)$$

at $pE°[w] = -4.14$. The $pE°[w]$ difference between CO_2 reduction to methane and carbohydrate oxidation to CO_2 (half-reaction (3.40)) is only 4.06, one-fifth the value for aerobic respiration.* Nevertheless, methanogens survive quite well and, in fact, are ubiquitous in sediments of all kinds, giving rise to a substantial worldwide methane production rate.

This sequence of microbial redox reactions is diagrammed in Figure 3.23. As oxidants are consumed in the conversion of reduced carbon to CO_2, the reduction potential falls to

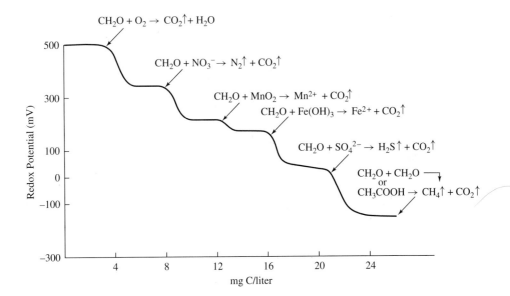

Figure 3.23 Sequence of redox reactions in aqueous environments. O_2 in natural waters at 20°C is sufficient to oxidize about 3.4 mg of organic carbon (shown here as CH_2O) per liter of water. When the rate of replenishment of O_2 from the atmosphere is slower than the rate of oxidation of CH_2O, oxygen is depleted and microbes will select the next most energetic oxidant in the sequence shown. For simplicity, only major products and their valence states are shown. See Table 3.7 for balanced equations. *Source:* W. M. Stigliani (1988). Changes in valued capacities of soils and sediments as indicators of nonlinear and time-delayed environmental effects. *Environmental Monitoring and Assessment* 10: 245–307. Copyright © 1988 by Kluwer, Academic Publishers. Reprinted with permission of Kluwer.

---

*Equation (3.46) plus two times equation (3.40) yields the overall reaction: $2CH_2O = CH_4 + CO_2$. Another important anaerobic alternative is acetate splitting: $CH_3COOH = CH_4 + CO_2$; the acetate is produced from cellulose by fermentative bacteria.

successively lower plateaus, corresponding to the successive $pE°[w]$ values of the reduction half-reactions. These plateaus are redox "buffer" regions of the carbon oxidation sequence. The reduction potential is buffered by the redox couples O_2/H_2O, NO_3^-/N_2, MnO_2/Mn^{2+}, $Fe(OH)_3/Fe^{2+}$, SO_4^{2-}/HS^-, and CO_2/CH_4 in the region of their respective $pE°[w]$ values. These couples, ordinarily, do not give reversible potentials at electrodes, but the metabolic activity of the vast array of microbes in soils and in water ensure that electron transfer is rapid, and that the couples do indeed behave reversibly on a time-scale of hours or days. Consequently, all redox-active materials respond to the reduction potential established by the microbial activity.

Bacteria also catalyze oxidation of reduced substances by molecular oxygen, even though such reactions can occur spontaneously in an aerobic environment. Thus HS^- oxidation to sulfate, the reverse of half-reaction (3.45), is catalyzed by sulfide oxidizers utilizing O_2 in half-reaction (3.39). These bacteria manage to extract energy from the HS^-/SO_4^{2-} and O_2/H_2O redox couples. Another important oxidation process is *nitrification*, the conversion of NH_4^+ to nitrate ion. Since plants take up and utilize nitrogen mainly in the form of nitrate, this is a key reaction in nature, especially in connection with the use of ammonium salts in fertilizers. The process actually occurs in two steps, ammonium to nitrite, NO_2^-, and nitrite to nitrate:

$$NH_4^+ + 2H_2O = NO_2^- + 8H^+ + 6e^- \tag{3.47}$$

$$NO_2^- + H_2O = NO_3^- + 2H^+ + 2e^- \tag{3.48}$$

These half reactions are catalyzed by two separate groups of bacteria, *Nitrosomonas* and *Nitrobacter*, each utilizing the oxidizing power of O_2 [see equation (3.39)] to extract energy from the process.*

In summary, redox potential can be considered as a kind of chemical switch in the aqueous environment, one that determines the sequence by which oxidants are utilized by microorganisms. Changes in redox potential can have important consequences for environmental pollution (Table 3.7).

2. Redox Potential and Its Influence on Water Quality

Many common water pollution problems result from redox chemistry. For example, conditions in subsurface waters are often sufficiently reducing, due to biological activity, to maintain iron in the soluble Fe^{2+} form. Exposure to air raises the redox potential enough to oxidize the Fe^{2+}, and the resulting Fe^{3+} hydrolizes, producing insoluble $Fe(OH)_3$. In many areas, the water drawn from wells runs brown from the precipitated $Fe(OH)_3$, which stains sinks, and can leave brown stains on laundry. (It also affects the taste of the water, but does not represent a health hazard, since $Fe(OH)_3$ is so insoluble that very little of it is absorbed in the intestines.)

a. Eutrophication in freshwater lakes.   Because the supply of oxygen is restricted, the species that inhabit an aquatic ecosystem are in a dynamic balance, one that is

---

*The sum of equations (3.47) and (3.48) plus two times equation (3.39) gives the overall reaction: $NH_4^+ + 2O_2 = NO_3^- + 2H^+ + H_2O$.

TABLE 3.7 REDOX REACTIONS, PRODUCTS, AND CONSEQUENCES

Redox Reaction	Reaction Products/Consequences
1. $O_2 + CH_2O$ $\rightarrow CO_2 + H_2O$	The aerobic condition, characterized by the highest redox potential, occurs when there is an abundance of O_2, and the relative absence of organic matter owing to oxic decomposition by aerobic microorganisms. Two examples are the aerobic digestion of sewage wastes, and the decomposition of organic matter near the surface of well-aerated soils. The end products, CO_2 and water, are nontoxic.
2. $(4/5)NO_3^- + CH_2O$ $+ (4/5)H^+ \rightarrow CO_2$ $+ (2/5)N_2 + (7/5)H_2O$	When molecular oxygen is depleted from the soil or water column, as would be the case, for example, in waterlogged soils and wetlands, available nitrate is the most efficient oxidant. Denitrifying bacteria consume nitrate and release N_2. N_2O, a greenhouse gas, is also released as a side-product. In agricultural soils, denitrification can lead to losses of nitrogen fertilizer amounting to as much as 20% of inputs. Denitrifying bacteria are also very active in heavily polluted rivers, or in stratified estuaries where organic matter accumulates. In some estuary systems, denitrification may significantly affect the transfer of nitrogen to the adjacent coastal waters and to the atmosphere.
3a. $2MnO_2 + CH_2O + 4H^+$ $\rightarrow 2Mn^{2+} + 3H_2O + CO_2$ **3b.** $4Fe(OH)_3 + CH_2O + 8H^+$ $\rightarrow 4Fe^{2+} + 11H_2O + CO_2$	In anaerobic environments where nitrates are in low concentration and manganese and ferric oxides are abundant, the metal oxides are a source of oxidant for microbial oxidation. This may be the case in natural soils, and in the sediments of lakes and rivers. The environmental significance of these metal oxides is that they serve a dual role. Not only are they a source of oxidants to microorganisms, they are also important for their capacity to bind toxic heavy metals, deleterious organic compounds, phosphates, and gases. When the metal oxides are reduced, they become water-soluble and lose their binding ability. This loss may result in the release of toxic materials.
4a. $(1/2)SO_4^{2-} + CH_2O$ $+ H^+ \rightarrow (1/2)H_2S + H_2O$ $+ CO_2$	Sulfidic conditions are brought about almost entirely by the bacterial reduction of sulfate to H_2S and HS^- accompanying organic matter decomposition. Sulfate reduction is very common in marine sediments because of the ubiquity of organic matter and the abundance of dissolved sulfate in seawater. In fresh water, such reactions are important in areas affected by acidic deposition in the form of sulfuric acid. H_2S is an extremely toxic gas. Sulfides are also important in scavenging heavy metals in bottom sediments.
4b. $MS_2 + (7/2)O_2 + H_2O$ $\rightarrow M^{2+} + 2SO_4^{2-} + 2H^+$	Conversion of a heavy-metal sulfide (MS_2) to sulfate may also occur when anaerobic sediments are exposed to the atmosphere, as in the case of the raising of dredge spoils. It may also occur when wetlands containing pyrites (FeS_2) are drained for agricultrue, or in coal mining areas as acid-mine drainage. One consequence may be an increase in acidification from the generation of sulfuric acid; another might be the release of toxic metals.
5. $CH_2O + CH_2O$ $\rightarrow CH_4 + CO_2$	Under anaerobic conditions at a redox potential of about -200 mV and in the presence of methogenic bacteria as may be found in swamps, flooded areas, rice paddies, and the sediments of enclosed bays and lakes, partially reduced carbon compounds can disproportionate to produce methane as well as CO_2. This reaction is more typical in freshwater systems because sulfate concentrations are much lower than in marine environments, averaging about one one-hundredth the concentration in seawater. Methane is a critical gas in the determination of global climate. Since the early 1970s, global atmospheric methane levels have been increasing at a rate of 1% per year. Although the reasons for this increase are still under investigation, the expansion of rice paddy cultivation in southeast Asia has been cited as a contributing cause. See discussion, Part II, pp. 123–125.

Source: W. M. Stigliani (1988). Changes in valued capacities of soils and sediments as indicators of nonlinear and time-delayed environmental effects. *Environmental Monitoring and Assessment* 10: 245–307.

easily disturbed by humans. In water, the O_2 concentration falls and the pE rises with increasing distance from the air-water interface. Thus, aerated soils support oxygen-utilizing microbes as well as higher life forms, while deeper in the soil, in the *saturated zone* where the soil pores are filled with water, anaerobic bacteria dominate and utilize progressively lower pE redox couples. Likewise in lakes, the sediments are generally oxygen-starved and rich in anaerobic microorganisms, while in the water column above, the O_2 concentration rises toward the surface. The concentration of O_2 at the surface is increased not only because the surface is in contact with air, but because the surface waters support the growth of vegetation and algae, which release O_2 as a product of photosynthesis. Because the sun's rays can penetrate only a certain distance into the water, photosynthesis is limited to the *euphotic zone*, whose depth depends on the water clarity. Below the euphotic zone, the O_2 concentration falls sharply.

The division of a lake into a euphotic zone and deep-water zone is stabilized, at least during the growing season, by *thermal stratification*. The surface waters are warmed by sunlight, but below the euphotic zone the water temperature falls sharply (Figure 3.24). Since warm water is less dense than cold water, a stable boundary is formed, called a *thermocline*; it is analogous to atmospheric inversion layers that can trap the air over cities under the right climatic conditions (pp. 143–144). The deep, cold waters are quiescent; physical mixing is largely confined to the euphotic zone. However, dead organisms and other debris from the surface layer sink, producing a steady flux of nutrients into the deep waters and sediments.

The biological productivity of a temperate lake varies annually in a cycle (Figures 3.24 and 3.25). The onset of winter diminishes the solar heating of the surface. The thermal stratification disappears and the water's density becomes uniform, allowing easy mixing by wind and waves, which brings nutrient-rich waters to the surface. In winter, the nutrient supply is high, but productivity is inhibited by low temperatures and light levels. Spring brings sunlight and warming, leading to a bloom of phytoplankton and other water plants. As plant growth increases, the nutrient supply diminishes and phytoplankton activity falls. Bacteria decompose the dead plant matter, gradually replenishing the nutrient supply, and a secondary peak of phytoplankton activity is observed in the autumn. Because the nutrient supply is limited in unpolluted waters, the BOD in the surface waters rarely outstrips the available oxygen.

This natural cycle can be disrupted, however, by excessive nutrient loading from human sources such as wastewaters or agricultural runoff. The added nutrients can support a higher population of phytoplankton, producing "algal blooms." When masses of algae die off, their decomposition can deplete the oxygen supply, killing fish and other life forms. If the oxygen supply is exhausted, the bacterial population may switch from predominantly aerobic bacteria to mainly anaerobic microorganisms that generate the noxious products (NH_3, CH_4, H_2S) of anaerobic metabolism. This process is called *eutrophication* or, more accurately, *cultural eutrophication*; eutrophication is the natural process whereby lakes are gradually filled in (Figure 3.26). Over time, an initially clear (*oligotrophic*) lake eutrophies gradually, filling with sediment and becoming a marsh eventually and then dry land. This process normally proceeds over thousands of years because biological growth and decomposition in the euphotic zone are closely balanced—the surface layers remain well oxygenated, and only a small fraction of biological production is deposited as sediment. When this balance is upset by overfertilization of the water, the eutrophication process accelerates greatly.

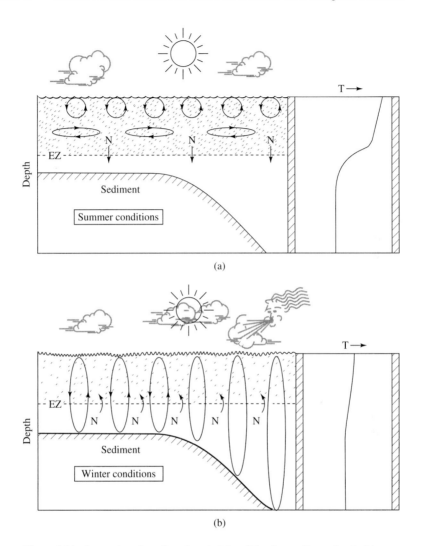

Figure 3.24   Seasonal cycling of nutrients in lakes. EZ = thermocline and end of the euphotic zone; stipple represents phytoplankton growth; N→ signifies direction of nutrient flow; enclosed arrows indicate circulation of waters. The solid line at the right is the temperature profile of the water column.

b. Nitrogen and phosphorus: the limiting nutrients.

The slow pace of natural eutrophication reflects the nutrient dynamics of an aquatic ecosystem (Figure 3.27). The nutrients are assimilated from the environment by the *primary producers*, the organisms that support photosynthesis and nitrogen fixation. The primary producers build the nutrients into their own tissues, which then serve as food for *secondary producers*, including fish. Dead plant and animal tissues are decomposed by bacteria, which restore the nutrients to the water. The growth of the primary producers is controlled by the *limiting* nutrient, the element that is least available in relation to its required abundance in the tissues. If the supply of the

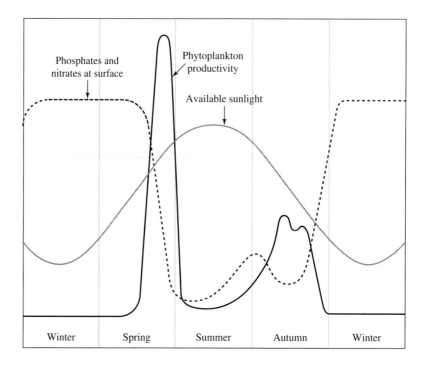

Figure 3.25 Seasonal phytoplankton productivity as a function of sunlight and nutrient concentration. *Source:* Adapted from W. D. Russel-Hunter (1970). *Aquatic Productivity* (New York: Macmillan Publishing Co., Inc.). Reprinted with permission from W. D. Russel-Hunter.

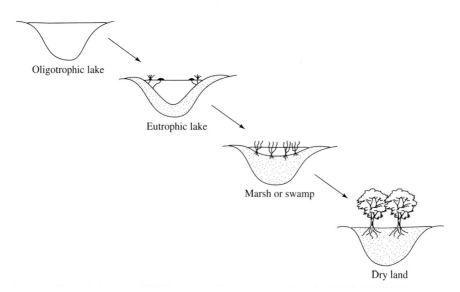

Figure 3.26 Eutrophication and the aging of a lake by accumulation of sediment.

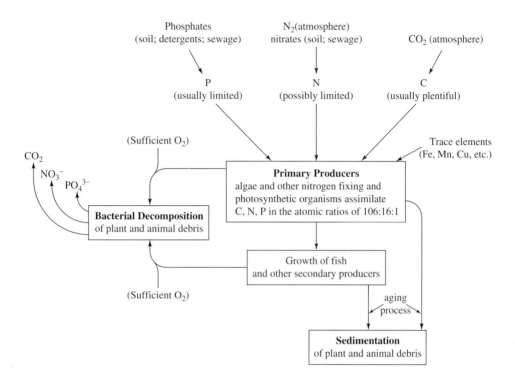

Figure 3.27   Nutrient cycling in an aquatic ecosystem.

limiting nutrient increases through overfertilization, the water can produce algal blooms, but not otherwise; conversely, management of the aquatic ecosystem requires that the supply of the limiting nutrient be restricted.

The major nutrient elements are carbon, nitrogen, and phosphorus, which are required in the atomic ratios 106:16:1, reflecting the average composition of the molecules in biological tissues. Numerous other elements are also required, including sulfur, silicon, chlorine, iodine, and many metallic elements. Because the minor elements are required in small amounts, they can usually be supplied at adequate rates in natural waters. On the other hand, carbon, the element required in the largest amounts, is plentifully supplied to phytoplankton from CO_2 in the atmosphere. Phytoplankton outrun the supply of CO_2 only under conditions of very rapid growth such as in some algal blooms. In these cases, the pH of the water can be driven as high as 9 or 10 through the required shift of the carbonate equilibrium

$$CO_3^{2-} + 2H^+ = H_2O + CO_2$$

The increase in pH can in turn alter the nature of the algal growth, selecting for varieties that are resistant to high pH.

Normally, the limiting nutrient element is either N or P. Although nitrogen makes up 80% of the atmosphere, it is unavailable except through the agency of N_2-fixing bacteria, living in symbiotic association with certain species of plants. On land, these species are rare enough to make nitrogen the limiting nutrient under most conditions. In water, however, N_2-fixing algal species are common, and nitrate ions are often abundant because of runoff from the land. Consequently, nitrogen is not usually limiting, although it may be in some regions, especially the oceans, where nitrate concentrations are low.

This leaves phosphorus as the element that is usually limiting to growth. Phosphorus has no atmospheric supply because there is no naturally occurring gaseous phosphorus compound. Moreover, the input of phosphorus in runoff from nonfertilized lands is usually low because phosphate ions, having multiple negative charges, are bound strongly to mineral particles in soils. In surface waters, most of the phosphorus is contained in the plankton biomass; the phosphorus availability depends on recycling of the biomass by bacteria.

Some of the phosphorus is lost to the deeper water and to the sediments when dead organisms sink. When the lake turns over in winter, the phosphorus in the deep waters is carried to the surface and supports the plankton bloom in the spring. Whether this phosphorus is available to the surface waters depends on conditions in the lake. At the bottom, phosphate ions may be adsorbed onto particles of iron and manganese oxide. However, when the sediment becomes anoxic, the metal ions are reduced to the divalent forms, the oxides dissolve, and the phosphate ions are released into solution (see notes on maganese and iron oxides in Table 3.7). Phosphate solubility is also increased through acidification since at successively lower pH values, HPO_4^{2-}, $H_2PO_4^-$, and H_3PO_4 are formed. (Their pK_a's are 12.4, 7.2, and 2.2, respectively.)

Under conditions of phosphorus limitation, human inputs of phosphate lead to enhanced biological production and the possibility of oxygen depletion. These inputs can arise from sewage, from agricultural runoff, especially where synthetic fertilizers (which contain phosphate salts) are applied intensively, and from polyphosphates in detergents. When phosphorus is added to lakes and rivers where the availability of phosphorus limits biochemical productivity, biomass production will increase in proportion to the amount of excess phosphorus added. The increased biomass raises the BOD of the water; if BOD increases to beyond 3.4 mg/l, oxygen will be depleted, leading to anoxia and anaerobic conditions. The most notorious instance of phosphate-induced eutrophication was in Lake Erie, which "died" in the 1960s. Excessive algal growth and decay killed most of the fish and fouled the shoreline. A concerted effort by the United States and Canada to reduce phosphate inputs was put into effect in the 1970s. Over $8 billion was spent in building sewage treatment plants to remove phosphates from wastewater, and the levels of phosphate in detergents were restricted. These efforts, along with other pollution control measures, succeeded in bringing the lake back to life. Commercial fisheries have revived, and the beaches are once again in use.

c. Anoxia and its effects on coastal marine waters.   Not only is anoxia a problem in freshwater lakes, it can be a problem in the intermediate or deep waters of an enclosed estuary, gulf, or fjord with restricted circulation between deep and surface waters. If inputs of organic carbon are high, dead biomass sinks into deeper waters where aerobic bacteria progressively consume the oxygen; if the deep layers fail to mix with surface layers, the oxygen is not replenished and anoxia sets in. Because seawater is rich in sulfate salts, the favored reaction under anaerobic conditions is sulfate reduction to hydrogen sulfide (H_2S), a chemical that is extremely toxic to fish and humans. Although H_2S is generally confined to the lower layers of seawater, during storms the deeper, anoxic layers can mix with surface layers, exposing aquatic life to the deadly gas.

Such events occurred in 1981 and 1983 in the enclosed marine areas off the east coast of Denmark. These events killed unprecedented numbers of fish through suffocation or poisoning by H_2S gas (Figure 3.28). The two episodes were triggered by sea storms, but the underlying cause was nutrient enrichment of the coastal waters. The source of the nutrient

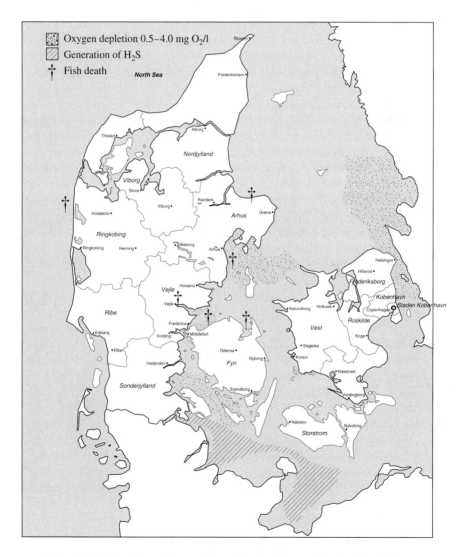

Figure 3.28   Coastal areas of eastern Denmark and southwestern Sweden affected by oxygen depletion, fish suffocation, and generation of H_2S.   *Source:* Miljostyrelsen (1984). *Oxygen depletion and fish kill in 1981: Extent and causes* (in Danish) (Copenhagen: Miljostyrelsen).

inputs was found to be primarily nitrogen in runoff from agricultural lands. The extra nitrogen increased the production of biomass (and hence organic carbon), which, upon bacterial decomposition, outstripped the oxygen available for aerobic degradation; anaerobic conditions produced H_2S, setting the stage for disaster.

Similar mechanisms have been at work in the United States, causing severe pollution of Chesapeake Bay. Because massive amounts of nutrients enter the bay from sewage, runoff, and atmospheric deposition, phytoplankton grow much faster than they can be foraged by consumer organisms such as oysters. The phytoplankton mass has clouded the waters; when these organisms sink to lower depths, they die from lack of light. In the depths, the dead plankton are consumed by bacteria that outstrip the supply of dissolved oxygen, making the bottom anoxic and allowing the production of H_2S. Without oxygen, the *benthic* (bottom-dwelling) organisms such as oysters and rooted plants cannot survive, and fish are displaced from their habitats.

The pollution of the Chesapeake appears to involve both nitrogen and phosphorus. Levels of these nutrients in the estuary rise and fall annually in seasonal patterns that are fairly complex (Figure 3.29). In winter, cold temperatures and lack of biochemical activity allow the concentration of O_2 to reach its annual maximum. At the same time, nitrogen enters in large amounts because winter is the period of maximum freshwater flow, with accompanying transport of sediment and runoff. Simultaneously, sedimentation is removing phosphorus from the water column, mainly through the precipitation of manganese and iron oxides, which absorb phosphorus efficiently and are insoluble under aerobic conditions. (Phosphorus is also removed during the settling of organic debris.) Beginning in the late spring and early summer, the oxygen levels decline due to increased biological activity. Nitrogen concentrations also decline for these reasons: 1) nitrogen is incorporated into biomass and sinks as the organisms die; 2) little new nitrogen is introduced in runoff; and 3) nitrogen is depleted as increasingly anoxic conditions force a switch from oxygen to nitrate as oxidant.

The opposite situation prevails for phosphorus. Under anaerobic conditions, phosphorus is liberated from the sediments, in large part due to the reduction of manganese and ferric oxides to Mn^{2+} and Fe^{2+}. In the 2+ valence states, the metals are soluble and release the bound phosphorus formerly adsorbed to the insoluble oxides of the metals. The phosphorus is readily mixed with the surface layers given the mechanical turbulence of estuarine environments. Thus, as conditions cycle from aerobic to anaerobic and back, the phosphorus is continuously recycled between the surface waters and the sediments. During anaerobic periods, phosphates are released to the water column to be taken up by microorganisms; during aerobic periods, phosphates are returned to the sediments. The amount of phosphate trapped in this cycle is vast, much greater than the annual quantities entering the estuaries from sewage effluents or other sources; it represents the cumulative inputs of many years. Thus, even though Maryland and Virginia banned detergents with phosphates in the 1980s, phytoplankton productivity is still excessive. Now the limiting nutrient may well be nitrogen, but nitrogen inputs are very difficult to control. Chesapeake Bay receives some of the highest atmospheric NO_x emissions in the world, mainly due to the density of traffic in the adjacent areas. Part of the strategy for cleaning up Chesapeake Bay might include reducing NO_x from vehicle exhausts, demonstrating once again the link between the atmosphere and the hydrosphere.

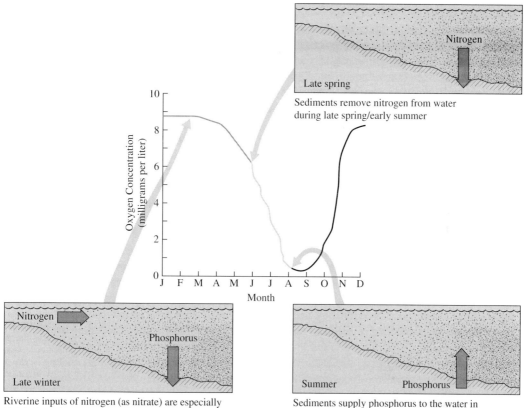

Figure 3.29   Oxygen concentration in water overlying the sediments with major seasonal net fluxes of nitrogen and phosphorous (insets) in the Patuxent River at the estuary of Chesapeake Bay.   *Source:* C. F. D'Elia (1987). Too much of a good thing: nutrient enrichment of the Chesapeake Bay. *Environment* 29(2): 6–11, 30–33. Copyright © 1987 by Environment. All rights reserved.

d. Wetlands as chemical sinks.   The episodes of fish-kills along the Danish coast might have been avoided if the original coastal wetlands had not been drained. Wetlands are typically anoxic and have large amounts of organic carbon; they create a natural buffer zone for nearby fresh or marine waters by trapping nitrates. The nitrates enter the wetlands in runoff, but are utilized by bacteria to oxidize stored carbon via the reduction of nitrate to N_2 or N_2O, which are vented to the atmosphere (Figure 3.30). By depleting the nitrates before they can enter the estuary, the surrounding wetlands limit the excessive growth of biomass and subsequent anoxic conditions in the estuary.

Furthermore, if the wetlands are of marine origin, they are likely to contain high concentrations of sulfur in the form of reduced sulfide minerals such as pyrite. Under the redox/ pH conditions prevalent in wetlands, these sulfides are highly insoluble and immobilized

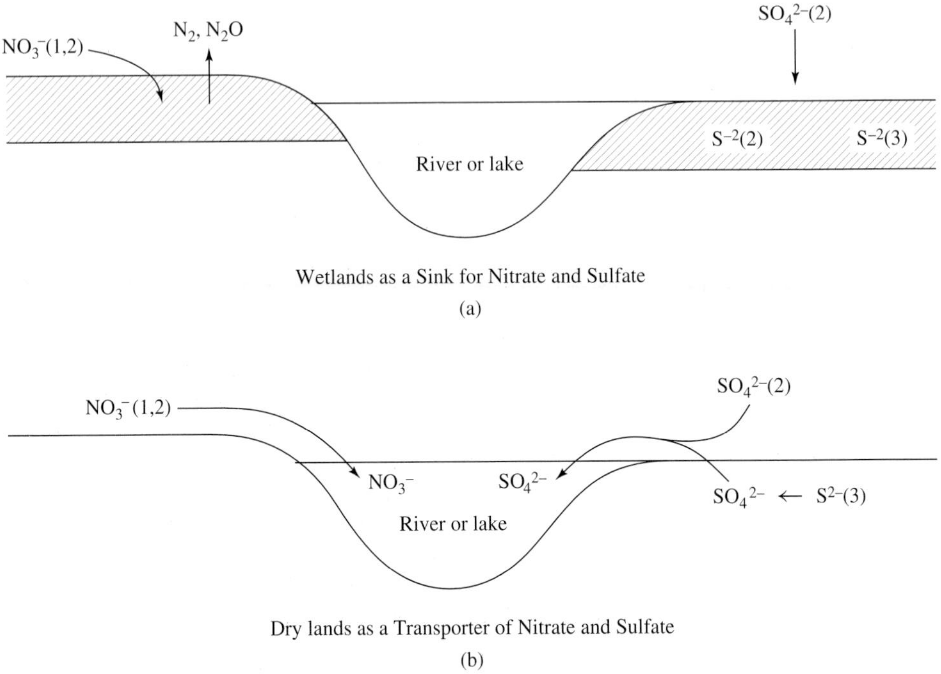

Wetlands as a Sink for Nitrate and Sulfate

(a)

Dry lands as a Transporter of Nitrate and Sulfate

(b)

1. Runoff of nitrogenous fertilizer
2. Input from acid deposition
3. Sulfide minerals from former marine sediments

Figure 3.30   (a) Ability of wetlands to buffer against nitrate and sulfate inputs to water bodies; (b) under conditions where wetlands become dry, none of the protective reducing reactions occur. In addition, accumulated sulfides may oxidize to sulfate as sulfuric acid, and leach into adjacent rivers or lakes.   *Source:* W. M. Stigliani (1988). Changes in valued capacities of soils and sediments as indicators of nonlinear and time-delayed environmental effects. *Environmental Monitoring and Assessment* 10: 245–307. Copyright © 1988 by Kluwer, Academic Publishers. Reprinted with permission of Kluwer.

(Figure 3.30). Draining the wetlands exposes these compounds to oxidizing conditions, producing a situation similar to acid mine drainage (pp. 222–223).

One example of this phenomenon occurred in a coastal area of Sweden near the Gulf of Bothnia, where wetlands were drained in the early 1900s for use as agricultural lands. As shown in Figure 3.31, draining the wetland shifted the Eh/pH conditions diagonally to the upper left, from the values typical of waterlogged soils to conditions close to those of acid mine drainage. The draining exposed sulfides to the atmosphere, and their oxidation to sulfuric acid acidified the soil and nearby lakes. The pH in one of these lakes, Lake Blamissusjon, dropped from 5.5 or higher in the last century to a current value of 3. Even though agricultural activities ceased in the 1960s, the lake has not recovered; it is widely known as the most acidic lake in Sweden.

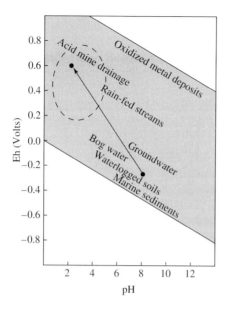

Figure 3.31   Eh/pH as a function of different aquatic environments. Oval enclosed by dashed line indicates region of highest solubility of heavy metals.   *Source:* Adapted from W. Salomons (1995). Long-term strategies for handling contaminated sites and large-scale areas. In *Biogeodynamics of Pollutants in Soils and Sediments,* W. Salomons and W. M. Stigliani, eds. (Berlin: Springer-Verlag). Copyright © 1994 by Springer-Verlag. All rights reserved.

e. Redox effects on metals pollution.   Changes in the redox potential can have important consequences for environmental pollution, especially with respect to metal ions such as cadmium, lead, and nickel. As shown in Figure 3.31, in general, the solubility of heavy metals is highest in oxidizing and acidic environments. At neutral to alkaline pHs in oxidizing environments, often these metals adsorb to the surface of insoluble $Fe(OH)_3$ and MnO_2 particles, especially when phosphate is present to act as a bridging ion. When the redox potential shifts to only slightly oxidizing or slightly reducing conditions as a result of microbial action, and the pH shifts toward the acidic range, $Fe(OH)_3$ and MnO_2 in soils and sediments are reduced and solubilized. The adsorbed metal ions can likewise become solubilized and move into groundwater (or into the water column of lakes when there is $Fe(OH)_3$ or MnO_2 in the sediment). Conversely, if sulfate is reduced microbially to HS^-, metal ions are immobilized as insoluble sulfides. But as we have seen, if sulfide rich sediments are exposed to air through drainage or dredging operations, then HS^- is oxidized back to sulfate, and the heavy metal ions are released.

A particularly important instance of biological redox mediation of heavy-metal pollution occurs in the case of mercury. Inorganic mercury, in any of its common valence states, Hg^0, Hg_2^{2+}, and Hg^{2+}, is not toxic when ingested; it tends to pass through the digestive system, although Hg^0 is highly toxic when inhaled. But the methylmercury ion $(CH_3)Hg^+$ is very toxic, regardless of the route of exposure. The environmental route to toxicity involves the same methanogenic bacteria that produce methane in anaerobic sediments. When exposed to mercury salts, they methylate the mercury, producing $(CH_3)Hg^+$ under acidic conditions; because methylmercury is soluble, it enters the aquatic food chain, where it is bio-accumulated in the protein-laden tissue of fish (see discussion in Part IV, pp. 314–319).

D. WATER POLLUTION AND WATER TREATMENT

The quality of surface and groundwater is of concern from two overlapping but distinct points of view: 1) human health and welfare, and 2) the health of aquatic ecosystems. Both aspects of water quality are enhanced by minimizing the impacts of human activities, but the specific issues and control measures are different.

1. Water Use and Water Quality

Water quality is as important an issue as water quantity. Although most of the water supply is returned to the stream flow after use, its quality is inevitably degraded. The effects are summarized in Table 3.8. The cooling of power plants by circulating water raises the temperature (thermal pollution), with adverse effects on the biota of the receiving waters. Discharge of sewage from homes and commercial establishments reduces the dissolved oxygen content, again upsetting the biological balance of surface waters. Industrial and mining activities contaminate water with a variety of toxic materials. Agriculture can foul surface and groundwaters with excess nutrients, and can lead to salinization of soil when irrigation waters evaporate, leaving salts behind.

In considering effects on water quality, it is useful to distinguish *point sources* and *nonpoint sources* of pollution. Point sources are factories and other industrial and commercial installations that release toxic substances into the water. In recent years toxic releases from point sources have been substantially reduced, especially in developed countries. The U.S. EPA, for example, targeted 17 toxic chemicals (the metals Cd, Cr, Hg, Ni and Pb, cyanide, and several chlorinated and nonchlorinated organic compounds—see Table 3.9 for the amounts released in 1993) for emissions reductions in a voluntary program involving the states, large industries, and community groups. Since 1988, when the program started, aqueous discharges of these chemicals have been reduced by 60–70%. These reductions affect water quality directly and indirectly, because they relieve stress on sewage treatment plants whose microbes (see later discussion) can be poisoned by these toxins.

Nonpoint sources include: emissions from transport vehicles; agricultural runoff, which can carry excess nutrients, pesticides, and silt into streams and groundwaters; and urban runoff, which can carry toxic metals and organics through storm drains into sewage treatment plants or directly into rivers and lakes (Figure 3.32). The progress made in con-

TABLE 3.8   EFFECTS ON WATER QUALITY FROM WATER USE

Water use	Effects on water quality
Domestic/Commercial	Decreases dissolved oxygen
Industrial/Mining	Decreases dissolved oxygen; pollutes water with toxic heavy metals and organics; causes acid mine drainage
Thermoelectric	Increases water temperature (thermal pollution)
Irrigation/Livestock	Causes salinization of surface and groundwaters; decreases dissolved oxygen (near feedlots)

TABLE 3.9 INDUSTRIAL RELEASES TO WATER OF EPA-TARGETED CHEMICALS (1993 IN METRIC TONS)

Chemical	To surface water	To public sewers	Total transfer
Cadmium and compounds	0.5	2.2	2.7
Chromium and compounds	113.6	201.2	314.7
Cyanide compounds	44.5	45.4	89.9
Lead and compounds	34.1	63.1	97.2
Mercury and compounds	0.2	0.0	0.2
Nickel and compounds	42.7	99.7	142.4
Benzene	8.5	140.0	148.5
Methyl ethyl ketone	89.5	343.2	432.6
Methyl isobutyl ketone	40.9	288.6	329.5
Toluene	60.4	439.4	499.8
Xylenes	25.7	338.1	363.8
Carbon tetrachloride	0.7	0.8	1.4
Chloroform	204.7	273.6	478.3
Dichloromethane	28.6	382.5	411.0
Tetrachloroethylene	4.6	50.4	55.0
1,1,1-Trichloroethane	4.9	27.4	32.0
Trichloroethylene	2.4	20.8	23.1
Totals	706.5	2,716.1	3,422.6

Source: U.S. Environmental Protection Agency (1995). *1993 Toxics Release Inventory,* Report 745-R-95-010 (Washington, DC: U.S. Environmental Protection Agency, Office of Pollution Prevention and Toxics).

trolling point sources of pollution has drawn attention to nonpoint sources, which account for an increasing fraction of the total pollutant load. For example, the relative contribution to the cadmium load to the Rhine River from point and nonpoint sources has changed over the past 30 years (Figure 3.33). While most of the Cd came from point sources in the 1970s, this is no longer the case, thanks to industrial controls. Now the greater share of Cd derives from nonpoint sources from urban and agricultural runoff (Cd is a contaminant in urban dust as well as in phosphate fertilizer).

2. Water and Sewage Treatment

Municipalities treat their water supplies for domestic and commercial uses to ensure freedom from disease and to eliminate odors and turbidity; they treat their sewage to reduce water pollution and eutrophication. In the former case, treatment begins with aeration to remove odors by purging dissolved gases and volatile organic compounds. Aeration also oxidizes any Fe^{2+} to Fe^{3+}, which forms $Fe(OH)_3$ and precipitates out. Additional Fe^{3+} or Al^{3+} ions are then added deliberately, usually as the sulfate salt, along with lime to adjust the pH. A voluminous $Fe(OH)_3$ or $Al(OH)_3$ precipitate is produced, which traps solid particles that may be suspended in the water supply. When the precipitate is collected and removed, the water is greatly clarified. If dissolved organic compounds require removal, the water can be passed through a filter of activated charcoal, although this is an expensive and uncommon step.

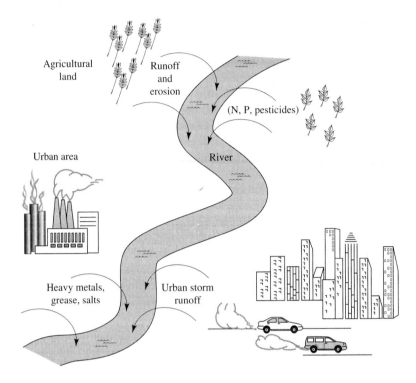

Figure 3.32   Nonpoint-source pollution from agricultural and urban areas.

Finally, disinfectant is added—chlorine, chlorine dioxide, or ozone—to kill microorganisms and provide safe drinking water.

Sewage treatment, similarly, relies to a large extent on settling and screening to remove solids; this physical separation step is called primary treatment. Most municipalities also carry out secondary treatment, which harnesses bacteria to metabolize organic compounds, converting them to CO_2. In this way the BOD is substantially decreased (Figure 3.34). If the sewage is not metabolized in this way, then the wastewater BOD can overwhelm the oxidizing capacity of receiving waters, leading to anoxic conditions. In secondary treatment, the wastewater is sprayed over a bed of sand or gravel that is covered by aerobic microorganisms, or else agitated with the microbes in a reactor. At the end of this process, the BOD is lowered by as much as 90%.

Although the microbes convert most of the organic matter to CO_2, they also incorporate some of it into new cells as the culture grows. These cells must be harvested from time to time, and are added to the sludge from the primary settling tank. Sludge disposal is a major issue in municipal sewage treatment. Since it is composed mainly of organic matter, sludge is an excellent fertilizer in principle. In practice, its application to cropland is restricted by the frequent presence of toxic metals that are flushed into the wastewater from domestic and industrial sources, or from urban runoff when storm drains are connected to sewer lines. Alternatively, the sludge can be incinerated and can provide energy for heating or electricity

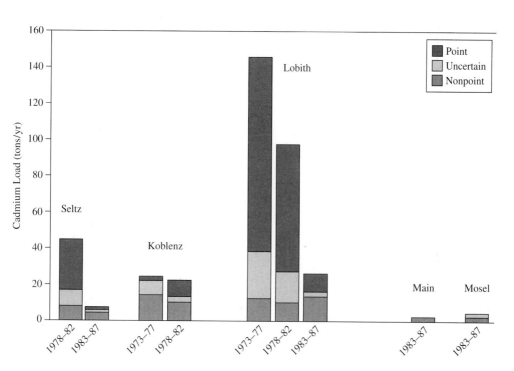

Figure 3.33   Estimated annual loads of cadmium at various stations on the Rhine River and its tributaries. The station at Lobith is at the German-Dutch border; the cadmium load there represents the sum of all upstream inputs in the Rhine Basin. The basin includes most of Switzerland, northeastern France, Luxembourg, a large part of southwestern Germany, and most of the Netherlands. *Source:* H. Behrendt (1993). *Point and Diffuse Loads of Selected Pollutants in the River Rhine and Its Main Tributaries,* Report RR-93-1 (Laxenburg, Austria: International Institute for Applied Systems Analysis). Copyright © 1993, International Institute for Applied Systems Analysis. Reprinted with permission.

production. Another alternative is to convert part of the sludge to methane, a high-quality fuel, by digesting it with anaerobic bacteria. However, poor economics and local opposition often militates against these options. Consequently, a good deal of sludge ends up in landfills. But as landfills become full, pressure for cropland application increases, so the issue of metals hazards and metals removal is again examined more closely. If the metals problem can be solved, sludge can become a valuable resource as fertilizer, instead of an environmental disposal problem.

Although secondary treatment is effective in reducing BOD, it does little to reduce the concentrations of inorganic ions, in particular, NH_4^+, NO_3^-, and PO_4^{3-}. These soluble ions are released with the wastewater, where they can eutrophy the receiving waters. Their removal requires tertiary treatment, in which additional chemical steps are added (Figure 3.35). Thus, phosphate is removed by precipitation with lime, producing the insoluble mineral hydroxyapetite, $Ca_5(PO_4)_3(OH)$. NH_4^+ can be converted to volatile NH_3 by adding lime to increase the pH, after which the pH is lowered again by CO_2 injection to reprecipitate the lime. Finally, the remaining organic compounds can be filtered out with activated charcoal, and

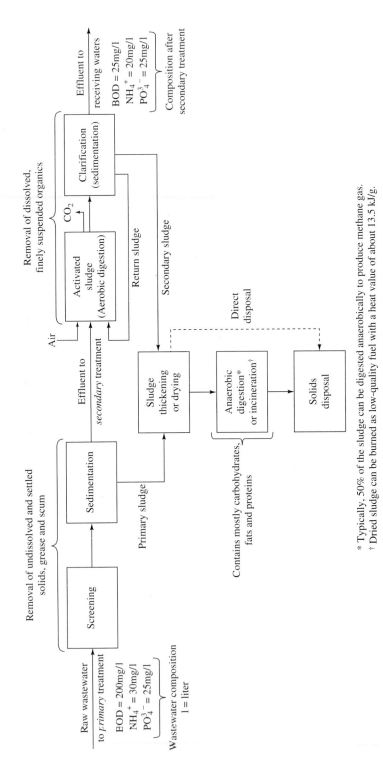

Figure 3.34 Primary and secondary treatment of municipal wastewater.

* Typically, 50% of the sludge can be digested anaerobically to produce methane gas.
† Dried sludge can be burned as low-quality fuel with a heat value of about 13.5 kJ/g.

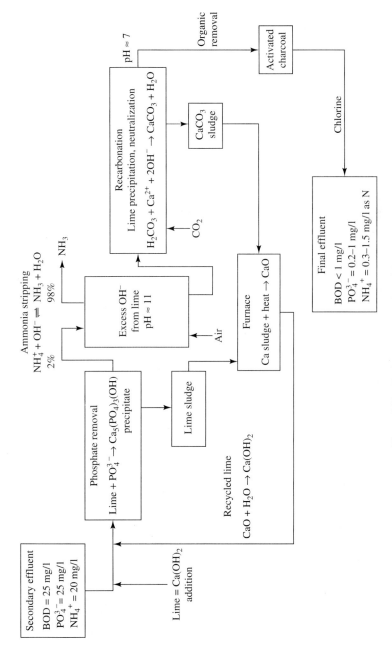

Figure 3.35 Tertiary treatment of municipal wastewater.

disinfectant can be added to produce quite pure water. These treatment steps add significantly to costs; ammonia stripping in particular is energy-intensive. An alternative process, requiring less energy, is to use nitrifying bacteria to convert NH_4^+ to NO_3^-, and then denitrifying bacteria to convert NO_3^- to N_2. However, these bacteria require careful control of growth conditions.

3. Health Hazards

a. Pathogens and disinfection.   The spread of pathogenic microorganisms through the water supply is the most serious pollution hazard to human health. Waterborne pathogens are ubiquitous throughout the world. Even waters that are untouched by humans can be contaminated by animal wastes. Hikers in the wilderness who drink untreated water from seemingly pristine streams often become infected by *Giardia*. The most serious problems, however, are created by the contamination of drinking water by human wastes. The failure to treat sewage and separate it from drinking water takes an enormous toll in disease. Unsanitary water is one of the most pervasive human problems, especially in developing countries.

Disinfection can take many different forms, one of the oldest of which relies on the removal of pathogens by the earth itself. Water filtered through soil and rock and into deep aquifers is generally free of microbes. Consequently, well water is generally pathogen-free. However, if the water table is shallow, or if the land is heavily laden with human or animal wastes, then the filtering capacity of the soil may be overwhelmed and wells contaminated.

The way to prevent microbial contamination is to keep the water supply as free from wastewater discharges as possible and to treat it with disinfectant. The disinfectants currently in use are ozone, chlorine dioxide, and chlorine. Of these, the most common is chlorine. When chlorine is added to water it disproportionates:

$$Cl_2 + H_2O = HOCl + H^+ + Cl^- \qquad (3.49)$$

The HOCl (hypochlorous acid) is an oxidant; the Cl is in the +1 oxidation state and is readily reduced to Cl^-. HOCl is the active ingredient of most bleaches, decolorizing fabrics by oxidizing colored molecules. (The colors generally reflect the presence of conjugated pi electrons, which are susceptible to oxidation.) Because it is a neutral molecule, HOCl passes through the cell walls of microorganisms and kills them by oxidizing vital molecules. Likewise, ozone and chlorine dioxide are powerful and diffusible oxidants, and kill microbes in the same way.

Chlorine is an effective and relatively inexpensive disinfectant with a proven record of success. Nevertheless, its use has become controversial because it can introduce organochlorine molecules into drinking water. HOCl is not only an oxidant but also a chlorinating agent. In particular, hydroxybenzenes are readily attacked by HOCl, and are converted to a variety of chlorinated compounds. Although hydroxybenzenes may be present from industrial wastes, they are also found naturally in surface waters as constituents of humic acids. Humic acids are complex molecules that derive from lignins (see pp. 16–17) and contain benzene rings with a variety of substituents including hydroxide. When plant matter decays, the lignins (the toughest molecules in plant tissue) persist the longest, and their breakdown products, the humic acids, are the main organic constituents of soils. Although polymeric, some humic acids dissolve in surface or groundwaters and enter the water supply. One of the more abundant reaction products of HOCl with humic acids is chloroform, $CHCl_3$. Most chlorinated water supplies, therefore, have trace levels of chloroform. Chloroform is a suspected liver carcinogen in humans, and there is some epidemiological evidence for a modest increase in the risk of bladder and rectal cancer from drinking chlorinated water.

The chloroform (and most other organic molecules) can be removed from the water supply by filtering it through activated charcoal, but this is an expensive measure. Instead, chlorine is being replaced in some communities by ozone or chlorine dioxide. (Even though the latter contains Cl, it is not an effective chlorinating agent.) These alternative disinfectants are coming into wide use, especially in Europe. They are more expensive than Cl_2, however, because they are too reactive to be stored or shipped, and must be generated on site. O_3 is generated by an electric discharge in air, while ClO_2 is made by oxidizing chlorite ion, ClO_2^- (often with Cl_2, which is reduced in the process to Cl^-). An additional disadvantage of both agents is that they are fast-acting and rapidly decomposed. In contrast, HOCl is less reactive; it acts more slowly and persists for some time in the water supply. Persistence is an advantage in many water systems where old and leaky pipes permit significant infiltration of water from the surroundings. HOCl offers a measure of protection against pathogens that might enter the supply through such leaks.

b. Organic and inorganic contaminants.

Surface and groundwaters can be contaminated by migration of chemicals from poorly maintained landfills, industrial waste sites, accidental spills, and leaks in storage tanks, especially those buried in the ground. Regulations for the cleanup of leaky underground oil and gasoline storage tanks have recently become major preoccupations of home- and gas-station owners in the United States. Drinking wells can be contaminated by trace amounts of petroleum fractions or chlorinated solvents or PCBs (polychlorinated biphenyls—previously used in transformers and pumps; see discussion, pp. 309–310, Part IV). These sparingly soluble organic compounds can escape confinement and migrate through the soil; they often accumulate in underground pools, from which they slowly enter the water table over a long period of time. Waste sites and spills can also leach metal ions into the water. Their transport is governed by binding equilibria, which are in turn influenced by soil pH and pE, as discussed in the preceding sections. When these ions enter streams and lakes, some metals such as cadmium and mercury, as well as most lipophillic organic compounds, bioaccumulate in the aquatic food chain and can make fish unsafe to eat (see Part IV, pp. 276–278, and discussion beginning on p. 314).

Remediation of contaminated soils and sediments is extremely difficult. Contaminated water can be extracted by specially dug wells and treated by aeration to volatilize lighter organic compounds, or by charcoal filtration to remove heavier organic compounds and metals. However, because of the volumes involved, this "pump-and-treat" approach is expensive; moreover, its effectiveness is often limited because of slow rates of contaminant transfer from soil reservoirs to the circulating water. Severely contaminated soils and sediments can be dug up or dredged and deposited elsewhere, presumably out of harm's way, but finding an appropriate dump site can be difficult. Moreover, the act of digging or dredging can stir up and transfer significant amounts of contaminated material to the surrounding air and water.

Agricultural areas can have water problems associated with the widespread application of fertilizers, herbicides, and pesticides. Although the trend in herbicide and pesticide use has for some time shifted from long-lived organic compounds to ones that break down fairly rapidly in the environment, some herbicides and pesticides can still accumulate in the groundwater, occasionally threatening farm wells.

Fertilizers can increase the level of nitrate ions in the groundwater. The main nitrate hazard is "blue baby syndrome," a condition of respiratory failure in babies having excessive nitrate in their diet. Some of the nitrate is reduced by anaerobic bacteria in the stomach to nitrite ion, NO_2^-. The nitrite oxidizes Fe^{2+} ions in hemoglobin to Fe^{3+}, which are unable to bind O_2. The Fe^{3+}– containing hemoglobin is called *methemoglobin*, and the condition is known as *methemoglobinemia*. In adults the methemoglobin is rapidly reduced back to the Fe^{2+} form, but in babies this process is slow. Nitrate-induced methemoglobinemia is now a rare condition in industrialized countries but remains a concern in developing countries.

Another worry about nitrate is the possibility that the nitrite produced in the stomach can react with amines in the diet to produce N-nitrosamines, R_2NNO (see p. 295). Nitrosamines have been shown to be carcinogenic in animal tests. However, the nitrate levels in drinking water are much lower than in cured meat products or cheeses, to which nitrates are added to inhibit the bacterium that causes botulism. Government agencies have instituted programs to reduce the levels of nitrates and nitrites in food products, and some manufacturers now add vitamins C and E in order to block nitrosamine formation.

E. SUMMARY

Water is available in abundance on planet Earth, and pure water is supplied continuously by the hydrological cycle, which is powered by the sun. The bounty of fresh water is distributed unevenly around the globe; even where it is abundant, the resource is often mismanaged through profligate use or through contamination from wastes. Access to unpolluted, pathogen-free water is a critical need for much of the world's population.

In soils, the major interactions of water are determined largely by acid-base equilibria established by buffering regimes that vary with soil type. Acidified water passes through the soil, where it is neutralized by limestone or clay minerals. This natural balance is upset when excessive, anthropogenically-derived acidification results in acid inputs that exceed the soil's buffering capacity. In this case, acids percolating through the soil are no longer neutralized, and receiving waters become acidified. Another outcome is that mobilization of toxic metals is greatly enhanced. Because buffering capacities diminish slowly, on a time scale of decades, long time lags may occur between the onset of pollution and its ultimate effects.

In aquatic ecosystems, water chemistry is determined by the redox potential. Reduced organic carbon is the major biochemical fuel of the biosphere, whose oxidation provides the primary energy upon which all life depends. Although O_2 is the most powerful oxidizing agent, it is only sparingly soluble in water, and is easily depleted when levels of organic carbon are excessive. In the absence of oxygen, microorganisms oxidize reduced carbon, utilizing other environmental oxidants. The choice of oxidant is determined in a sequence of reactions regulated by the redox potential. Water pollution problems arise because some of these reactions cause the release of harmful by-products.

The problem of water pollution is thus one largely of capacity depletion: acid buffer capacity in soil-water interactions, and oxidation capacity in receiving waters. Reduction of water pollution thus requires strategies that focus not only on reductions in emissions of specific pollutants having a direct impact on water quality, but also on maintenance and replenishment of the vital capacities. In some cases this may be achieved by redirecting misplaced resources. For example, provided they are free of toxic chemicals, sewage wastes, instead of being discharged to water bodies, can be applied as fertilizer to the land, which has a much higher oxidizing capacity.

PROBLEM SET III: HYDROSPHERE

1. The global hydrological cycle is driven by solar evaporation of water on land and sea. Calculate the solar energy required to drive the global hydrological cycle using the data in Figure 3.1, and assuming that heat of evaporation of water (both salt and fresh) at 15°C (the average global temperature) is 44.3 kJ per mole. Compare your answer with the value given in Figure 1.1. Compare it to the global anthropogenic primary energy consumption in 1990 (3.7×10^{17} kJ/year).

2. **(a)** Using data from Table 3.1, calculate the global supply of water per capita in 2025, when the world population is projected to be 8.5 billion. In 1990 the population was 5.3 billion.

   (b) With respect to the demand for water, which amounted to 3,240 km^3 in 1990 (Table 3.1), irrigation accounted for 69% of global water withdrawals. The global efficiency of irrigation is around 37%. Calculate what the demand would have been if global efficiency of irrigation had been 70%.

3. Consider a corn yield of 7,400 kg per hectare (equivalent to 120 bushels per acre). If 25 kg (one bushel) of corn consumes about 20 m^3 of water during the growing season, what is the ratio of the weight of corn to the weight of water consumed? Where does most of the water end up? Assuming a rainfall of 30 cm/yr, calculate the minimum quantity of irrigation water required per hectare to grow the corn (1 hectare = 10^4 m^2).

4. Due to increasing diversion of the water supply for other purposes, and to frequent droughts, many U.S. cities experience chronic shortages of water for domestic uses. Assume you are the mayor of a city plagued by water shortages, and you decide to focus your efforts on decreasing the *demand* for water rather than increasing the *supply*. Devise a strategy for this "demand management" approach. First estimate the gallons per capita per day of domestic water use, given that household water use in the United States in 1990 was about 3.50×10^{13} liters/yr for a population of around 250 million (1 gallon = 3.785 liters). Secondly, you learn from a public survey that total domestic water use comprises toilet flushing (38%), bathing (31%), laundry and dishes (20%), drinking and cooking (6%), brushing teeth and other miscellaneous uses (5%). Flush-toilets in your town use about 20 liters per flush, but water-conserving varieties recommended by the Plumbing Manufacturers Institute average about 13 liters per flush; the average household shower uses about 25 liters of water per minute, but water-saving shower heads are available that spray about 10 liters of water per minute; water-efficient dishwashers and washing machines are available that reduce water use by 25%. Devise a long-term plan for reducing water consumption, and calculate a

reasonable goal (in terms of gallons of water per capita per day) to be achieved in the next five to ten years.

5. H_2S boils at $-61°C$, H_2Se at $-42°C$, and H_2Te at $-2°C$. Based on this trend, at what temperature would one expect water to boil? Why does water boil at a much higher temperature?

6. If the concentration of atmospheric CO_2 were to double from its current value of 350 ppm, what would be the calculated pH of rainwater (assuming that CO_2 were the only acidic input)? With respect to rising CO_2 levels, do we have to be concerned about enhanced acidity of rain in addition to potential climate warming?

7. Describe the three major buffer ranges for neutralizing acidic inputs to soils. For each of the ranges, include in your description: (1) the pH range over which the buffer operates; (2) the major chemical component(s) that participates in the buffering reactions; and (3) the chemical reaction by which H^+ is neutralized.

8. Why is the process of acid deposition beneficial from the point of view of air quality? How does this process transfer a short-term air-pollution problem into long-term problems of soil and water pollution?

9. Why is carbonic acid effective in weathering parent soil rock, but not an important factor in lake acidification?

10. **(a)** With reference to Figure 3.21, assume that the average annual deposition of sulfur in the watershed of Big Moose Lake between 1880 and 1920 was 0.8 g S/m^2, and from 1921 to 1950 it was 2.5 g S/m^2. Calculate the cumulative acid equivalents (eq) per m^2 that were deposited in the watershed over the entire period 1880 to 1950.

(b) Assume that the watershed soils buffered acidity via exchange reactions with base cations on clay mineral surfaces. Assume further that base saturation (β) declined from 50% in 1880 to 5% in 1950. Calculate the total Cation Exchange Capacity (CEC_{tot}), and the buffering capacity ($CEC_{tot} \times \beta$) in 1880 (in units of eq/m^2). Also assume that the silicate buffer rate (br_{Si}), which replenishes the base cations in soil from weathering of silicate, was negligible over this period.

(c) Do the same calculation as in 2(b), but assume br_{Si} was equal to 0.02 eq m^{-2}yr^{-1} over the time scale from 1880 to 1950. What would you expect the pH of the soil water to be when the base saturation approaches 5%?

11. Write the equations for the oxidation of pyrite (FeS_2) and give two examples of anthropogenic activities that can initiate the oxidation. What is the environmental effect manifested by pyrite oxidation?

12. **(a)** Show how the numbers given in Table 3.5 were calculated for natural sources. Make the following assumptions: the pH of unpolluted rainfall is 5.7, and annual global precipitation is 496,000 km^3/yr; production of NO_2 from lighting is 20×10^{12} g N/yr; production of SO_2 from volcanoes is 20×10^{12} g S/yr; biogenic production of dimethylsulfide (DMS) and H_2S is 65×10^{12} g S/yr.

(b) Calculate moles of hydrogen ions per year generated from coal burning and smelting, assuming this activity generates about 93×10^{12} g S/yr. Assume combustion processes from all activities generate about 20×10^{12} g N/yr.

(c) Calculate the acid-neutralizing capacity of the atmosphere due to generation of ammonia (NH_3). Assume that natural and anthropogenic sources each generate about 60×10^{12} gN/yr. Calculate the total *net* generation of H^+/yr from both natural and anthropogenic sources.

(d) Calculate the average moles $H^+m^{-2}yr^{-1}$ from natural sources. (The area of the globe is 510×10^{12} m^2.) Calculate the average moles $H^+m^2yr^{-1}$ in industrialized regions of the globe. (Assume that the area of industrialized regions is $11.75 \times 10^{12}m^2$.)

13. (a) What class of molecules is responsible for most of the reducing power in aqueous environments?

(b) What parameter is a measure of reducing power?

14. Five hundred kg of n-propanol ($CH_3CH_2CH_2OH$) are accidentally discharged into a body of water containing 10^8 liters of H_2O. By how much is the BOD (in milligrams per liter) of this water increased? Assume the following reaction:

$$C_3H_8O + 9/2\,O_2 \rightarrow 3CO_2 + 4H_2O$$

15. A lake with a cross-sectional area of 1 km^2 and a depth of 50 meters has a euphotic zone that extends 15 meters below the surface. What is the maximum weight of the biomass (in grams of carbon) that can be decomposed by aerobic bacteria in the water column of the lake below the euphotic zone during the summer when there is no circulation with the upper layer? The bacterial decomposition reaction is:

$$(CH_2O)_n + nO_2 \rightarrow nCO_2 + nH_2O$$

The solubility of oxygen in pure water saturated with air at 20°C is 8.9 mg/l; 1 m^3 = 1,000 liters.

16. Assume that algae need carbon, nitrogen, and phosphorus in the atomic ratios 106:16:1. What is the limiting nutrient in a lake that contains the following concentrations: total C = 20 mg/l, total N = 0.80 mg/l, and total P = 0.16 mg/l? If it is known that half the phosphorus in the lake originates from the use of phosphate detergents, will banning phosphate builders slow down eutrophication?

17. Name the six most important oxidants in the aquatic environment, and how the redox potential regulates their reactivity.

18. (a) If a lake contains high concentrations of dissolved Mn^{2+} and Fe^{2+}, what would be the concentration of dissolved NO_3^- and why?

(b) What environmental effect may accompany reduction of MnO_2 and $Fe(OH)_3$?

19. In anaerobic marine environments, what toxic gas can be generated and by which reaction (name reactants and products)?

20. Explain the "phosphate trap" in the estuary of Chesapeake Bay. Why was a local ban on phosphorus in detergents not particularly helpful in mitigating eutrophication in the estuary?

21. (a) Explain why anaerobic freshwater wetlands with high concentrations of organic carbon can serve as natural buffers against sulfates and nitrogen oxides (give reactions).

(b) When other oxidants are absent from such wetlands, which redox reaction is likely to predominate, and which products will be emitted?

22. An estuarine creek in New Jersey contains large amounts of mercury bound as sulfide (with K = 10^{-52}) under the prevailing environmental conditions (pH = 6.8; Eh = –230 mV). Environmental scientists have been asked to assess the potential impacts of the polluted sediments. They conclude that the mercury poses no danger in its current state. However, they caution against any action that would expose it to air and increase its redox potential. Explain why the scientists come to this conclusion?

23. Characterize nonpoint-source pollution, and give two examples, one in urban areas and one in agricultural areas. Why is it more difficult to control this type of pollution than point-source pollution?

SUGGESTED READINGS III: HYDROSPHERE

J. W. Maurits la Riviere (1989). Threats to the world's water. *Scientific American* 261(3): 48–55.

S. Postel (1986). Increasing water efficiency. In *State of the World* (Annual Worldwatch Institute Report on Progress Toward a Sustainable Society). L. Brown, Project Director, (New York: W. W. Norton & Company).

S. Postel (1990). Saving water for agriculture. In *State of the World 1990* (Annual Worldwatch Institute Report on Progress Toward a Sustainable Society). L. Brown, Project Director, (New York: W. W. Norton & Company).

Council for Agricultural Science and Technology (CAST) (1988). *Effective Use of Water in Irrigated Agriculture,* Report 113 (Ames, Iowa: Council for Agricultural Science and Technology).

W. B. Solley, R. R. Pierce, and H. A. Perlman (1993). *Estimated Use of Water in the United States in 1990,* U.S. Geological Survey Circular 1081 (Washington, DC: Department of the Interior).

National Acid Precipitation Assessment Program (NAPAP) (1993). *1992 Report to Congress* (Washington, DC: National Acid Precipitation Assessment Program).

National Research Council (1986). *Acid Deposition, Long-Term Trends* (Washington, DC: National Academy Press).

J. Alcamo, R. Shaw, and L. Hordijk, eds. (1990). *The RAINS Model of Acidification, Science and Strategies in Europe* (Dordrecht, The Netherlands: Kluwer Academic Publishers for International Institute for Applied Systems Analysis, Laxenburg, Austria).

W. M. Stigliani and R. W. Shaw (1990). Energy use and acid deposition: the view from Europe. *Annual Review of Energy* 15: 201–216.

W. M. Stigliani (1988). Changes in valued capacities of soils and sediments as indicators of nonlinear and time-delayed environmental effects. *Environmental Monitoring and Assessment* 10: 245–307.

C. F. D'Elia (1987). Too much of a good thing: nutrient enrichment of the Chesapeake Bay. *Environment* 29(2): 6–11, 30–33.

W. Salomons and W. M. Stigliani, eds. (1995). *Biogeodynamics of Pollutants in Soils and Sediments, Risk Assessment of Delayed and Non-Linear Responses* (Berlin: Springer-Verlag).

Council for Agricultural Science and Technology (CAST) (1992). *Water Quality, Agriculture's Role,* Report 120 (Ames, Iowa: Council for Agricultural Science and Technology).

D. S. Knopman and R. A. Smith (1993). Twenty years of the clean water act: has U.S. water quality improved? *Environment* 35(1): 16–20, 34–41.

W. M. Stigliani, P. R. Jaffé, and S. Anderberg (1993). Heavy metal pollution in the Rhine Basin. *Environmental Science and Technology* 27(5): 786–793.

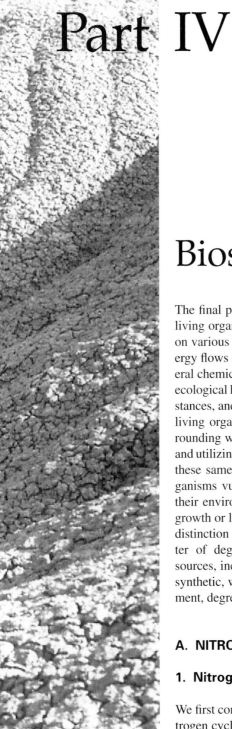

Part IV

Biosphere

The final part of this book concerns the biosphere—the realm of living organisms and their interactions. We have already touched on various aspects of the biological world in connection with energy flows and air and water chemistry, but now we focus on several chemical and biological topics that directly affect human and ecological health: food production, nutrition, pesticides, toxic substances, and carcinogenesis. What links these topics is the need of living organisms to take in and process materials from the surrounding world. All organisms have developed ways of ingesting and utilizing substances essential for growth and development. But these same processes of intake and metabolism make living organisms vulnerable to the effects of non-nutritive substances in their environment. When these substances interfere with normal growth or life, we describe them as toxic. But as we shall see, the distinction between essential and toxic substances is often a matter of degree. Moreover, toxic substances come from many sources, including natural ones. Whether a chemical is natural or synthetic, what counts is its reactivity, persistence in the environment, degree of exposure, and influence on biochemistry.

A. NITROGEN AND FOOD PRODUCTION

1. Nitrogen Cycle

We first consider food production and the important role of the nitrogen cycle. The central process in biological energy production is the photosynthetic conversion of CO_2 to reduced carbon compounds, which are then used as fuel by all manner of life (see discussions of the carbon cycle on pp. 12–14 and of photosynthesis on pp. 68–69). However, organisms need more than carbon; as we discussed in the context of aquatic ecosystems (pp. 231–234), they also require nitrogen, phosphorus, and several other elements in smaller amounts. Most of these elements are available in soil in sufficient quantities to support an adequate level of natural plant growth.

Possibilities for increasing productivity beyond natural limits, however, are often constrained by the supply of nitrogen available to the plant.* While 80% of the atmosphere consists of molecular nitrogen, N_2 is an extremely stable and unreactive form. In order to participate in biological reactions, nitrogen must be *fixed;* that is, it must be combined with other elements. The cycling of nitrogen through the environment is illustrated in Figure 4.1. Some N_2 is fixed nonbiologically through reaction with O_2 at sufficiently high temperatures in combustion or lightning. The nitrogen oxides formed in the atmosphere are converted to nitric acid and washed out in rain, thus providing the soil with a supply of nitrate. Plants can utilize nitrate in the production of protein and other essential organic nitrogen compounds.

But the amount of nitrogen available through the pathway of nitrogen oxides and nitrate is insufficient to support the abundant plant life we know. The majority of naturally oc-

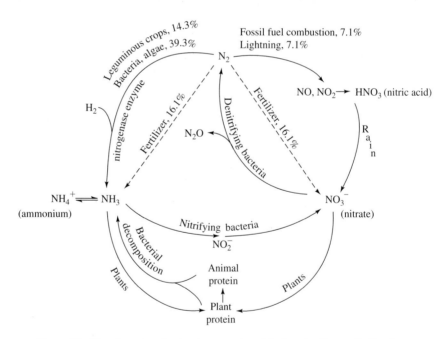

Figure 4.1 The nitrogen cycle, showing percent contributions to nitrogen fixation from natural and human sources. (Total contribution of synthetic fertilizers, assuming equal shares of ammonia and nitrate, is 32.2%.)

*Phosphorus can also be a limiting nutrient, although plants require 16 atoms of N for every atom of P (p. 233). At specific times and places, availability of light and water can also be limiting factors.

curring nitrogen fixation is accomplished by certain bacteria and blue-green algae (*cyanobacteria*) that are able to reduce N_2 to NH_3. These organisms possess a specialized biochemical apparatus, the *nitrogenase* enzyme complex, whose chemistry is illustrated in Figure 4.2. The complex consists of two proteins, the "Fe protein," which contains a polynuclear iron-sulfur complex, and a "Mo-Fe protein," which contains a polynuclear molybdenum-iron-sulfur complex (Figure 4.3). The reduction of N_2 is carried out at the active site of the Mo-Fe protein, where six electrons and six protons are added to N_2, producing two molecules of NH_3. The role of the Fe protein is to transfer electrons to the Mo-Fe protein in coordination with the hydrolysis of MgATP (ATP = adenosine triphosphate) to MgADP (ADP = adenosine diphosphate), a process that releases energy. Even though the overall reaction

$$N_2 + 3H_2 = 2NH_3 \quad \Delta G = -94 \text{ kJ/mole} \tag{4.1}$$

is downhill in energy, the N_2 bond is so strong (941 kJ/mol) that additional energy must be provided by the organism in the form of MgATP to overcome the activation barrier.

Plants can use ammonia directly for their nitrogen source; animals gain their nitrogen by eating plants. When plants and animals die, the reduced nitrogen in their tissues is converted to ammonia by bacterial decomposition and added to the ammonia pool. The ammonia can be used as fuel by other bacteria (*Nitrosomonas*), which convert NH_3 to NO_2^- (nitrite), using O_2 as oxidant. Still other bacteria (*Nitrobacter*) oxidize the nitrite further, to NO_3^- (nitrate). (The overall process of nitrogen oxidation is called *nitrification*.) Plants can also utilize nitrate as a nitrogen source. Thus, there is a continual cycling between oxidized and reduced forms of fixed nitrogen in soils.

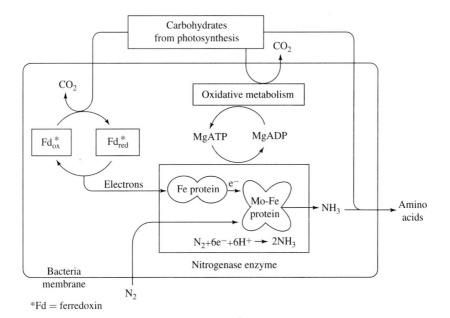

Figure 4.2   The chemistry of the nitrogenase enzyme system in algae and bacteria.

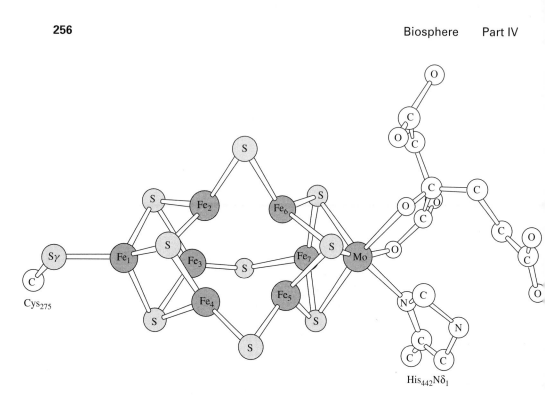

Figure 4.3 Structure of the iron-molybdenum cofactor of the molybdenum-iron protein of nitrogenase.* Seven iron and one molybdenum atoms form a cluster with bridging sulfur atoms. The cluster is ligated to the protein by two side chains from the cysteine 275 and histidine 442 residues. In addition, an organic acid, homicitrate, is bound to the molybdenum atom. The cavity in the center is a possible site of binding and activation of N_2. *Source:* S. J. Lippard and J. M. Berg (1994). *Principles of Bioinorganic Chemistry* (Sausalito, California: University Science Books). Reprinted with permission, copyright © 1994, University Science Books. All rights reserved.

If the fixed nitrogen were not returned to the atmosphere, the atmospheric pool of N_2 eventually would be depleted. But the nitrogen cycle is closed by *denitrifying* bacteria, which use NO_3^- instead of O_2 as the oxidant in metabolic reactions, reducing the nitrate back to N_2 (see Part III, p. 226). Both nitrification and denitrification processes release some N_2O as a side-product. N_2O is a greenhouse gas and the main source of stratospheric NO, a principal actor in O_3 destruction chemistry (see pp. 121–125, 153–154).

2. Agriculture

The major natural sources of nitrogen for agriculture are the nitrogenase-containing bacteria, some of which grow in symbiosis with a limited variety of crop plants, most notably those of the legume family. These symbiotic bacteria are contained in the nodules of such legumes as beans, peas, alfalfa, and clover. When the plants die, most of the nitrogen is returned to the soil in fixed form, where it is available for other kinds of plants; a small fraction is re-

---

*Figure was drawn from structural data provided by J. Kim and D. C. Reese (1992). Structural models for the metal centers in the nitrogenase molybdenum-iron protein. *Science* 257: 1677–1682.

turned to the atmosphere via the denitrification reaction. The fertilizing ability of legumes is the reason behind crop rotation, an ancient agricultural practice in which legumes are planted alternately with cereals, grains, and other vegetables to maintain the productivity of the non-legume plants. In the absence of crop rotation, plants that do not fix nitrogen deplete the soil's nitrogen stores quickly, unless these store are replenished by the addition of fertilizer.

One traditional fertilizer is animal manure, but increasingly in recent decades it has been supplanted by artificial fertilizers produced industrially. The industrial production of nitrogen fertilizer (dashed lines in Figure 4.1) is accomplished via the Haber process, in which reaction (4.1) is carried out over an iron catalyst. Even with a catalyst, the reaction requires high pressures (100 atm) and temperatures (500°C). (In contrast, the nitrogenase enzyme operates at ambient pressure and temperature, but energy input in the form of MgATP—see Figure 4.2—is required.) The resulting ammonia can be injected directly into crop-bearing soils or, more conveniently, it can be added as ammonium nitrate salt, produced by air oxidation of half the ammonia to HNO_3, which is recombined with the remaining ammonia:

$$NH_3 + 2O_2 = HNO_3 + H_2O \tag{4.2}$$

$$NH_3 + HNO_3 = NH_4NO_3 \tag{4.3}$$

The Haber process [equation (4.1)] requires a source of hydrogen gas. Currently, the most economic process for obtaining hydrogen is from methane:

$$CH_4 + 2H_2O = 4H_2 + CO_2 \tag{4.4}$$

The global production of nitrogen fertilizers has increased dramatically in the past three decades (Figure 4.4), and is now more than double the estimated N_2 fixation rate of leguminous crops (90 versus 40 Tg/yr of fixed N). The total of these two anthropogenic sources roughly equals the estimated preindustrial N_2 fixation rate, and the nitrogen oxide

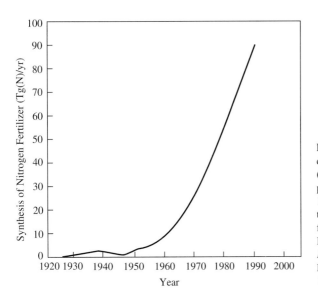

Figure 4.4 Global nitrogen fertilizer production from 1920 to 1990 in Tg N/yr (Tg = teragrams = 10^{12} g); cumulative production over this 70-year period was 1420 Tg N, with half of that occurring between 1982 and 1990. *Source:* Adapted from A. P. Kinsig and R. H. Socolow (1994). Human impacts on the nitrogen cycle. *Physics Today* November, 1994: 24–31. Reprinted with permission, copyright © 1994, Physics Today. All rights reserved.

contribution from fossil-fuel combustion adds another 20 Tg/yr. Thus, human activities now appear to dominate the global nitrogen cycle, although this was not the case as recently as 1970. The relationship of these fixation rates to the global fluxes and pools of nitrogen is illustrated in Figure 4.5. Human activity has increased the rates of both nitrogen fixation and denitrification. The sum of the increased fixation on land and increased runoff of fixed nitrogen from land to the oceans, however, does not appear to be entirely balanced by the increase in denitrification on land and in the oceans. This result suggests that fixed nitrogen is now accumulating in the terrestrial pool (see problem 2, Part IV).

Most of the global fixed nitrogen is in the ocean, where it is divided nearly equally between organic and inorganic forms. The inorganic ions, NH_4^+ and NO_3^-, do not accumulate in soils; because they are soluble, they are quickly taken up by plants and bacteria or are washed out of the soil and transported to the oceans. The terrestrial fixed nitrogen is mainly in organic matter.

The application of fertilizer can dramatically improve agricultural yields (Figure 4.6). The quadrupling of yields for corn and other grains between 1940 and 1990 made the U.S. and Canadian Midwest the breadbasket of much of the world. Moreover, crop yields have increased everywhere (Figure 4.7) as a result of fertilizer availability and improved crop strains.

Improved yields have been particularly important for poor countries, many of which have made substantial strides in expanding agricultural production to meet the nutritional

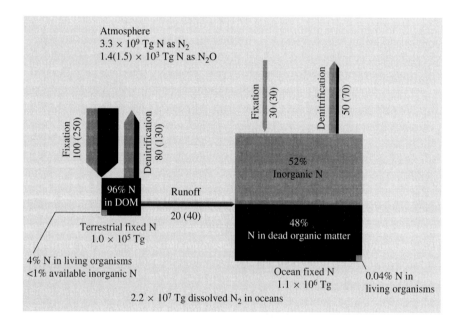

Figure 4.5   Global fluxes (Tg N/yr) and pools of nitrogen (Tg N). Light portions of arrows indicate preindustrial flows; dark portions and numbers in parentheses indicate current flows. DOM signifies *dead organic matter. Source:* Adapted from A. P. Kinsig and R. H. Socolow (1994). Human impacts on the nitrogen cycle. *Physics Today* November, 1994: 24–31. Reprinted with permission, copyright © 1994, Physics Today. All rights reserved.

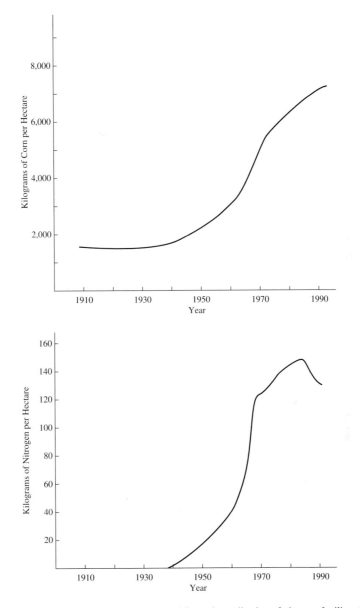

Figure 4.6   Historical trends for the United States in application of nitrogen fertilizer (in kg N per hectare) and yield of corn (in kg corn per hectare).   *Sources:* Fertilizer use rates are from D. Pimentel et al. (1973). Food production and the energy crisis. *Science* 182: 443–449; U.S. Department of Agriculture (1987). *Fertilizer Use and Price Statistics, 1960–85,* Statistical Bulletin No. 750 (Washington, DC: Economic Research Service); World Resources Institute (1994). Agricultural inputs, 1979–91 (Table 18.2). In *World Resources, 1994–95* (Oxford, UK: Oxford University Press). Corn yields are adapted from L. R. Brown (1994). Facing food insecurity. In *State of the World,* L. R. Brown et al. (New York: W. W. Norton). Reprinted with permission, copyright © 1994, Worldwatch Institute. All rights reserved.

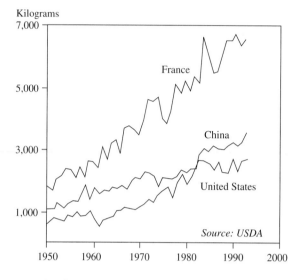

Kilograms

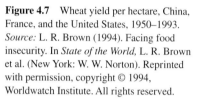

Figure 4.7 Wheat yield per hectare, China, France, and the United States, 1950–1993. *Source:* L. R. Brown (1994). Facing food insecurity. In *State of the World*, L. R. Brown et al. (New York: W. W. Norton). Reprinted with permission, copyright © 1994, Worldwatch Institute. All rights reserved.

needs of their growing populations. Much was accomplished in the 1960s through the "Green Revolution," when plant scientists working in research institutes in Mexico and the Philippines developed new strains of wheat and rice that produce high yields when fertilized. Traditional varieties of grain indigenous to tropical and subtropical countries do not respond well to fertilizer (Figure 4.8). They grow too tall and fall over in winds and heavy rain. The new strains are dwarf plants with stalks that remain short while the grain matures. An additional benefit of dwarf varieties is that a larger number of dwarf plants can be grown per unit-area than traditional, less compact varieties, which have large leaves and extended root networks to absorb the small amounts of nutrients and water often found in tropical soils. However, the higher plant density has increased the required amount of water and made farms more dependent on irrigation (see Figure 3.2, p. 195). In addition, the traditional plants were more resistant to indigenous pests, so the new plants had to be protected with pesticides (see discussion beginning on p. 274). Thus, the Green Revolution had substantial costs as well as benefits.

Despite the progress of recent decades, the steadily rising curves of world food production are beginning to level off, and it seems likely that per capita production will decline as the population continues to rise (Figure 4.9). This decline is due partly to a slowdown in the development of new plant varieties subsequent to the breakthroughs of the 1960s, and partly to the fact that crop yields can be improved only so far by fertilization. When sufficient fertilizer has been added to support plant growth at its maximum rate, then further additions either have no effect or actually inhibit growth. The curves of diminishing returns for corn in various soils are shown in Figure 4.10. It is clear that increasing the rate of fertilization from current levels in the United States would not increase corn yields significantly. In some areas rice yields have actually declined since about 1985, and there are indications from long-term experiments that higher yields induced by commercial fertilizers may deplete soils of other essential nutrients.

Intensive fertilization brings other problems as well. One is pollution of estuaries (see pp. 234–237) and aquifers as the ammonium and nitrate ions that are not taken up by the

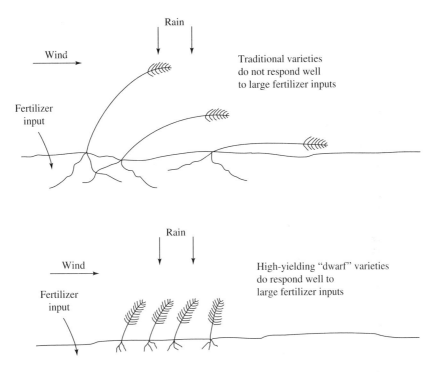

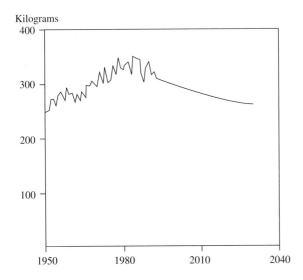

Figure 4.8 Comparison in response to fertilizer inputs between traditional varieties of grain and the dwarf varieties developed during the Green Revolution.

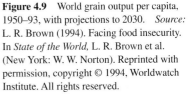

Figure 4.9 World grain output per capita, 1950–93, with projections to 2030. *Source:* L. R. Brown (1994). Facing food insecurity. In *State of the World,* L. R. Brown et al. (New York: W. W. Norton). Reprinted with permission, copyright © 1994, Worldwatch Institute. All rights reserved.

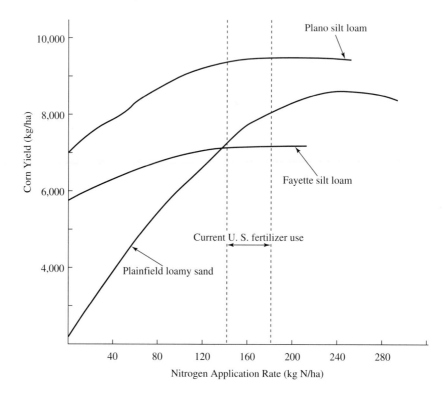

Figure 4.10   Yield response of corn to nitrogen inputs applied to three soils.   *Source:* S. L. Oberle and D. R. Keeney (1990). A case for agricultural systems research. *Journal of Environmental Quality* 20:4–7. Reprinted with permission, copyright © 1990, American Society for Agronomy, Crop Science Society of America, and Soil Science Society of America.

plants are washed away. Another disadvantage is the energy intensity of fertilizer production. The cost of commercial fertilizer is tied directly to that of natural gas, the source of the hydrogen required in the Haber process. During the petroleum crisis of 1973, gas prices soared and fertilizer became too expensive for third-world farmers. There were significant production declines and widespread famine.

Relative to the traditional fertilization methods of crop rotation and the application of manure, commercial fertilizers increase soil erosion. The traditional practices add complex organic matter to the soil, maintaining its structure; the soluble commercial fertilizers add no organic matter, making the soil more susceptible to being blown and washed away once the crop is harvested.

Erosion is not caused solely by commercial fertilizers, of course; the main factor is poor land-management practices. A big part of the problem is the cultivation of marginal lands that are not suited for agriculture, especially in poor countries with rapidly growing populations. In countries with population pressures, a vicious cycle ensues in which the newly cultivated land is rapidly degraded, thereby forcing the cultivation of still more marginal lands. Eventually, the soil can be re-formed by accumulation of dead organic matter and the weathering of minerals, but this takes a long time. Soil-erosion rates on U.S. croplands are estimated to be about 18 times faster than natural soil re-formation, while in Africa, Asia, and South America, the ratio of erosion to re-formation may be as high as 30 or 40 to 1.

Erosion and land degradation are serious problems. Erosion lowers crop productivity because of the loss of nutrients, soil biota, and organic matter, and especially because of the lowered availability of water, which is held less well by eroded soils. In addition, erosion greatly increases the sedimentation of rivers, reservoirs, and harbors. Sedimented stream bottoms prevent several species of fish (including salmon and trout) from reproducing. Severe salmon losses in the Northwest region of the United States have been attributed, in part, to stream sedimentation from agriculture and logging.

Improved land management can stem erosion losses substantially. For example, ridge planting and crop rotation combined with the use of winter cover crops can dramatically reduce erosion, when compared to standard monoculture with intensive fertilization. There is currently much interest in "low-input" or "organic" farming, in which the need for synthetic fertilizer is diminished or eliminated by crop rotation and the application of manure.

Improving the rates of natural nitrogen fixation and photosynthesis is a major topic of current agricultural research. One avenue of investigation involves experiments in plant breeding and genetic engineering designed to produce symbiotic association between nitrogen-fixing bacteria and cereal plants such as wheat and rice. If this work is successful, it opens the door to freeing the production of important cereal crops from the current dependence on nitrogen supplementation.

3. Nutrition

We now turn our attention to the biochemical pathways through which the food we eat keeps us functioning. This is the chemistry of biological metabolism.

The major nutritional categories are carbohydrates, fats, and proteins (Figure 4.11). Carbohydrates are sugar molecules linked together in a long chain. Fats are triglycerides of fatty acids, which have long hydrocarbon chains bonded to a glycerol unit. Proteins are composed of strings of amino acids joined by peptide bonds; each amino acid contains a characteristic side chain, R.

a. Energy and calories. A simplified flowchart for the main biochemical processes is given in Figure 4.12. Carbohydrates are the immediate product of photosynthesis and also the immediate source of biological energy in the process of respiration. Most of the energy that we need to keep our various bodily functions going is obtained from the oxidation of carbohydrates. Fats represent a form of biological energy storage. In comparison with carbohydrates, fats contain less oxygen and more carbon and hydrogen, whose oxidation is the source of our energy. The energy content of fats is 9 Calories/g, compared with 4 Calories/g for carbohydrates. Also, fats are immiscible with water, whereas carbohydrates are hydrophilic. Carbohydrates are usually found in association with a quantity of water that is approximately four times their weight. Consequently, the conversion of carbohydrates to fats represents a concentration of energy in lightweight portable form. A person who weighs 70 kg (154 pounds) contains about 16% fat. This represents enough energy for bodily requirements for 30 days. If this amount of excess energy were stored as carbohydrate with its

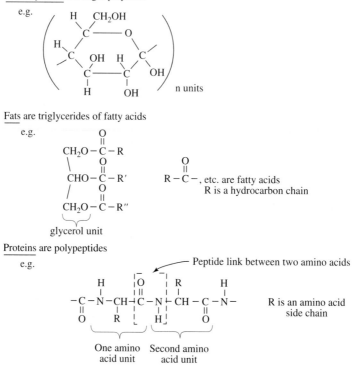

Figure 4.11   Chemical structures of carbohydrates, fats, and proteins.

associated water, the body weight would have to be 185 kg (406 pounds). When we eat more food than we require for our energy needs, the excess calories are stored as fat. However, the biological oxidation of fat is slower than that of carbohydrates. When we need energy, we burn off our carbohydrate supply first and then call upon the fat stores.

b. Protein.   Aside from an energy supply, we need to maintain the biochemical machinery of the body itself. This is the realm of protein chemistry. Proteins make up most of the body's structural tissues and also the myriad enzymes, the biological catalysts that carry out the thousands of reactions that are necessary for the maintenance of life. There are 20 different kinds of amino acids, each with a different chemical side chain (Figure 4.13). Each kind of protein molecule is made up of a fixed sequence of these amino acids. Most of the amino acids can be synthesized by the body from a variety of starting materials as long as there is an adequate supply of protein nitrogen in the diet. There are, however, eight amino acids (Figure 4.13 box) that the body cannot synthesize: valine, leucine, isoleucine, threonine, lysine, methionine, phenylalanine, and tryptophan.

These essential amino acids must be obtained directly from the diet and in amounts that allow us to maintain a proper balance of amino acids overall. For example, the ratio of

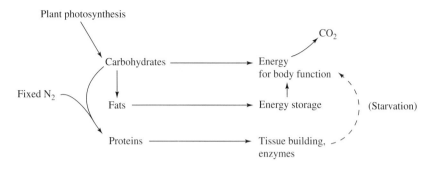

Figure 4.12   Flow chart of biochemical functions of carbohydrates, fats, and proteins.

tryptophan to lysine needs to be 0.6/2.6 = 0.23, because the average human protein contains 0.6 tryptophans and 2.6 lysines per 100 amino acids (Table 4.1). If this ratio is exceeded, the amount of protein that can be made by the body is limited by the lysine; the extra tryptophan is simply burned up. The notion of essential amino acid balance is similar to the concept of limiting nutrients, discussed earlier in connection with the growth of algae in water (pp. 231–234). Whichever amino acid is present in the lowest amount relative to the frequency with which it is used is the one that limits the total amount of protein produced.

Different food sources contain different amounts of protein; but they also vary in the amino acid composition of their protein. As might be expected, the amino acid composition of animal protein is fairly close to human protein. For this reason, milk and meat provide a rather close approximation to the amino acid balance we need (Table 4.1). Plant protein, on the other hand, is farther from the human composition. The cereal plants, particularly wheat, are quite deficient in lysine. The lysine frequency in wheat is less than half of that of the human average, which means more than twice as much wheat protein as milk protein is needed to supply human needs. Even though the remaining amino acids are closer to the correct proportions, they cannot be utilized without sufficient lysine. For this reason, wheat protein is often referred to as low-quality protein in comparison with the high-quality protein contained in meat, eggs, and milk.

Different plants show different patterns of deviation from the ideal amino acid balance. While wheat is deficient in lysine, beans have lysine in abundance. However, beans are deficient in methionine, which wheat has in abundance. In this respect, wheat and beans are complementary. If the two are mixed in equal amounts, the amino acid balance is greatly improved. It takes less total vegetable protein for the mixture than for either one alone to provide human requirements. A diet of half wheat protein and half bean protein provides a protein mixture that is only 10% less efficient than milk protein in providing the right balance. It is no accident that beans and rice or wheat products are traditionally eaten together in many parts of the world. In order to be effective complements, the different proteins must be mixed in the same meal, so that they are digested together. Thus, vegetarians must pay attention to balancing different protein sources.

Although meat is an important component in the diets of relatively prosperous nations, most people in the world cannot afford to buy it, and it is possible to do without meat altogether and still remain healthy. Moreover, although meat is a high-quality source of

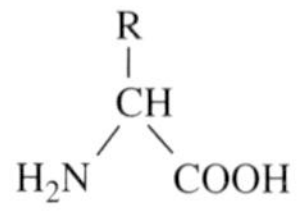

General formula

Amino acid*	R (side chain)
1. Glycine	—H
2. Alanine	—CH$_3$
3. Serine	—CH$_2$OH
4. Aspartic acid	—CH$_2$COOH
5. Glutamic acid	—CH$_2$CH$_2$COOH
6. Asparagine	—CH$_2$CONH$_2$
7. Glutamine	—CH$_2$CH$_2$CONH$_2$
8. Arginine	—CH$_2$CH$_2$CH$_2$NHC(NH)NH$_2$
9. Cysteine	—CH$_2$SH
10. Tyrosine	—CH$_2$ ◯ OH

| 11. Proline | 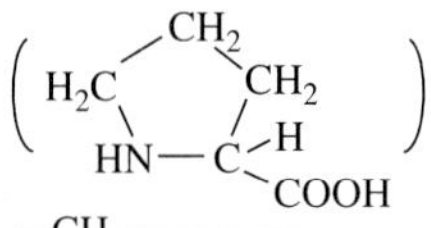 (structure) } General formula including R group |

| 12. Histidine | 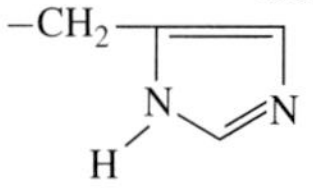 —CH$_2$ (imidazole ring) |

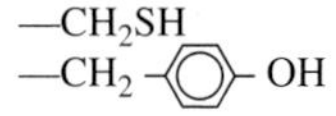

Essential amino acids	
13. Valine	—CH(CH$_3$)$_2$
14. Leucine	—CH$_2$CH(CH$_3$)$_2$
15. Isoleucine	—CH(CH$_3$)CH$_2$CH$_3$
16. Threonine	—CH(OH)CH$_3$
17. Lysine	—CH$_2$CH$_2$CH$_2$CH$_2$NH$_2$
18. Methionine	—CH$_2$CH$_2$-S-CH$_3$
19. Phenylalanine	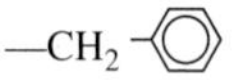 —CH$_2$ ◯
20. Tryptophan	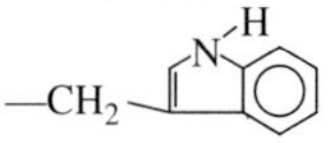 —CH$_2$ (indole ring)

*1–12 can be synthesized by the human body if sufficient protein nitrogen is present in the diet. 13–20 cannot be synthesized and must be obtained directly from food.

Figure 4.13   The amino acids and their chemical structures.

protein, it represents a considerable waste of biological energy because animals typically store a rather small fraction of the food they consume in the form of meat. It takes 10 g of plant protein to produce 1 g of beef protein. Most countries do not have enough plant protein to spare for meat production on a large scale. If there is sufficient grass to support grazing animals, they can make a net contribution to the human food supply. However, in meat-producing countries, it is common to feed animals high-quality plant food such as cereals and soybeans to fatten them more quickly.

TABLE 4.1   ESSENTIAL AMINO ACID CONTENT IN COMMON FOODS

Essential amino acid	Average frequency of occurrence in human protein (out of 100 amino acids)	Out of 100 amino acids number present in:			
		Cow's milk	Meat	Beans	Wheat
Tryptophan	0.6	0.5	0.5	0.4	0.6
Phenylalanine	3.1	3.8	2.7	2.8	2.8
Lysine	2.6	3.3	3.4	2.8	1.1
Threonine	2.0	2.0	2.0	1.6	1.3
Methionine	3.5	2.8	3.3	2.2	3.2
Leucine	4.0	4.2	3.3	3.1	3.1
Isoleucine	2.5	2.8	2.3	2.2	1.6
Valine	3.2	3.6	2.7	2.5	2.0

Sources: Data from F. E. Deathrage (1975). *Food for Life* (New York: Plenum Press); President's Science Advisory Committee (1967). *The World Food Problem, Report of the Panel on World Food Supply,* Vol. II (Washington, DC: President's Science Advisory Committee).

c. Minerals and vitamins.

The human diet must contain a balanced assortment of building materials for maintaining the biochemical machinery. The needed elements must be present in the correct proportions. They are H, C, N, O, P, S, Na, K, Mg, Ca, Fe, Zn, Cu, Co, Cr, Mo, Se, I, and perhaps other elements in exceedingly small amounts. Moreover the elements must be in assimilable chemical forms. We are unable to use carbon as CO_2 or nitrogen as NO_3^-, although plants do so. Also, ferric hydroxide is useless as a dietary source of iron because of its insolubility, and many soluble ferric salts are also ineffective, probably because they are converted to ferric hydroxide in the alkaline milieu of the intestine. Microbes and plants produce chelating agents, which extract iron from the ferric hydroxide in their vicinity, but these are not present in animal biochemistry.

Once the elements are assimilated, they are incorporated into the required biological molecules through myriad biosynthetic pathways. There are, however, some needed molecules that the body is incapable of synthesizing. These include the essential amino acids, which must be provided in the proper portions by dietary protein, and the vitamins.

The vitamins have been discovered through the disease states induced when they have been deficient in the diet. In 1747 James Lind discovered that citrus fruit was effective in treating British sailors who suffered from scurvy. In 1932 the active ingredient in citrus fruit was found by Albert Szent-Györgyi and by Charles King to be ascorbic acid, vitamin C. Thirteen vitamins are known today. The last one (vitamin B_{12}) was discovered 45 years ago, and it is unlikely that any more will be found. Many people have lived for years on intravenous solutions that contain only the known vitamins and other nutrients.

The vitamins fall into two classes: fat-soluble and water-soluble (Figure 4.14). The water-soluble vitamins either are enzyme cofactors or are required in cofactor synthesis. For example, niacin provides the pyridine end of NAD (nicotinamide adenine dinucleotide), while riboflavin is incorporated into FAD (flavin adenine dinucleotide). Both of these cofactors are used to catalyze many biological redox reactions. The roles of the fat-soluble vitamins are more complex. Vitamin A, or retinol, is incorporated into the visual pigment,

Vitamin	Structural formula	Dietary sources	Deficiency symptoms
Fat-soluble vitamins			
Vitamin A	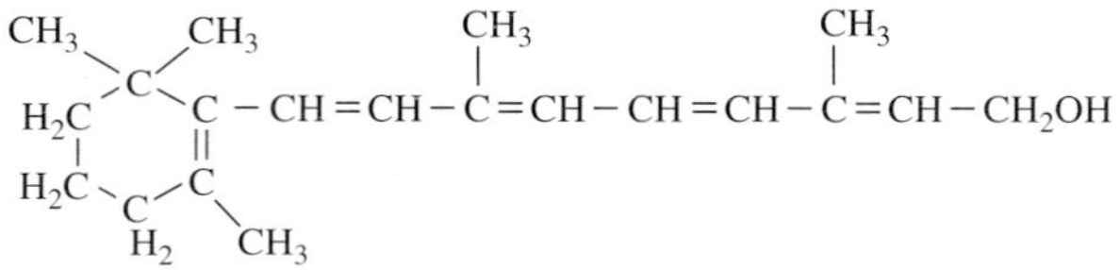 Retinol	Fish-liver oils, liver, eggs, fish, butter, cheese, milk. A precursor, β-carotene, is present in green vegetables, carrots, tomatoes, squash	Night blindness, eye inflammation
Vitamin D	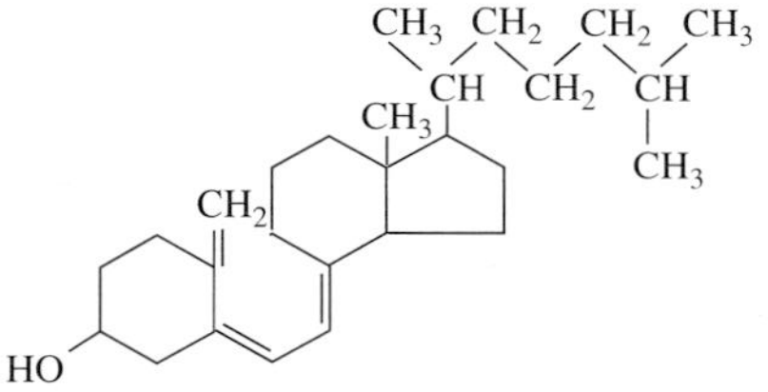 Vitamin D$_3$	Fish-liver oils, butter, vitamin-fortified milk, sardines, salmon. Body also obtains this compound when ultraviolet light converts 7-dehydrocholesterol in the skin to vitamin D	Rickets, osteomalacia, hypoparathyroidism
Vitamin E	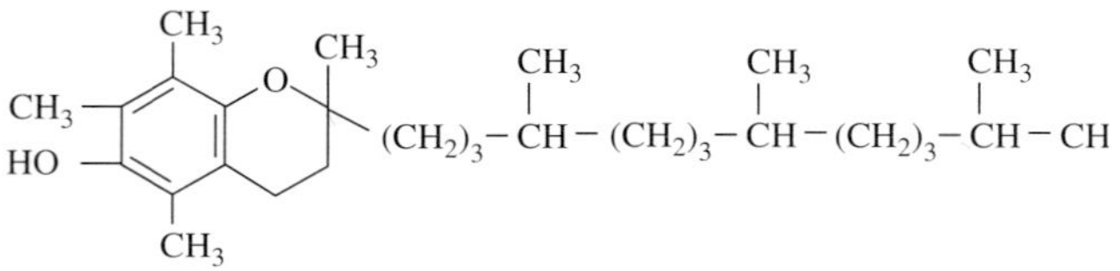 α-Tocopherol	Vegetable oils, margarine, green leafy vegetables, grains, fish, meat, eggs, milk	Anemia in premature babies fed inadequate infant formulas
Vitamin K	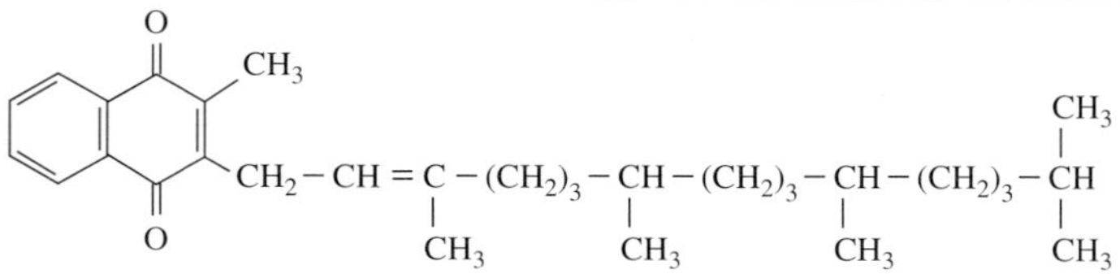 Vitamin K$_1$	Spinach and other green leafy vegetables, tomatoes, vegetable oils	Increased clotting time of blood, bleeding under skin and in muscles

Figure 4.14 Vitamins: structures, dietary sources, and deficiency symptoms. *Source:* H. J. Sanders (1979). Nutrition and health. *Chemical and Engineering News* 57(13):27–46. Reprinted with permission, copyright © 1979, American Chemical Society.

Water-soluble vitamins

	Structure	Sources	Deficiency
Thiamine (vitamin B_1)	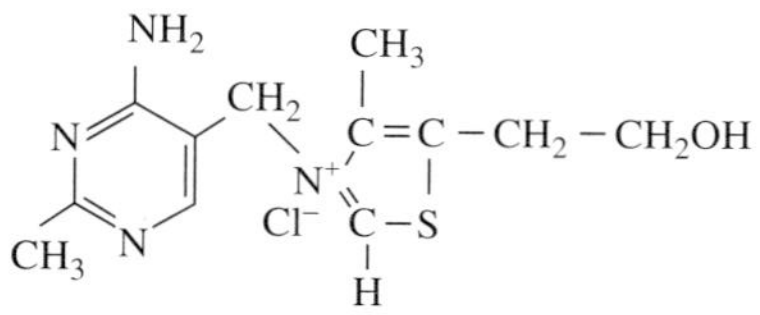 Thiamine chloride	Cereal grains, legumes, nuts, milk, beef, pork	Beriberi
Niacin (nicotinic acid)	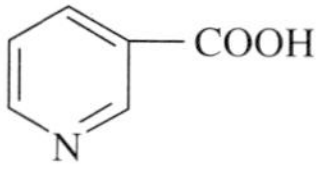 —COOH	Red meat, liver, turnip greens, fish, eggs, peanuts	Pellagra
Riboflavin (vitamin B_2)	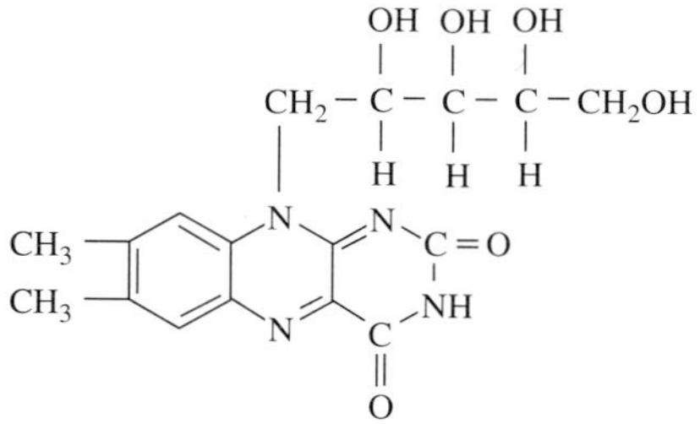	Milk, red meat, liver, green vegetables, whole wheat flour, fish, eggs	Dermatitis, glossitis (tongue inflammation), anemia
Pyridoxine (vitamin B_6)	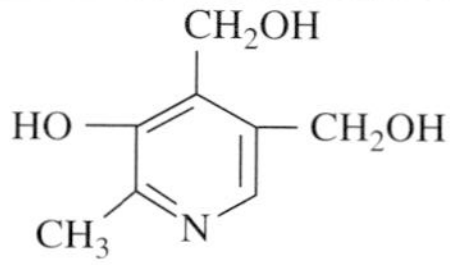 Pyridoxol	Eggs, meat, liver, peas, beans, milk	Dermatitis, glossitis, increased susceptibility to infections, irritability, convulsions in infants

Figure 4.14 (*continued*)

Vitamin	Structural formula	Dietary sources	Deficiency symptoms
Pantothenic acid	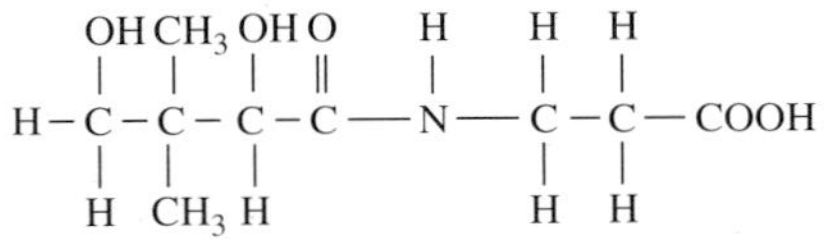 $\text{HO—CH}_2\text{—C(CH}_3)_2\text{—CH(OH)—CO—NH—CH}_2\text{—CH}_2\text{—COOH}$	Liver, beef, milk, eggs, molasses, peas, cabbage	Gastrointestinal disturbances, depression, mental confusion
Folic acid (pteroylglutamic acid)	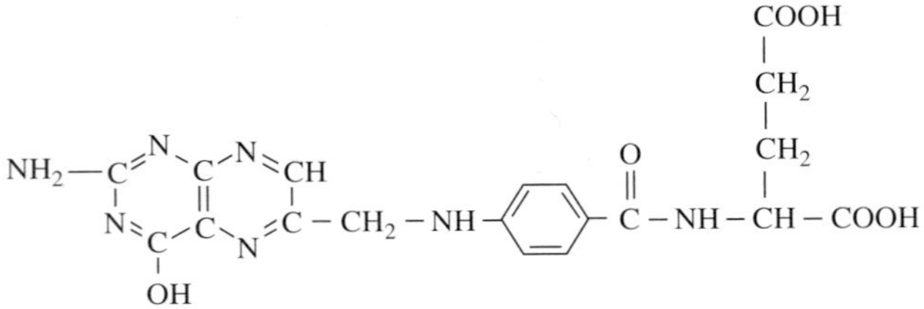 (pteridine ring)—CH_2—NH—C_6H_4—CO—NH—CH(—CH_2—CH_2—COOH)—COOH	Liver, mushrooms, green leafy vegetables, wheat bran	Anemias, gastro-intestinal disturbances
Biotin	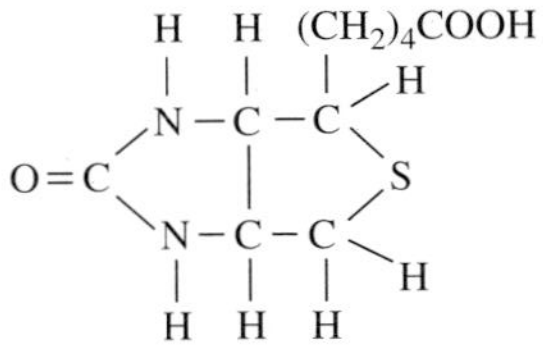 (bicyclic ureido-thiophene ring)—$(CH_2)_4$—COOH	Beef liver, kidney, peanuts, eggs, milk, molasses	Dermatitis

Figure 4.14 *(continued)*

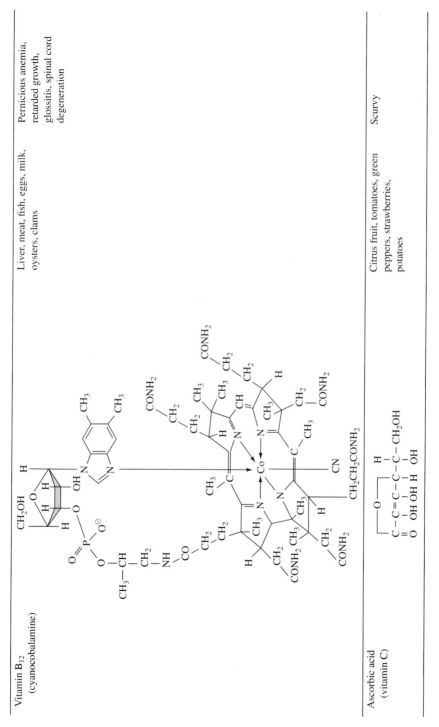

Vitamin B$_{12}$
(cyanocobalamine)

Liver, meat, fish, eggs, milk,
oysters, clams

Pernicious anemia,
retarded growth,
glossitis, spinal cord
degeneration

Ascorbic acid
(vitamin C)

Citrus fruit, tomatoes, green
peppers, strawberries,
potatoes

Scurvy

Figure 4.14 *(continued)*

rhodopsin. Vitamin D is required for the proper deposition of calcium in bones. Vitamin K is involved in the blood clotting mechanism.

d. Antioxidants. Vitamins E and A are natural antioxidants, protecting membranes from damage by molecular oxygen. So is β-carotene, the orange pigment of carrots and other vegetables, which is converted to vitamin A in the body (it is two molecules of retinol linked by a double bond, which replaces the OH group and an H atom on the terminal carbon atom). Vitamin C also has antioxidant properties and protects molecules in biological fluids. All of these antioxidants have double bonds, which readily react with free radicals (recall the role of olefins in the chemistry of smog, pp. 174–175).

There is much evidence implicating free-radical damage in the aging process and as a causative factor in a number of diseases, including cancer and heart disease. Radicals are generated in the body as side-products of O_2 reduction. The release of partially reduced O_2, superoxide and peroxide, can lead to the production of hydroxyl radicals by transition metal-catalyzed reactions of the kind discussed earlier in connection with transition metal activation of O_2 in raindrops (pp. 139–141). The extremely reactive hydroxyl radicals attack biological molecules in their vicinity, generating organic radicals, which react further in radical chain reactions. These reactions can lead to mutations and cancer, through damage to DNA, and to other disease states through damage to membranes or other critical parts of the biochemical machinery.

Because of the deleterious effects of radicals, organisms have evolved a variety of molecular defenses. These include enzymes that destroy the partially reduced O_2 species before they can generate hydroxyl radicals. These are *superoxide dismutases,* which catalyze superoxide disproportionation

$$2O_2^- + 2H^+ = O_2 + H_2O_2 \qquad (4.5)$$

and *catalases,* which catalyze peroxide disproportionation

$$2H_2O_2 = O_2 + 2H_2O \qquad (4.6)$$

Together, these two classes of enzymes ensure that partially reduced forms of oxygen are converted to O_2 and H_2O. This conversion is possible because these partially reduced forms are unstable with respect to O_2 and water; it takes more energy per electron to add one electron to O_2 (superoxide), or two of them (peroxide), than to add four of them (water). Another strategy for avoiding radicals is to make sure that redox-active metals, such as iron, copper, and manganese, are tied up in *chelates* (see discussion of chelating agents and detergents, p. 217), so they cannot bind and activate O_2. In biological systems, transition metals do not circulate as free ions, but are bound and chelated to proteins and to smaller organic molecules. This sequestering of metals has been demonstrated to be important for avoiding radicals; in experimental systems, overwhelming the natural chelation system by adding excess metal salts to biological tissues can provoke free-radical damage.

The last line of defense against radicals are the antioxidants, like vitamins E, A and C, which react sacrificially with radicals before they can reach biological targets. Not all antioxidants in the diet are vitamins, since not all antioxidants are specifically required for essential biological functions. These nonvitamin antioxidants include both naturally occurring molecules and synthetic molecules added to processed foods to preserve them from rancidity caused by oxidative damage. Synthetic antioxidants include butylated hydroxytoluene (BHT) and butylated hydroxyanisole (BHA) (Figure 4.15). These phenolic compounds react readily with radicals by donating a hydrogen atom; the resulting phenoxyl radical is stabilized by the electron-donating alkyl and methoxy substituents on the benzene rings.

Minimum required doses of the vitamins have been set by examining the levels below which deficiency diseases set in. However, the optimum level of dietary vitamins is still intensely debated. There have been numerous claims that supplements of vitamins higher than the minimum doses produce beneficial health effects. The most illustrious proponent of vitamin therapy was Linus Pauling, who argued forcefully that the optimum level of vitamin C is much higher than the minimum level, and that large doses can prevent colds and even cancer. Clinical studies, however, have been equivocal. In general, there is a growing body of evidence that antioxidants may decrease the incidence of various diseases, including cancer and heart disease. But even this conclusion has been clouded by results reported in 1994 of a study on 29,000 male smokers in Finland, which showed an 18% higher incidence of lung cancer in those who took β-carotene supplements than in those who did not. And a study published in 1995 reported that excessive amounts of vitamin A consumed during the early months of pregnancy increase the risk of birth defects (although β-carotene consumption was not associated with this risk).

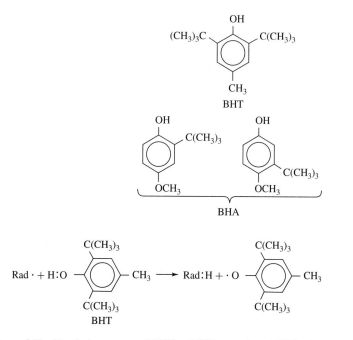

Figure 4.15   Chemical structures of BHT and BHA; reaction of BHT with a radical. *Source:* J. W. Hill and D. K. Kolb (1992). *Chemistry for Changing Times,* 7th Edition (Upper Saddle River, New Jersey: Prentice Hall). Copyright © 1994 by Prentice Hall, reprinted with permission.

These cautionary results have led health authorities to withhold recommendations on antioxidants. At present they are able to conclude only that the diet should contain adequate fresh fruit and vegetables, which supply needed vitamins. Fresh produce also supplies many other natural chemicals that are likely to have beneficial effects; it was recently discovered, for example, that a compound in broccoli (sulforophane) can block tumor growth in experimental rats.

B. INSECTICIDES AND HERBICIDES

1. Insect Control

A serious limitation on the human food supply is that we must share plant foods with insects. It has been estimated that the weight of the world's insect population exceeds that of its human inhabitants by a factor of 12. Only a small fraction of the insect species, about 500 species out of the world total of five million, actually feed on human crops, but these have the potential of doing enormous damage. Indeed, it is estimated that 30% of agricultural crops are consumed by insects worldwide.

Moreover, some insect species, including mosquitoes, fleas, and tsetse flies, are carriers of devastating human diseases. The death toll through the ages from such insect-borne diseases as malaria, yellow fever, bubonic plague, and sleeping sickness has been much larger than from warfare. Insect invasions, sometimes on a massive scale, have been a recurrent part of human history.

a. Persistent insecticides: organochlorines.   Attempts to combat insect pests were relatively ineffective until the development of modern chemical pesticides. The first of these was para-dichlorodiphenyltrichloroethane, DDT (Figure 4.16). Introduced by the Allies during World War II to control typhus and malaria outbreaks, DDT has since saved millions of additional lives through disease-vector control. Its discoverer, the Swiss chemist Paul Muller, won the Nobel Prize for Medicine and Physiology in 1948. After the war, DDT was the first pesticide to come into widespread agricultural use.

In many respects DDT is an ideal insecticide. It is chemically stable and degrades only slowly under environmental conditions. Having low volatility, it also evaporates slowly, and it is not readily washed away because of its low solubility in water. These three characteristics makes it a persistent insecticide. Each application is effective for a long time.

Because it is hydrophobic, DDT readily penetrates the waxy outer coating of insects and, once inside, quickly paralyzes the insect. DDT acts by binding to the nerve cells of insects in a way that holds open the molecular channels that admit sodium ions, which in turn leads to uncontrolled firing of the nerves. DDT's toxicity to animals, including humans, is low, because animals absorb much less of the chemical in their tissues. It is this combination of persistence and selective toxicity to insects that made DDT such a successful insecticide.

However, it was not long before DDT began to lose its effectiveness because of the buildup of insect resistance; its use had begun to decline by 1960, over a decade before it was banned for most uses in the United States and other industrialized countries (see below).

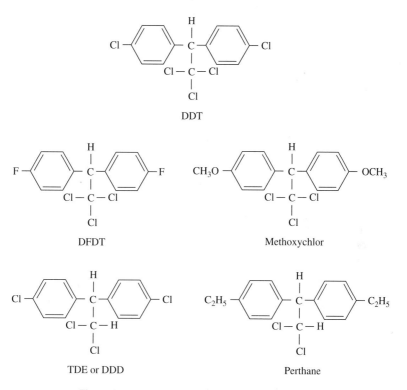

Figure 4.16   Chemical structures of DDT and its analogs.

Under heavy DDT application, individual insects that are relatively resistant to it are more likely to survive than those that are susceptible, and new generations have a steadily higher incidence of resistant characteristics. The main resistance factor for DDT is an enzyme called *DDT-ase,* which catalyzes the *dehydrochlorination* (loss of H and Cl atoms) of DDT to form DDE (Figure 4.17), which has a new double bond at the central carbon atom. Because this atom is now trigonal instead of tetrahedral, DDE has a distinctly different shape than DDT and no longer binds strongly to insect nerve cells. Thus, insects that have evolved the ability to make high levels of DDT-ase can transform DDT into an innocuous molecule.

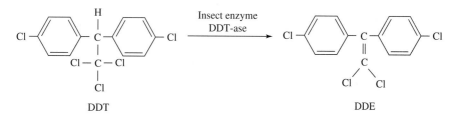

Figure 4.17   Mechanism for insect resistance to DDT.

Because of insect resistance, new insecticides were developed to supplement and replace DDT in various applications. The first of these were variations on the DDT theme (Figure 4.16), molecules that retain the DDT framework but have different chemical substituents that are not as rapidly attacked by resistant insects' DDT-ase. Some substitutions worked while others did not; for example, replacing the chlorine atoms at the ends of the molecule with hydrogen atoms greatly decreased insecticidal activity. Finding effective variations of DDT was a matter of trial and error, because the nature of the binding site on the insect nerve cells was unknown. The more successful variants were brought into widespread use, until resistance to them also built up in the insect population. The nature of resistance means that no insecticide can remain effective for long, and it spurs the drive to develop new insecticides.

It was soon discovered that other organochlorine molecules, quite different from DDT, were also insect neurotoxins. Several of these were products of an addition reaction (called a *Diels-Alder* addition) between perchlorocyclopentadiene and an olefinic molecule (Figure 4.18). These "cyclodiene" insecticides also came into widespread use, as did toxaphene, a complex mixture resulting from the reaction of the naturally occurring hydrocarbon camphene with chlorine.

b. Ecosytem effects: bioaccumulation. DDT's success came at a price. Because insects are part of a very complex web of predator-prey relationships, a broad-spectrum insecticide like DDT is bound to have effects on the entire ecosystem. For example, when DDT was introduced in Borneo, as part of a malaria eradication campaign by the World Health Organization in the 1960s, mosquitoes were suppressed, but so were other insect species, including a wasp that preyed on the caterpillars that lived in the thatched roofs of village houses. Without the wasps, the caterpillar population exploded, and the thatch roofs were consumed. In addition, the dead mosquitoes were eaten by gecko lizards, which became sick and were easy prey for village cats. As a result of eating all the sick lizards, the cats in turn sickened and died, resulting in an explosion of the rat population. The rats ate the local crops and threatened an outbreak of bubonic plague. The Borneo government had to reintroduce cats into the affected region.*

The mosquito-lizard-cat connection illustrates the important principle of *bioaccumulation* in *food chains*. Chemicals in the prey become concentrated in the predator if the chemicals are not broken down and excreted rapidly. This is the case for persistent hydrophobic chemicals, which accumulate in the fat tissues of the predator. When the predator is eaten in turn, the chemical concentrates further in the fat of *its* predator. Each link in the food chain provides successive concentration. Figure 4.19 illustrates the accumulation of DDT in a typical aquatic food chain. If the plankton contain 0.04 ppm DDT, then the clams feeding on the plankton have 10 times as much, while fish can have five times more if they feed on the clams. Finally, the fish-eating birds, at the top of the food chain, can build up quite high levels, up to 75 ppm in their fat tissue.

DDT can have biochemical effects other than neurotoxicity. One of them seems to be disruption of the avian hormonal system that controls calcium deposition during egg forma-

---

*P. R. Ehrlich and A. H. Ehrlich (1981). *The Causes and Consequences of the Disappearance of Species* (New York: Random House).

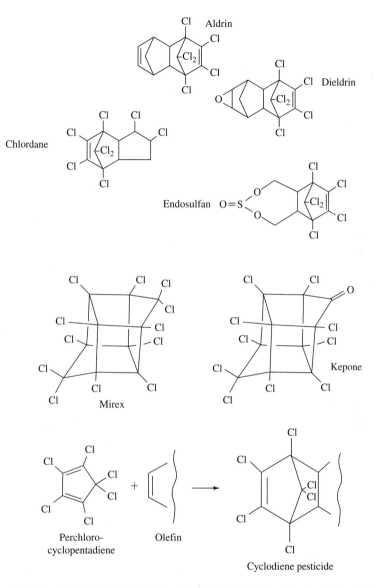

Figure 4.18   Cyclodiene insecticides, formed via the Diels-Alder condensation (bottom) of perchlorocyclopentadiene with an olefinic molecule.

tion. As a result, birds having high levels of DDT (or its metabolic product, DDE, which has the same effect) lay eggs with shells that are too thin to endure until hatching. The population of the peregrine falcon and of other birds of prey fell precipitously in the years following the widespread use of DDT.

In 1962 Rachel Carson's book *Silent Spring* brought the ecological dangers of the uncontrolled use of insecticides sharply into public view, and subsequent legal and political

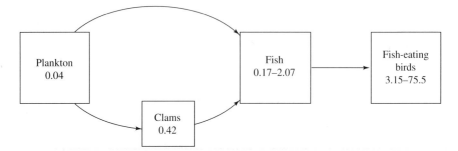

Figure 4.19   Accumulation of DDT in the aquatic food chain (in units of parts per million). *Source:* Adapted from C. A. Edwards (1973). *Persistent Pesticides in the Environment,* 2nd edition (Cleveland, Ohio: CRC Press). Reprinted with permission, copyright © 1973 by CRC Press.

actions led to severe restrictions on the use of DDT and other persistent insecticides in many countries during the 1970s. Since then the threatened bird populations have recovered significantly in these areas. However, many other countries continue to use DDT, particularly in areas where malaria remains endemic. DDT is relatively cheap, and remains reasonably effective in mosquito control. Indeed, stopping DDT use can have dire consequences, as in the case of Sri Lanka, where a DDT-based mosquito control program had reduced the number of reported cases of malaria from 2,800,000 in 1948 to just 17 in 1963. After spraying was stopped in 1964, largely because of the political response to *Silent Spring,* malaria quickly returned and reached 2,500,000 cases by 1969. A new control program brought some improvement, but malaria remains a serious problem.*

To get around the bioaccumulation problem, it is possible to engineer a molecule that has DDT's effectiveness in some insecticidal applications, but is less persistent. One example is methoxychlor (Figure 4.16), in which the para-Cl atoms of DDT are replaced by methoxy groups, $-OCH_3$. The methoxy groups increase water solubility and susceptibility to degradation reactions. Consequently, methoxychlor does not bioaccumulate significantly and continues to be approved for use against flies and mosquitoes. However, the nonpersistent character of methoxychlor makes it more expensive to use since a single application is not as long-acting. As usual, there is no "free lunch" in environmental protection, and difficult choices often have to be made.

c.   Nonpersistent insecticides: organophosphates and carbamates. Other insecticides have been developed that are nonpersistent because they break down rapidly into harmless and water-soluble products, once released into the environment. Because they do not last long, they must be highly potent. The two classes of widely used nonpersistent insecticides—organophosphates and carbamates (Figure 4.20)—are in fact powerful neurotoxins. Indeed, the organophosphate insecticides are in the same chemical family as the nerve-gas chemical-warfare agents that were developed during and after World War II.

*K. Mellanby (1992). *The DDT Story* (Farnham, Surrey, UK: British Crop Protection Council).

(a)
Organophosphates

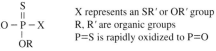

$$
\begin{array}{c}
\quad\quad S \\
\quad\quad \| \\
RO - P - X \\
\quad\quad | \\
\quad\quad OR
\end{array}
$$

General formula

X represents an SR′ or OR′ group
R, R′ are organic groups
P=S is rapidly oxidized to P=O

examples:

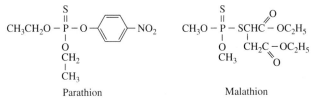

Parathion   Malathion

(b)
Carbamates

$$
\begin{array}{c}
\quad\quad O \\
\quad\quad \| \\
RO - CNHR′
\end{array}
$$

General formula

examples:

Carbaryl   Aldicarb

Figure 4.20 Chemical structures of organophosphate and carbamate insecticides.

The organophosphates and carbamates both work by inhibiting the enzyme *cholinesterase,* which hydrolyzes the neurotransmitter *acetylcholine.* Neurotransmitters are molecules that are released by a nerve cell in order to fire an adjacent nerve cell (Figure 4.21) by diffusing across the gap between the cells, called the *synapse,* and binding to receptors on the second cell. There are many kinds of neurotransmitter molecules, but the one responsible for firing motor nerve cells in higher life forms is acetylcholine. Once acetylcholine binds to its receptors, a motor nerve cell will continue to fire until the acetylcholine is hydrolyzed by cholinesterase, which is present in the synapse. If the enzyme is inhibited, then nerve firing continues uncontrollably, leading to paralysis and death.

How do the insecticides inhibit cholinesterase? The enzyme works by binding acetylcholine, and then carrying out a displacement reaction on the acetyl group, using the OH group of a serine amino acid residue that is located at the enzyme active site (Figure 4.22). The choline part of the molecule is released in this reaction, leaving the acetyl group bound to the enzyme. Next, the enzyme induces a water molecule to carry out a second

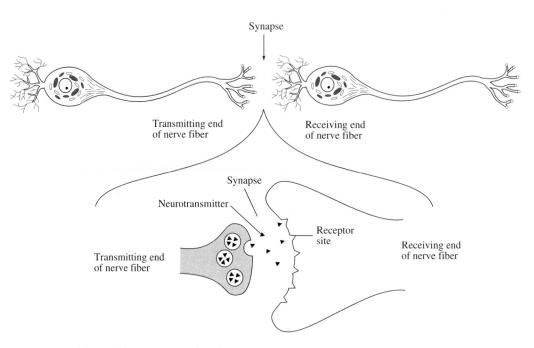

Figure 4.21 The transmission of a nerve impulse by release of neurotransmitter molecules across the synapse. *Source:* P. Buell and J. Gerard (1994). *Chemistry in Environmental Perspective* (Upper Saddle River, New Jersey: Prentice Hall). Copyright © 1994 by Prentice Hall, reprinted with permission.

displacement at the acetyl group, resulting in the release of acetic acid. The enzyme is then ready to carry out a second round of catalysis. The insecticide molecules trick the enzyme by mimicking acetylcholine. They bind to the active site and induce the serine residue to carry out a displacement reaction, just as acetylcholine does. But instead of an acetyl group, the enzyme ends up with a bound phosphoryl group in the case of an organophosphate insecticide, or a bound carbamyl group in the case of a carbamate insecticide (Figure 4.22). These groups are much less susceptible to attack by water than is the acetyl group. In both cases the atom being attacked, carbon in the case of carbamyl and phosphorus in the case of phosphoryl, is attached to additional electronegative groups that inhibit the nucleophilic attack by the incoming water molecule. Consequently, the enzyme is blocked by the carbamyl or phosphoryl group from hydrolyzing acetylcholine. The second step in the enzymatic reaction is greatly slowed by these groups, resulting in a much lower catalytic rate.

The potency of the inhibitor depends on the rate of the first reaction, in which the enzyme's active-site serine is captured. The better the *leaving group* (X for the organophosphates, OR for the carbamates—Figure 4.22), the faster the reaction. For example, fluoride is an excellent leaving group because the fluoride ion is quite stable in water (as opposed to methoxide, for example, which is a strong base and is harder to displace). Organophosphate molecules with fluoride substituents are extremely powerful cholinesterase inhibitors, and

Normal mode of action

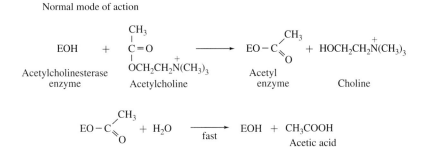

Inhibition by organophosphate insecticide

$$EOH + X-\underset{\underset{O}{\|}}{\overset{\overset{OR}{|}}{P}}-OR' \longrightarrow EO-\underset{\underset{O}{\|}}{\overset{\overset{OR}{|}}{P}}-OR' + HX$$

Organophosphate   Phosphoryl
                                                  enzyme

$$EO-\underset{\underset{O}{\|}}{\overset{\overset{OR}{|}}{P}}-OR' + H_2O \xrightarrow{slow} EOH + HO-\underset{\underset{O}{\|}}{\overset{\overset{OR}{|}}{P}}-OR'$$

Inhibition by carbamate insecticide

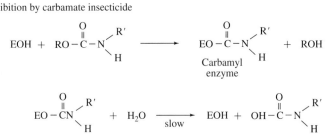

Figure 4.22   Mechanisms of cholinesterase inhibition by organophosphate and carbamate insecticides.

the nerve gases are molecules of this class (an example is sarin, methylisopropoxyfluorophosphate).

In order to use organophosphates as insecticides without massively poisoning people and other animals, the reactivity of the organophosphate must be toned down. An effective way to do this is to replace the P=O group with the P=S group (producing a *phosphorothioate*). The S atom significantly deactivates the P atom toward nucleophilic attack and slows the rate of reaction with cholinesterase. However, once inside the insect, the S atom is rapidly attacked by oxidative enzymes, and the molecule is converted back to a potent organophosphate neurotoxin. Animals also have oxidative enzymes, but at levels much lower than in insects. Thus, the phosphorothioate turns the insect's biochemistry on itself and significantly decreases toxicity toward other species.

Nevertheless, the organophosphate and carbamate insecticides are generally much more toxic than organochlorine insecticides. Some of the most widely used ones, such as parathion and aldicarb, are highly toxic and have caused death and injury to many agricultural workers. Thus, the environmental advantage of these nonpersistent agents is counterbalanced by their health impacts on agricultural workers. In the case of Sri Lanka, mentioned above, DDT spraying for malaria control was replaced by parathion spraying, which resulted in many deaths among spraying crews, whereas none had been caused by DDT.

It is sometimes possible to introduce added safety by clever molecular engineering. A good example is malathion (Figure 4.20), whose toxicity is several hundredfold less than parathion. Its diesterthiomethyl substituent is a good leaving group, making malathion an effective neurotoxin once the P=S group is replaced by P=O. But the ester groups are rapidly hydrolyzed by *carboxylase* enzymes, and the resulting dicarboxylthiomethyl substituent is a poor leaving group because of its negative charge. These enzymes are abundant in animals but not in insects. Thus, the biochemical differences among organisms again are exploited to improve the selectivity of the toxin.

d. Natural insecticides: pyrethroids.   Many plants have evolved their own chemical defenses against insects, and there is interest in using these natural and environmentally benign insecticides. Usually the plants' molecules are too difficult to extract and/or too complicated to manufacture on a commercial scale. But they may inspire chemists to make new kinds of insecticides. Pyrethroids (Figure 4.23) are an example. Used by humans for centuries, pyrethrin is obtained from *pyrethrum,* a daisy-like flower. It is available commercially but is of limited use because it is unstable to sunlight; pyrethrins are *too* nonpersistent to be effective. Recently, however, pyrethrin-like insecticides have been developed that are chemically modified to improve their stability in the environment.

e. Pheromones: biological controls, and integrated pest management. Insect resistance remains a problem for all insecticides. Not only do the chemicals become progressively less effective, but they also sometimes remain more effective against natural insect enemies of the target pest, thereby actually making the problem worse than it was at the beginning. This is because the predator species are usually slower to reproduce than the prey species, so resistance takes longer to develop among them. A fairly common pattern is

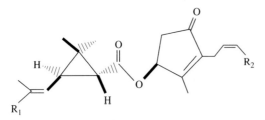

Figure 4.23   General structure of pyrethroids (R_1 and R_2 are alkyl groups).

that the introduction of a new insecticide causes an immediate decline in the population of the insect to be controlled, followed a few years later by a population explosion of a strain of the same insect against which the insecticide is no longer effective.

For this reason, increasing attention has been given to the development of insect-control methods that operate more selectively. For example, genetic engineering has been pressed into service to develop crop plants which harbor genes from bacteria that produce natural toxins harmful to insects. These toxins effectively deter insect pests from feeding on the crop. Although insect resistance can still develop, since resistant insects will have better access to a food supply, this is likely to happen more slowly than if the crop area were sprayed with the toxin. Moreover, ecosystem effects should be minimized.

Research on the biochemistry of insects has led to the discovery of hormones that control their growth and sexual behavior. If applied at the right time, these chemicals can disrupt the insect's life cycle. For example, juvenile hormones (Figure 4.24) regulate growth and can disrupt the synchronization of metamorphosis if they are applied externally.

Many insects rely on *pheromones,* molecules that act as messengers between individuals, guiding them to each other or to their food supply (Figure 4.24). Application of sex attractants can confuse the insects and prevent them from finding mates. It is also feasible to use pheromones to bait insect traps that contain high concentrations of toxic chemicals and are therefore much more effective in killing the insects than is broad-scale spraying.

Another technique of insect control is using chemicals or radiation to sterilize large numbers of male insects, which are bred for the purpose. When these sterile insects are released, they mate with the native population but produce no offspring. As a result, the total insect population is substantially reduced. For example, this strategy has been used to control the Mediterranean fruit fly in California. It is also possible to introduce predators artificially to control the population of pests and insects. This solution requires caution, however, because the predator can become so well adapted that it becomes a bigger pest than the target insect.

These new methods of controlling specific insect species are still under development. They require careful planning and timing and are much more sophisticated than simply spraying a field with a chemical insecticide. Because they are effective against only one species at a time, they are costlier than applying the broad-spectrum insecticides. Nevertheless, with the increasing environmental costs and decreasing effectiveness of the traditional insecticides, pressure is mounting for their development and use.

These pressures have actually produced a significant decline in insecticide use in the United States since 1975 (Figure 4.25). Massive spraying has in many cases been replaced by integrated pest management (IPM) strategies, which utilize a combination of tactics, including crop rotation (which helps to forestall infestation by resistant species), tillage practices, water management, biological controls, and insecticides. Central to IPM is the concept of an optimal treatment threshold; the aim is not to completely eradicate the pest species, but rather to hold its population below a given damage level. IPM has been successfully applied to major crops, including cotton. In the mid-1970s cotton accounted for over 40% of all insecticides used in the United States, but by 1982 the application rate had declined fourfold, from about 6.5 to 1.7 kg insecticide per hectare.

Juvenile hormones: natural insect growth regulators

Example:

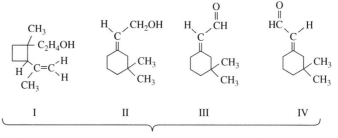

General formula

Specific formula: Controls stages of growth in:

R = H , R′ = OCH$_2$CH$_3$ Potato aphid, cockroach, grain-eating beetles

R = H , R′ = OCH$_2$C ≡ CH   Green peach aphid, pea aphid, citrus mealybug

R′ = OCH$_3$, R′ = OCH(CH$_3$)$_2$ Mosquito, apple maggot, Mediterranean fruit fly

(a)

Pheromones: natural insect sex attractants

Example:

I   II   III   IV

Mixture of molecules emitted by male boll weevil as sex attractant

(b)

Figure 4.24   Use of (a) juvenile hormones and (b) pheromones in insect control.   *Sources:* (a) C. Hendrik et al. (1976). Insect juvenile hormone activity of alkyl (2E,4E)-3,7,11-trimethyl-2,4-dodecadienoates. Variations in the ester function and in the carbon chain. *Journal of Agriculture and Food Chemistry* 24: 207–218. (b) R. D. Henson et al. (1976). Identification of oxidative decomposition products of the boll weevil pheromone, grandlure, and the determination of the fate of grandlure in soil and water. *Journal of Agriculture and Food Chemistry* 24: 228–231. Copyright © 1976 by American Chemical Society. Reprinted with permission.

2. Herbicides

Weeding out undesirable plants from crops is an essential aspect of food production. In commercial agriculture, manual and mechanical weeding has been widely replaced by using herbicides. In addition to saving labor, herbicides are applied in "no-till" agriculture, which

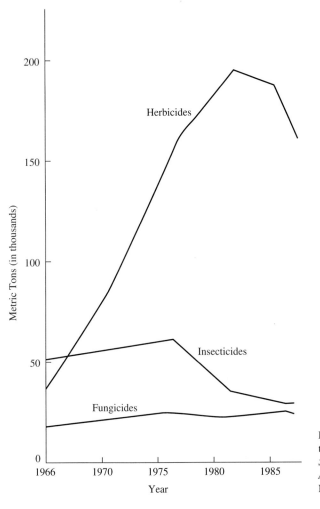

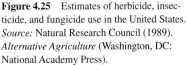

Figure 4.25   Estimates of herbicide, insecticide, and fungicide use in the United States. *Source:* Natural Research Council (1989). *Alternative Agriculture* (Washington, DC: National Academy Press).

minimizes erosion by reducing disturbance of the soil. In this practice, fields are planted without plowing by injecting seeds directly into the ground, after weeds have been controlled with herbicides. Broad application of no-till agriculture since the mid-1960s largely explains the dramatic expansion of herbicide use in the United States (Figure 4.25). The lion's share of this increase (80% by the mid-1980s) was attributable to just two crops, corn and soybeans (Figure 4.26). No-till agriculture requires high levels of herbicides because weeds must be thoroughly eradicated; even a small percentage of weeds remaining after herbicide application produces enough seeds to restore the weed population and choke off the crop. Nevertheless, herbicide use has been declining since the early 1980s, partly because of concerns about health and ecosystem effects similar to those surrounding insecticides.

The largest class of herbicides are the *triazines* (Figure 4.27), of which the best known is *atrazine,* the main agent used in cornfields. Atrazine does not bioaccumulate significantly,

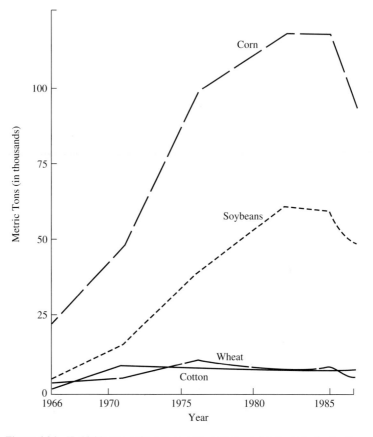

Figure 4.26   Herbicide use estimates on corn, cotton, soybeans, and wheat. *Source:* Natural Research Council (1989). *Alternative Agriculture* (Washington, DC: National Academy Press).

being moderately soluble in water, but it is fairly persistent and is often detectable in farm wells. Although it is not very toxic (LD_{50} = 1870 mg/kg—see next section and Table 4.2), there is some concern about correlations of high well-water concentrations with cancer and birth defects. Atrazine is being replaced in some areas by *metolachlor* (Figure 4.27), which degrades more rapidly in the field.

Another herbicide, *paraquat* (Figure 4.27), has been used extensively in marijuana-eradication campaigns. It is quite toxic and is suspected of having produced lung damage in some marijuana smokers. It is also used in developing countries to combat louse infestations and has produced cases of paraquat poisoning there.

The chlorophenoxy compounds 2,4-D and 2,4,5-T (Figure 4.27) are effective herbicides, 2,4-D for weeds in lawns and 2,4,5-T for clearing brush. 2,4-D is still used in large quantities, but 2,4,5-T was banned in the 1970s because of dioxin contamination (see below, pp. 302–309). A mixture of the two herbicides, the infamous Agent Orange, was used ex-

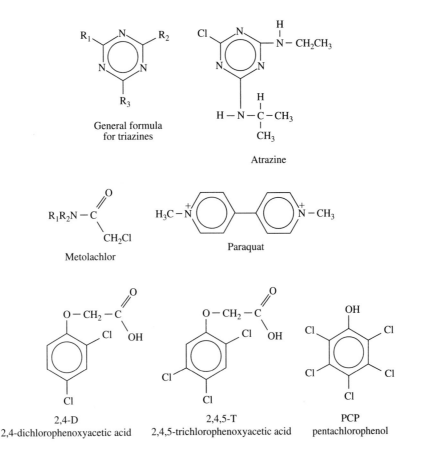

Figure 4.27 Chemical structures of selected herbicides, and the wood preservative pentachlorophenol (PCP).

tensively as a defoliant during the Vietnam War. Another member of this class of chemical, pentachlorophenol (PCP) (Figure 4.27), is widely used as a wood preservative.

A notable development is the recent introduction of a new, apparently environmentally friendly compound, glyphosate, (N-(phosphonomethyl)-glycine, $^-HO_3PCH_2N^+H_2CH_2COOH$). A simple amino acid derivative, glyphosate is water-soluble but clings to soil. Because of its ionic structure, it is attracted to ion-exchange sites on soil particles, and is therefore not washed into groundwater. Glyphosate is readily metabolized by soil microorganisms and has a half-life in soils of about 60 days, on average. It does not bioaccumulate to any significant extent.

When sprayed on plants, glyphosate enters the leaves and inhibits an enzyme that is required for the synthesis of the aromatic amino acids, phenylalanine, tyrosine, and tryptophan. Blockage of this essential biosynthetic pathway kills the plant. The pathway is different in animals and plants, and animals are unaffected by glyphosate, which they rapidly eliminate. However, all plants are affected, and glyphosate cannot be used to kill weeds selectively among crops. It is effective in clearing all vegetation from an area, and it is increasingly used to prepare fields for no-till agriculture. Once glyphosate has been applied, the field can safely be planted because new plants do not absorb glyphosate from the soil; the molecules are held too tenaciously by the soil particles.

Glyphosate is also useful in combating exotic species of plants that invade a habitat and crowd out native plants. For example, glyphosate was used to restore Wingham Brush, a nine-hectare remnant of Australia's rainforest, in the Manning River Valley of New South Wales. Wingham Brush is an important habitat of the flying fox, a native species of bat. The flying fox population had been threatened by foreign weeds that were smothering the trees where the bats live. These weeds were cleared by the application of glyphosate, allowing the indigenous plants to recover.

C. TOXIC CHEMICALS

Toxic chemical species have been discussed at many points in this book, and we now direct our attention to the many ways that chemicals can be harmful to living things.

1. Acute and Chronic Toxicity

It is useful to distinguish between an acute effect, in which there is a rapid and serious response to a high but short-lived dose of toxic chemical, and a chronic effect, in which the dose is relatively low but prolonged, and a time lag occurs between initial exposure and the full manifestation of the effect. Acute poisons interfere with essential physiological processes, leading to a variety of symptoms of distress and, if the interference is sufficiently severe, to death. Chronic toxins have more subtle effects, often setting in motion a chain of biochemical events that leads to disease states, including cancer.

Sorting out these effects, the province of toxicology and epidemiology, is not an easy matter. The body's biochemistry is extremely complex, and it changes all the time in response to diet, activity, stress, and a variety of environmental factors. There are large differences among individuals, based on variations in genetics and in life's circumstances. Moreover, there are stringent limits on the use of humans as experimental subjects, so that most of the available data are from experimental animals or from studies of adventitious exposure in the workplace or in the environment. Consequently, conclusions about toxic effects are seldom hard and fast, and are frequently modified in light of new studies.

Acute toxicity is relatively easy to gauge. At high-enough levels, the effects of toxins on bodily function are obvious and fairly consistent across individuals and species. These levels vary enormously for different chemicals. Almost everything is toxic at some level, and the difference between toxic and nontoxic chemicals is a matter of degree.

The most widely used index of acute toxicity is LD_{50}, the lethal dose for 50% of a population. This number is obtained by graphing the number of deaths among a group of experimental animals, usually rats, at various levels of exposure to the chemical, and interpolating the resulting *dose-response* curve to the dose at which half the animals die (Figure 4.28). The dose is generally expressed as the weight of the chemical per kilogram of body weight,

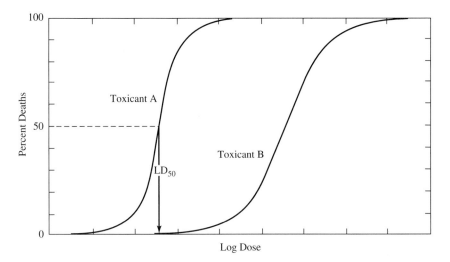

Figure 4.28   Illustration of a dose-response curve in which the response is the death of the organism; the cumulative percentage of deaths of organisms is plotted on the y-axis. *Source:* S. E. Manahan (1991). *Environmental Chemistry,* Fifth Edition (Boca Raton, Florida: Lewis Publishers, an imprint of CRC Press). Reprinted with permission, copyright © 1991 by CRC Press.

on the assumption that toxicity scales inversely with the size of the animal. Table 4.2 lists LD_{50} values for several substances, showing nine orders of magnitude variation between the most toxic (*botulin* toxin, the agent responsible for botulism) and the least toxic (sugar). Among insecticides, we see that DDT is about 30 times less toxic than parathion but 12 times more toxic than malathion, at least when measured in rats or mice.

Chronic effects are much more difficult to evaluate, especially at the low exposure levels that are likely to be encountered in the environment. In an experimental setting, the lower the dose, the fewer the animals that show any particular effect. To obtain statistically significant results, a study might have to include a prohibitively large number of animals. The only available recourse is to evaluate effects of a series of high doses, and then to extrapolate the dose-response curve to the expected incidence at low doses. But extrapolation may have to extend over several orders of magnitude, and there is no assurance that the actual dose-response function is linear. The biochemical mechanisms that control effects may be different at high and low doses. The controversy over this issue is especially heated in the context of animal testing for cancer (see below—pp. 298–299).

Toxicologists are increasingly turning to biochemical studies, using all the techniques of molecular biology, in order to elucidate the effects of toxicants at the molecular level. The expectation is that a more thorough understanding will provide a better basis for evaluating toxicity risks. Great strides have been made in probing the mechanism of action of various classes of toxicants (the dioxins are a good example; see pp. 303–304), but it is not yet possible to translate this understanding into a quantitative estimate of physiological effects.

The other approach to evaluating health risks is epidemiology, the study of human exposure to chemicals in the workplace or in the environment and the effect on the health of a

TABLE 4.2 LD$_{50}$ VALUES OF SELECTED CHEMICALS

Chemical	LD$_{50}$(mg/kg)*
Sugar	29,700
Ethyl alcohol	14,000
Vinegar	3,310
Sodium chloride	3,000
Atrazine	1,870
Malathion (insecticide)	1,200
Aspirin	1,000
Caffeine	130
DDT (insecticide)	100
Arsenic	48
Parathion (insecticide)	3.6
Strychnine	2
Nicotine	1
Aflatoxin-B	0.009
Dioxin (TCDD)	0.001
Botulin toxin	0.00001

*For rats or mice.

Source: P. Buell and J. Gerard (1994). *Chemistry in Environmental Perspective* (Upper Saddle River, New Jersey: Prentice Hall).

population. Epidemiology, in principle, can provide data that is most directly relevant to risk estimation. The problem is that the variables in epidemiological studies are difficult to control; despite sophisticated statistical analysis, it may be hard to ascertain that a particular effect is not influenced by some other factor, such as smoking or poor diet, rather than by the chemical under study. It is also hard to select a control population without bias, which might skew the estimation of risk. For example, the frequently used method of selecting controls via randomized phone numbers has recently been shown to underrepresent poor people (perhaps because they are less likely to respond to a phone request to participate in a study). Obtaining statistically significant results depends greatly on the sample size, just as in animal studies. Larger numbers are needed when relatively small risks are being evaluated, as is usually the case in environmental exposures. Not surprisingly, results are more reliable when the risk is large than when it is small. For example, the smoking-related risk of lung cancer (see p. 296) is easy to demonstrate statistically, because lung cancer incidence is 10–20 times higher in smokers than in nonsmokers. But the breast cancer risk associated with hormone replacement therapy has been much more difficult to establish, despite laboratory evidence for a connection (see pp. 300–301). Two major studies appeared in 1995, one showing a 1.3- to 1.7-fold increase in the breast cancer risk for women taking estrogen and/or progesterone, and the other showing no added risk.

Both experimental and epidemiological approaches are important for examining a special class of toxic effect that is of increasing concern: prenatal effects on the fetus. The tragedy of birth defects resulting from the introduction of the drug thalidomide in the 1960s sensitized everyone to the possibility of *teratogenic* effects of environmental chemicals in addition to those of drugs. Screening for such effects with experimental animals is now routine. In addition to obvious birth defects, the possibility of developmental deficits resulting from prenatal exposure to toxins is of increasing concern. The occurrence of fetal alcohol syndrome is a flagrant example. But there may be more subtle effects from environmental exposures. For example, a study of families living on the Lake Michigan shore who regularly ate fish caught in the lake found that verbal test scores of four-year-olds decreased noticeably for those with the highest exposure to PCBs (see below, pp. 309–310) at birth (Figure 4.29).

2. Cancer

Of all the possible effects of chemicals in the environment, none is more feared than cancer, and none has generated more controversy. The public's fear of cancer has driven regulatory agencies to set very low tolerances for many chemicals in various environmental settings, from food to drinking water to toxic waste sites. These standards continue to stir debate; they are claimed to be too lenient by many environmental activists and too strict by manufacturers and by others who might have to pay for required cleanups. Because of the uncertainties associated with the available data, as discussed in the preceding section, it is very hard to establish the truth of the matter. In the absence of hard evidence, there is a great deal of room for subjective factors that influence our perceptions of risk.

      a. Mechanisms.   Cancers occur when cells divide uncontrollably, eventually consuming vital tissues. The normal mechanisms that limit cell growth and division are disrupted. This can happen in many different ways, but the common thread is that mutations occur in the cell's DNA at positions which specify the synthesis of key regulatory proteins. It has been shown that several such mutations are required to *transform* a normal cell into a cancerous one. This requirement explains why there is a long *latency* period, often 20 years or more, between exposure to a cancer-causing substance and the actual occurrence of cancer. Because of the probabilistic nature of mutations, the risk of cancer increases with age. Although children and young adults can and do develop cancers, most cancers are primarily

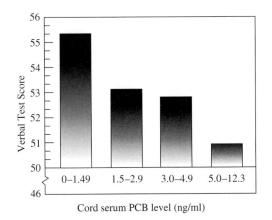

Figure 4.29   Test outcomes (McCarthy verbal test scores) of the 1990 Lake Michigan case study of four-year-old children; the children's scores are graphed versus the PCB concentrations in the umbilical cord serum at birth.   *Source:* J. L. Jacobson et al. (1990). Effects of *in utero* exposure to polychlorinated biphenyls and related contaminants on cognitive functioning in young children. *Journal of Pediatrics* 116: 38–44. Reprinted with permission, copyright © 1990 by Journal of Pediatrics. All rights reserved.

diseases of old age. One of the causes of increasing cancer incidence is simply the increase
in life expectancy during the last century

A mutation occurs when DNA is mistranscribed during cell division. Maintaining the
genetic code requires the correct pairing of bases via complementary H-bonding (Figure 4.30), when a new DNA strand is copied from an old one. If an incorrect base is some-
how incorporated into the sequence, then the error will be propagated in succeeding generations of the cell. If the incorrect base is part of a gene, then an error is introduced into the
protein for which the gene codes, and the protein may misfunction. Mutations occur all the
time because the fidelity of DNA transcription cannot be perfect. The normal error rate is
very low (about one in 100 million), but it is not zero. Although the mutations are more or
less random, there is some probability that they will occur at sites coding for regulatory proteins, and an additional, much smaller probability that enough critical mutations will accumulate to transform a given cell. Since our bodies contain billions of cells, and because we
live through many cycles of cell division, it is likely that all of us harbor potentially cancerous cells. But they lead to cancer only rarely because the body has several lines of defense.

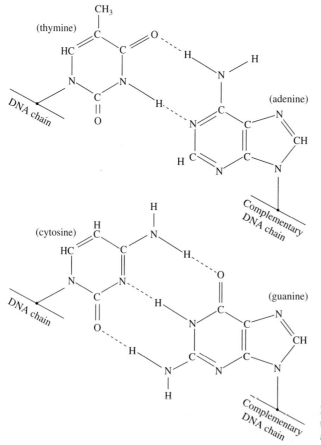

Figure 4.30   Base pairing in DNA between
thymine and adenine, and between cytosine
and guanine.

The cell itself has a variety of repair enzymes whose job is to detect incorrect base pairs and correct them. These enzymes greatly reduce the probability of accumulating enough critical mutations to produce cancer. In addition, the immune system provides powerful protection: cancer cells can be detected and destroyed by virtue of characteristic changes in their surface molecules. Finally, the development of full-blown cancer may require additional biochemical or physiological events. For example, solid tumors require a blood supply in order to grow, and must induce the body to provide a network of blood vessels.

Once in a while, all of these impediments are overcome, and a cancer results. The normally low probability of this happening can be increased by a variety of factors. An important one is genetics. Individuals may inherit a genetic defect that increases the cancer risk. The defect may involve a faulty repair enzyme, so that mutations survive more readily. Or there may be a pre-existing mutation in a gene for one of the regulatory proteins, which increases the odds of accumulating the remaining required mutations. Current genetic research is uncovering a wide range of genes in which mutations increase the risk of developing specific cancers.

Other factors involve exposure to cancer-inducing chemicals (*carcinogens*) or to dietary components that affect this exposure. For example, there is evidence that roughage in the diet protects against colon cancer, probably because the undigested fibers absorb carcinogenic molecules and sweep them out of the colon. Carcinogens can operate in two ways: they can be mutagens, inducing mutations by attacking the DNA bases, or they can be promoters, which increase the cancer probability indirectly. For example, promoters can act by increasing the rate of cell division. The more often cells divide, the greater the probability that cancerous mutations will accumulate. Thus, alcohol is a promoter of liver cancer because its consumption in excessive amounts causes cell proliferation in the liver, which is the organ that handles alcohol metabolism.

There are two requirements for mutagens: 1) they must react with the DNA bases in ways that alter their hydrogen bonding with a complementary base; since the bases are electron-rich, the mutagens tend to be *electrophiles;* 2) they must gain access to the nucleus where the DNA is located. Many electrophiles are not mutagenic because they react with other molecules and are deactivated before they can reach the nucleus. For this reason, most mutagenic chemicals are not themselves reactive but are converted into reactive metabolites by the body's own biochemistry. The body has a variety of ways of ridding itself of foreign chemicals (*xenobiotics*). One of the most important is *hydroxylation* of lipophilic organic compounds. For example, when benzanthracene is hydroxylated (Figure 4.31), not only does a hydroxy group increase water solubility, but it also serves as a point of attachment for other hydrophilic groups such as glucuronide sulfate, which increase the water solubility further and promote excretion by the kidneys. Hydroxylation is accomplished by inserting one of the oxygen atoms of O_2 into a C–H bond, the remaining oxygen atom being reduced to water by supplying two electrons from a biological reductant:

$$O_2 + -C-H + 2e^- + 2H^+ = -C-O-H + H_2O \tag{4.7}$$

This is a tricky reaction because the highly reactive oxygen atom must be generated exactly where it is needed; otherwise it will attack any molecule in its vicinity, adding to the supply of free radicals. The reaction is carried out by a class of enzymes, cytochrome P450, which contain a heme group (Figure 4.32) to bind the O_2 (just as hemoglobin does; see

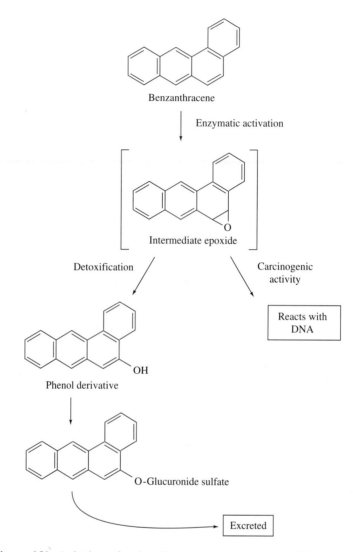

Benzanthracene

Enzymatic activation

Intermediate epoxide

Detoxification

Carcinogenic
activity

Reacts with
DNA

Phenol derivative

OH

O-Glucuronide sulfate

Excreted

Figure 4.31 Activation of polycyclic aromatic hydrocarbons (PAHs). *Source:* C. Heidelberger (1975). Chemical carcinogenesis. *Annual Review of Biochemistry* 44: 79–121. Copyright © 1975 by Annual Reviews Inc. Reproduced with permission.

Figure 2.40) and an adjacent binding site for the xenobiotic molecule. Despite this juxtaposition of the reactants, the immediate product is sometimes not the hydroxylated molecule but rather an epoxide precursor (Figure 4.31), which is a potent electrophile. Since this precursor is generated inside the cell, it has a chance of diffusing into the nucleus and reacting with the DNA before it rearranges to the hydroxylated product. This is the reason that PAH compounds (see pp. 166–167) like benzanthracene are carcinogenic. Another possibility is

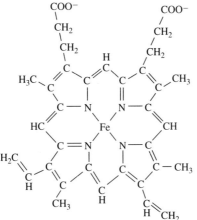

Figure 4.32   The structure of heme.

that the hydroxylated product can itself be a precursor to a reactive agent. For example, hydroxylation of dimethylnitrosamine (Figure 4.33), another carcinogen, releases formaldehyde (CH_2O), leaving an unstable intermediate that is a source of methyl carbonium ion (CH_3^+), a powerful electrophile, which can react readily with DNA if generated nearby.

PAHs and nitrosamines are anthropogenic carcinogens, but there are many natural ones as well. Aflatoxins, which are complex products of a mold that infests peanuts and other crops, are powerful carcinogens. Biochemist Bruce Ames, developer of the Ames mutagenicity test (see next section), points out that the plants we eat contain natural pesticides, many of which are turning out to be mutagenic when tested. He has estimated that the average American eats 1.5 g per day of natural pesticides, about 10,000 times more than the amount of synthetic pesticide residues. Ames and others have also drawn attention to the high natural level of mutagenesis due to oxidative damage to DNA from the side-products of normal O_2 metabolism (see discussion of antioxidants, pp. 272–273). This research puts the damage caused by synthetic chemicals in the context of the natural background level of DNA damage and repair.

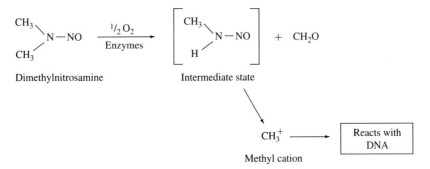

Figure 4.33   Activation of dimethylnitrosamine in the body.

b. Cancer incidence and testing.   Despite extensive epidemiological studies, it is not easy to tease out the contribution of environmental chemicals to cancer incidence, for the reasons mentioned above (pp. 289–291). For example, even though radon is thought to be a more serious cancer hazard than any other environmental chemical, the studies on radon in houses do not agree as to whether the cancer incidence is elevated when the radon levels are higher than the U.S. EPA's guideline of 4 pCi/l (see pp. 36–38).

However, some cancer causes are firmly established by epidemiological data. The most striking evidence is the historical data on lung cancer and smoking (Figure 4.34). A manyfold rise in U.S. lung cancer mortality tracked the increase in cigarette smoking, with a lag of several decades, and this happened in different historical periods for men and for women. Smoking accounts for 30% of all U.S. cancer deaths (along with 25% of fatal heart attacks). Similarly clear is diet's role in cancer, as is strongly suggested by data showing marked changes in the pattern of cancer incidence when people migrate from one part of the world to another (Figure 4.35). The rates and types of cancers contracted by migrating ethnic groups change when their diets change. It is thought that high levels of salt or smoked fish in the Japanese diet may account for excess stomach cancers, while high fat in the U.S. diet might be responsible for a higher rate of colon cancer. However, the actual contributions of dietary components to cancer incidence (or to protection from cancer) have been hard to pin down.

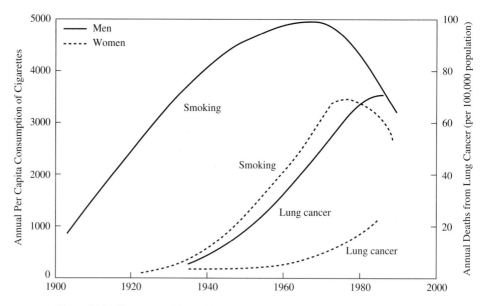

Figure 4.34   Cigarette smoking and lung cancer in the United States; death rates are averages for all ages. *Sources:* Drawn from data given in: L. Garfinkel and E. Silverberg (1990). Lung cancer and smoking trends in the United States over the past 25 years. In *Trends in Cancer Mortality in Industrial Countries*, D. L. Davis and D. Hoel, eds. (New York: The New York Academy of Sciences); J. A. Meyer (1990). *Lung Cancer Chronicles* (New Brunswick, New Jersey: Rutgers University Press).

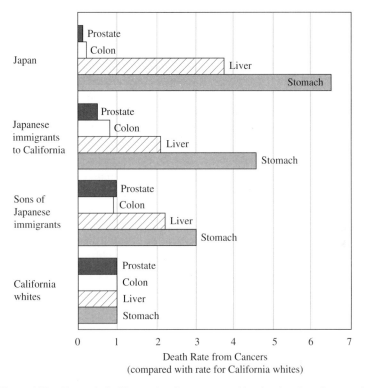

Figure 4.35 Change in incidence of various cancers with migration from Japan to the United States.

Data on occupational exposure have firmly implicated several industrial chemicals. For example, vinyl chloride causes liver cancer, benzene causes leukemia, and asbestos causes mesothelioma, a cancer of the lining of the lung. However, exposure of people at large to these chemicals is far lower than in an occupational setting, and the hazard at these lower levels can only be guessed by extrapolation.

Alternatively, carcinogenic risk can be estimated from test data. Since many carcinogens are mutagens, carcinogens can be screened by using the Ames bacterial test. The suspect carcinogen is administered to mutant bacteria that are unable to grow in the absence of the amino acid histidine in the culture medium. Certain additional mutations will produce a *revertant* organism, capable of growing again in the histidine-deficient medium. The stronger the mutagen, the greater the number of revertant organisms produced. Thus, the number of revertant colonies is a measure of the mutation rate, which can be determined at various concentrations of the test substance. (The test can also be used to monitor complex mixtures for mutagenic activity in order to separate and identify the active ingredient.) Since some chemical compounds are not mutagens until they are metabolically activated, in order to assay carcinogenicity the Ames test requires adding a rat-liver extract, which contains the cytochrome P450 enzymes responsible for oxidative activation of the carcinogen by the hydroxylation mechanisms described in the preceding section. Because bacteria are very different from people, the test cannot distinguish reliably all human carcinogens or evaluate their potency. It is, however, an inexpensive and useful screening method.

The main source of carcinogenicity data has been animal tests, usually involving rats. Cancers are counted over the lifetime of the animal at various doses, and the results are extrapolated to typical exposure levels in order to obtain an estimate of the cancer risk. Because of the need to obtain statistically significant results on a limited number of animals, most of the data are at the *maximum tolerated dose* (MTD), above which acute toxicity symptoms occur.

The use of the MTD has been criticized on the grounds that even in the absence of overt toxic symptoms, there may be significant organ damage and resulting cell proliferation, which increases the mutation probability. There may be additional reasons why conditions at the MTD in animals may have little relevance for human exposure. For example, saccharin carries a carcinogenicity warning when marketed as an artificial sweetener because it was found to cause bladder cancer in male rats at high doses. But subsequent mechanistic research established that these cancers are associated with a protein, α_{2u}-globulin, which is specific to male rats and is not present in humans (or even female rats). The tumors occur when the bladder lining regenerates after erosion by a precipitate of the protein with saccharin in the urine. This mechanism would not occur in humans, even at high doses, and indeed the epidemiological evidence on saccharin is negative.

Despite their deficiencies, animal tests can serve as a rough guide for comparing carcinogenic risk from different substances. Ames has proposed an index for this purpose, HERP (human exposure dose/rodent potency). It is calculated by estimating the lifetime ex-

TABLE 4.3   COMPARISON OF RISK FROM EXPOSURE TO CARCINOGENS

HERP%*	Risk agent
0.001	Tap water, 32 oz (chloroform)
0.003	DDT, from remaining residues
0.004	EDB,[†] before it was banned
0.03	Peanut butter, 1 sandwich (aflatoxin)
0.03	Comfrey herb tea, 1 cup (symphytine)
0.07	Brown mustard, 1 teaspoon (allyl isothiocyanate)
0.1	Mushroom, 1 raw (hydrazines)
0.1	Basil, 1 g of dried leaf (estragole)
2.8	Beer, 12 oz (ethyl alcohol)
4.7	Wine, 8 oz (ethyl alcohol)

*HERP: Human exposure dose/rodent-potency. Cancer hazard based on a typical person's average daily exposure to the substances over a lifetime. Calculated as a percentage of the LD_{50} for rats (the dose that causes cancer in half the exposed rats).

[†]EDB stands for ethylene dibromide, a fungicide and fuel additive.

Source: P. Buell and J. Gerard (1994). *Chemistry in Environmental Perspective* (Upper Saddle River, New Jersey: Prentice Hall).

posure for an average person and dividing by the rodent LD_{50} for death by cancer. The results (Table 4.3) suggest that exposure to pesticide residues or to tap water are much weaker hazards than are such common items in our diet as wine, beer, mushrooms, and peanut butter.

Because there are currently no viable alternatives to animal tests, they will no doubt continue to be used as a factor in assessing cancer risks. As additional mechanistic insights emerge from biochemical research, they can be factored into the evaluation of the significance of particular tests and may alter experimental protocols.

3. Hormonal Effects

Recently, increasing concern has focused on the biochemical role of environmental chemicals that mimic hormone functions. Hormones are messenger molecules, excreted by various glands, that circulate in the bloodstream and powerfully influence the biochemistry of specific tissues. Hormone activity is initiated by binding to receptor proteins in the target cells.

There are two kinds of hormones, water-soluble and lipid-soluble, with entirely different mechanisms of actions. Water-soluble hormones, such as insulin, are peptides and proteins. They bind to receptor proteins embedded in the target cell membrane, analogous to the neurotransmitter receptors (Figure 4.21). This binding induces the activation of enzymes inside the cell, which catalyze the synthesis of interior messenger molecules; in turn, these *second messengers* bind and activate proteins that control metabolic processes.

The lipid-soluble hormones are *steroids,* derivatives of cholesterol (Figure 4.36). They diffuse through cell membranes and are picked up at the inside surface by specific receptor proteins that are dissolved in the interior fluid (*cytosol*) of the target cell. Hormone binding changes the shape of the receptor protein and enables it, after diffusion to the nucleus, to turn on specific genes (Figure 4.37). Thus, the steroid hormones act by inducing the synthesis of enzymes and regulatory proteins.

Environmental chemicals can also bind to hormone receptors if they have the proper shape and distribution of electrical charges. This is unlikely to be a problem for peptide hormones because water-soluble xenobiotics are quickly excreted. But lipophilic xenobiotics, which are stored in the fat tissue, might bind to steroid hormone receptors. If the resemblance to the hormone is close enough, the xenobiotic can turn on the same biochemical machinery, but if the resemblance is only partial, then binding may not activate the receptor. In that case, the xenobiotic blocks the hormone and depresses its activity; it is an antihormone. Either way, there is potential for upsetting the biochemical balance controlled by the hormone. A mechanism of this kind is probably responsible for DDT's disruption of calcium deposition in birds' eggs (pp. 276–278).

The sex hormones are in the steroid class; estrogens and androgens induce and maintain the female and male sexual systems. They have become a focus of attention because of reports of malformed sex organs in wildlife. In particular, alligators in a Florida lake were found to have impaired reproductive systems (abnormally small penises in the males), low rates of hatching, and high levels of DDE, the breakdown product of DDT, in their tissues. DDE contamination resulted from spills of a DDT-containing pesticide along the lakeshore. Subsequent tests showed that DDE binds to androgen receptors and blocks their activity. This evidence of environmental demasculinization has fueled speculation that something similar

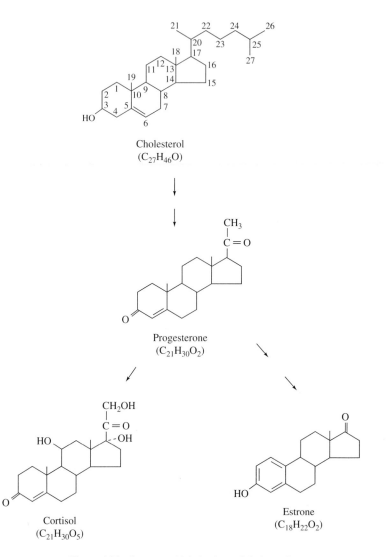

Figure 4.36 Some steroid derivatives of cholesterol.

may be going on in human males, because of reports from a number of clinics that sperm counts among men have been going down for a number of years. But the validity of these data as indicators of male fertility has been questioned.

Estrogenic xenobiotics have provoked more concern because of the association of estrogen with breast cancer in women. Estrogen binding to receptors in the breast stimulates the proliferation of breast cells; as we have seen in the preceding section, cell proliferation promotes mutagenesis and cancer. A link between breast cancer and estrogen has been established in laboratory animals, and there is a statistical association, albeit equivocal (see discussion on epidemiology above, pp. 289–290) between estrogen therapy and breast can-

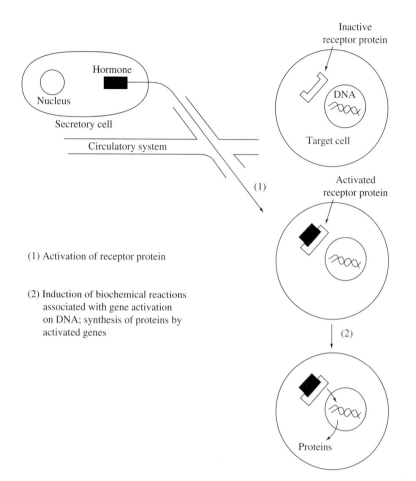

Figure 4.37   Mechanism of steroid hormone function.

cer incidence. (There is stronger evidence that estrogen protects against heart disease and osteoporosis.) The incidence of breast cancer has been rising, and many environmental chemicals have recently been found in laboratory assays to be estrogenic. These include DDT, the antioxidant BHA, and a variety of organic chemicals that are either used as plasticizers or are products of high-temperature treatment of plastics. Putting these facts together, many people worry that exposure to these chemicals may put women at risk for breast cancer. Skeptics point out, however, that the levels of exposure are low, and that the xenobiotics should be swamped out by the body's own estrogen (although the endogenous estrogen level fluctuates cyclically, while the xenobiotics do not). This is currently an area of active research.

4. Organic Pollutants: Dioxins and PCBs

There are many organic compounds in the environment, some of them toxic. We have already dealt with oil spills in Part I (pp. 23–27), air pollutants in Part II (pp. 162–175), and pesticides and herbicides in preceding sections. There are numerous organic waste products from industrial operations or from the use and disposal of manufactured products that may contaminate air, land, and water from effluents, from leakage of waste dumps, or from accidental spills and fires.

We focus on two classes of pollutants that have often been in the news in recent years: dioxins and polychlorinated biphenyls (PCBs). These are chlorinated aromatic compounds that are persistent and bioaccumulative and may cause cancer and other health problems.

a. Dioxins. The term dioxins is shorthand for a family of polychlorinated dibenzodioxins (Figure 4.38), sometimes abbreviated PCDDs. They are not made intentionally but are formed as contaminants in several large-scale processes, including 1) combustion, 2) paper pulp bleaching with chlorine, and 3) manufacture of certain chlorophenol chemi-

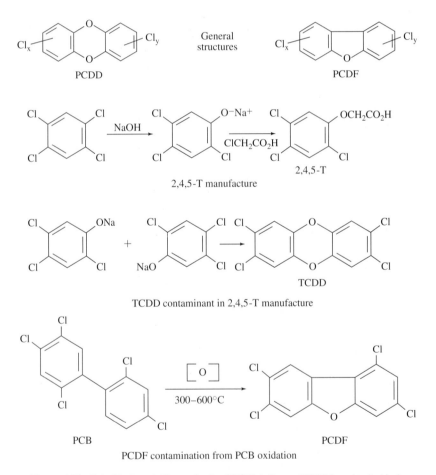

Figure 4.38 Polychlorinated dibenzodioxins (PCDDs), furans (PCDFs), and polychlorinated biphenyls (PCBs); chemical structures and reactions.

cals. It was this last process that brought dioxin its initial notoriety as a contaminant of the herbicide 2,4,5-T, a component of Agent Orange. The herbicide was made by reacting chloroacetic acid with 2,4,5-trichlorophenol, which was itself formed by reacting 1,2,4,5-tetrachlorobenzene (Figure 4.38). During this prior reaction, which was carried out at a high temperature, some of the trichlorophenoxide condensed with itself (Figure 4.38) to form 2,3,7,8-tetrachlorodibenzodioxin (TCDD); Agent Orange contained about 10 ppm of this material. It was subsequently shown that control of the reaction temperature and of the trichlorophenoxide concentration could keep the TCDD contamination to 0.1 ppm, but it was too late to save the herbicide, which was banned in the United States in 1972.

1) Toxicity.   TCDD turned out to be extraordinarily toxic to laboratory animals, producing birth defects, cancer, skin disorders, liver damage, suppression of the immune system, and death from undefined causes. The LD_{50} in male guinea pigs was only 0.6 $\mu g/kg$. Consequently, there was great alarm when TCDD was found at a number of industrial waste sites. And in 1983 the U.S. government offered to purchase houses in the town of Times Beach, Missouri, after its roads were found to have been contaminated by dioxin in waste oil from a 2,4,5-T manufacturer, oil that had been sprayed by a hauler to control dust. The most serious instance of environmental contamination occurred in 1976 when an explosion in a factory in Seveso, Italy that manufactured 2,4,5-T released a few kg of TCDD into the town and its surroundings.

As evidence on toxicity has accumulated, however, the risk to humans from dioxin has become less clear. The variation in toxicity among species turned out to be large, with LD_{50} values that were orders of magnitude higher in other animals than in guinea pigs (Table 4.4).

TABLE 4.4   ACUTE TOXICITIES OF 2,3,7,8-TETRACHLORO-
DIBENZODIOXIN IN EXPERIMENTAL ANIMALS

Species	Route	LD_{50} (micrograms per kilogram)
Guinea pig (male)	Oral	0.6
Guinea pig (female)	Oral	2.1
Rabbit (male, female)	Oral	115
Rabbit (male, female)	Dermal	275
Rabbit (male, female)	Intraperitoneal	252–500
Monkey (female)	Oral	< 70
Rat (male)	Oral	22
Rat (female)	Oral	45–500
Mouse (male)	Oral	< 150
Mouse (male)	Intraperitoneal	120
Dog (male)	Oral	30–300
Dog (female)	Oral	> 100
Frog	Oral	1,000
Hamster (male, female)	Oral	1,157
Hamster (male, female)	Intraperitoneal	3,000

Source: Data from F. H. Tschirley (1986). Dioxin. *Scientific American* 254(2): 29–35.

Moreover, although the Seveso contamination produced many wildlife deaths and exposed many people, no serious human health effects were found for many years; nor have any been tied to the Times Beach contamination. Recently, however, rates for a number of cancers were found to be elevated in the exposed Seveso population, although the small numbers make the statistics uncertain. The U.S. EPA has carried out a dioxin reassessment (1995), which concludes that dioxin is likely to increase cancer incidence in humans. At high levels, PCDDs cause chloracne, a painful skin inflammation, but these levels have only been encountered in accidental industrial exposures. In laboratory animals, low doses have been found to be teratogenic and to lead to developmental abnormalities.

Rapid strides are being made in understanding TCDD's complex biochemistry. The molecule binds strongly to a receptor protein that is present in all animal species. This receptor, called Ah (for aryl hydrocarbon) is activated by a number of planar aromatic molecules (its natural substrate is still unknown); the binding of TCDD is particularly strong, with an equilibrium constant for dissociation of 10^{-11} molar. Like a hormone receptor (Figure 4.37), the Ah receptor interacts in complex ways with the cell's DNA. One effect is the induction of a cytochrome P450 enzyme (a variant labeled 1A1), which is responsible for hydroxylating a number of xenobiotics, including PAHs (but not TCDD itself, since its chlorine atoms deactivate the ring toward oxidation). There are additional effects on a variety of biochemical pathways, which are currently under study. It remains uncertain, however, whether all of TCDD's toxic effects originate in its binding to the Ah receptor.

2) Paper Bleaching and Combustion Sources.   A variety of PCDDs are formed in small quantities when chlorine is used to bleach paper pulp, probably via chlorination of the phenolic groups in lignin (see lignin structure, p. 17). There has been concern about trace dioxin contamination of paper products, and about bioaccumulation of dioxin in waters receiving paper mill effluents. Dioxin emissions are being reduced by switching from chlorine to chlorine dioxide, which is an oxidant but not a chlorinating agent (see discussion of water disinfection, pp. 246–247).

The main source of dioxin in the environment, however, is combustion. When material containing chlorine is combusted, dioxin is produced in traces; because the volume of material combusted annually is huge, these traces add up to a substantial aggregate environmental load. As might be expected, the dioxin emission rate correlates roughly with the chlorine content of the combustion feed,* although dioxin formation is highly dependent on combustion conditions and on the type of pollution controls, if any. It appears that the chlorine need not be organically bound, since wood stoves have been found to produce dioxins, and the chlorine in wood is mostly sodium chloride. The main mechanism of dioxin formation appears to involve reaction of organic fragments in the combustion zone with HCl and O_2. The HCl formation rate might be expected to depend on the form of the chlorine in the combusted material, but this question has not been resolved. Dioxin formation is catalyzed on the surface of fly ash, probably by transition metal ions, and is favored at moderate temperatures, with a maximum at about 400°C. At lower temperatures, the reaction slows down and

---

*V. M. Thomas and T. G. Spiro (1995). An estimation of dioxin emissions in the United States. *Toxicology and Environmental Chemistry* 50: 1–37.

the products remain adsorbed on the fly ash, whereas at higher temperatures the dioxins are oxidized further.

Combustion produces a wide range of PCDD *congeners* (molecules with the same structure, but with varying numbers and positions of chlorine substituents). Also formed are the structurally similar polychlorinated dibenzofurans, PCDFs (Figure 4.38). In the context of combustion products, "dioxin" means the aggregate of PCDDs and PCDFs, also abbreviated to PCDD/Fs. These terms are used interchangeably in the ensuing discussion. Both classes of molecules are toxic, but the toxicity varies among the congeners. Toxicity is assumed to be roughly proportional to the strength of binding to the Ah receptor. TCDD is the most toxic of the dioxins; toxicity decreases progressively when chlorine atoms are removed from the 2,3,7 and 8 positions, or when they are added to the remaining positions on the rings. These alterations reduce the "fit" of the molecule to the binding site on the Ah receptor. A similar toxicity pattern is observed for the PCDF congeners, but the toxicity is about an order of magnitude lower for the PCDFs than for the PCDDs. In order to gauge the effects of exposure to these chemicals, a scale of international toxicity equivalence factors (I-TEFs) has been established based on toxicity relative to TCDD, which is assigned a value of 1 (Table 4.5). This factor is 0.1 for 2,3,7,8-PCDF, for example, and 0.001 for the octachloro congeners of either series. With these factors one can convert the distribution of both classes of molecules (PCDD/Fs) into a single toxicity equivalent quantity (TEQ), expressed in grams of TCDD-equivalents. For example, 1.0 g each of TCDD and 2,3,7,8-PCDF would have a TEQ value of 1.1 g.

Dioxin inventories have been estimated for a number of countries from data on emission rates for various kinds of combustion and on the total amount of material combusted. Because emission rates are highly variable and the data are few, the estimates are not very certain, but they do provide a rough comparison of the various sources (Figure 4.39). It seems clear that municipal- and hospital-waste incinerators are the major sources, at least in developed countries. But pollution-control devices can cut the incinerator emission rates dramatically. A combination of spray dryers and fabric filters can be effective in removing

TABLE 4.5   INTERNATIONAL TOXICITY EQUIVALENCY
FACTORS FOR PCDDs AND PCDFs

Congener	PCDD series	PCDF series
2,3,7,8	1 (defined)	0.1
1,2,3,7,8	0.5	0.05
2,3,4,7,8		0.5
1,2,3,4,7,8	0.1*	0.1[†]
1,2,3,4,6,7,8	0.01	0.01[‡]
octachloro	0.001	0.001

*same value for 1,2,3,6,7,8- and 1,2,3,7,8,9- congeners
[†]same value for 1,2,3,6,7,8-, 1,2,3,7,8,9-, and 2,3,4,6,7,8- congeners
[‡]same value for 1,2,3,4,7,8,9- congener

Source: N. J. Bunce (1994). *Environmental Chemistry,* Second Edition (Winnipeg, Canada: Wuerz Publishing, Ltd.).

Sources

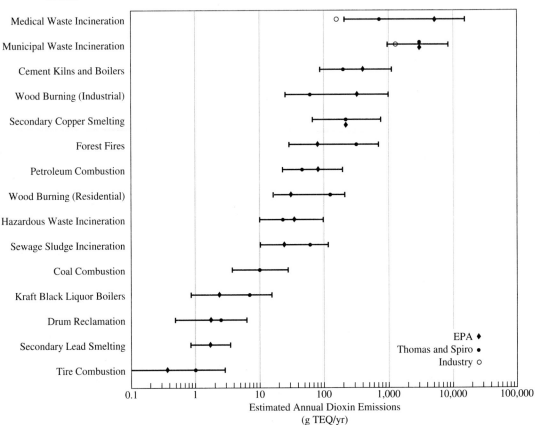

Figure 4.39   Estimated annual U.S. TEQ (toxic equivalent) dioxin emissions from combustion sources, circa 1990. The length of horizontal lines indicates the estimated uncertainty.   *Source:* V. Thomas and T. Spiro (1996). The U.S. dioxin inventory: are there missing sources? *Environmental Science and Technology* 30(2): 82A–85A. Copyright © 1996 by American Chemical Society. Reprinted with permission.

PCDD/Fs from the incinerator exhaust gases. Ironically, electrostatic precipitators, which were installed on older incinerators to reduce particle emissions, can actually increase dioxin formation, apparently through catalysis by particles lodged on the precipitator surfaces. Under pending regulations of the U.S. EPA, incinerator emissions of PCDD/Fs are expected to diminish to 1% of current emissions.

Incinerators account for about 80% of the U.S. dioxin inventory, while a host of other combustion sources account for most of the remainder. Some dioxin is released by evaporation from pentachlorophenol-treated wood (Figure 4.27), but this is mostly the octachloro congener, which has low toxicity. The spraying of 2,4,5-T herbicide, which used to be a major contributor to the inventory, has now been eliminated. Among the numerous other combustion sources, cement kilns are significant because they often burn organic waste materials as a fuel source. Wood furnaces and stoves are collectively a significant source; the emission rate is low, but the amount of wood burned annually is quite large.

3) Natural Sources?   Wood-derived dioxins raise the interesting issue of natural versus anthropogenic sources.* It has been suggested that forest fires are a major source of dioxins. The amount of wood burned in U.S. forest fires is actually about the same as the amount of wood burned in domestic and industrial stoves and furnaces; if the emission rates are about the same, the dioxin production should also be comparable. However, there are no data on dioxin production in uncontained fires. Likewise, we know very little about other possible sources in nature. Contrary to popular opinion, organochlorines are not exclusively synthetic but are widely produced as natural products by a variety of microorganisms. Soil organisms produce peroxidase enzymes to break down lignin, and these are capable of incorporating chloride ions into carbon-chlorine bonds. It is not known to what extent PCDD/Fs might result from this natural chemistry, but dioxins have been found in compost piles.

Could natural sources of dioxins outweigh anthropogenic ones? The sedimentary record suggests not. It is possible to establish the trend in the deposition rate over time by analyzing the dioxin profile of cores extracted from lake bottoms. Sediment from Siskiwit Lake has been examined in this way (Figure 4.40); the lake is located on an island in Lake Superior, far from any pollution source. Dioxins must have reached Siskiwit Lake by long-range transport through the atmosphere. The dioxin deposition rate is found to have increased eightfold between 1940 and 1970, the period of great expansion in the industrial use of chlorine. In contrast, forest fires in the U.S. actually diminished by more than a factor of four in the same period, thanks to more effective fire-control measures. Since 1970 the dioxin deposition rate has declined by about 30% (Figure 4.40), in parallel with the phaseout of 2,4,5-T spraying, and with improved incinerator technology. These trends seem to rule out predominantly natural sources. It is interesting that the dioxin in sediment is mainly the octachloro congeners, probably because the less chlorinated congeners are selectively volatilized from dioxin-bearing particles during long-range transport.[†]

4) Exposure.   The total U.S. PCDD/F emission rate from combustion is estimated to be 6–9 kg/yr TEQ. But this total is spread out over an enormous area, and the atmospheric concentrations are very low. Exposure from breathing dioxin-laden air is minimal, even if one lives next to an incinerator. As with other hydrophobic materials, exposure to dioxins is determined by bioaccumulation mechanisms (see pp. 276–278). The main exposure route for humans is dietary: meat, dairy products, and fish (Table 4.6). The dioxins deposit on hay and feed crops consumed by cows, which concentrate the dioxins in their fat tissues. Likewise,

*G. Grible (1994). The natural production of chlorinated compounds. *Environmental Science and Technology* 25 (7): 310A–318A.

[†]R. A. Hites (1990). Environmental behavior of chlorinated dioxins and furans. *Accounts of Chemical Research* 23: 194–201.

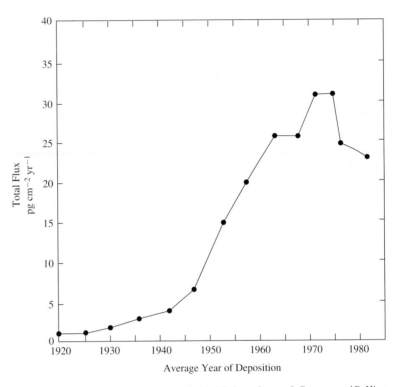

Figure 4.40   Flux of PCDD and PCDF to Siskiwit Lake.   *Source:* J. Czuczwa and R. Hites
(1986). Airborne dioxins and dibenzofurans: sources and fates. *Environmental Science and
Technology* 20(2): 195–200. Copyright © 1986 by American Chemical Society. Reprinted
with permission.

TABLE 4.6   AVERAGE CONTENT OF 2,3,7,8-TETRACHLOROBENZODIOXIN IN THE AMERICAN
FOOD SUPPLY

Food	TCDD concentration in picograms per gram (pg/g)*	Average TCDD intake (pg/person/day)*
Ocean fish	500	8.6
Meat	35	6.6
Cheese	16	0.31
Milk	1.8	0.20
Coffee	0.1	0.04
Ice cream	5.5	0.04
Cream	7.2	0.01
Sour cream	10	0.01
Cottage cheese	2.1	0.01
Orange juice	0.2	0.01
Total		15.9

* 1 *picogram* (pg) = 10^{-12} gram.
Source: Data from S. Henry et al. (1992). Exposures and risks of dioxin in the U.S. food supply. *Chemosphere*
25: 235–238.

fish concentrate dioxins from algae, which absorb dioxins from fallout, and also from local pollution sources such as pulp bleaching plants or sewage and wastes. As a result, we all have detectable concentrations of dioxins in our fat tissue, although the most prevalent one is the octachloro congener, which is not very toxic (Table 4.7). The average daily dose of PCDD/Fs is estimated to be roughly 0.1 ng (*nanogram* = 10^{-9}g) TEQ/day in the United States. This dose is not far from levels at which biochemical effects can be detected in laboratory animals. If the dioxin deposition rate is declining, as the sedimentary record indicates, then the average exposure should also decline.

b. Polychlorinated biphenyls.   As the name implies, polychlorinated biphenyls (PCBs) are made by chlorinating the aromatic compound biphenyl (see Figure 4.38 for the molecular structure of a specific PCB). A complex mixture results, with variable numbers of chlorine atoms substituted at various positions of the rings; a total of 209 congeners are possible. PCBs were manufactured in the United States from 1929 to 1977, with a peak production of about 100,000 tons a year in 1970. They were used mainly as the coolant in power transformers and capacitors because they are excellent insulators, are chemically stable, and have low flammability and vapor pressure. In later years they were also used as heat-transfer fluids in other machinery and as plasticizers for polyvinylchloride and other polymers; they found additional uses in carbonless copy paper, as de-inking agents for recycled newsprint, and as weatherproofing agents. As a result of industrial discharges and the disposal of all these products, PCBs were spread widely in the environment.

Because PCBs are chemically stable, they persist in the environment, and because they are lipophilic, they are subject to bioaccumulation, as DDT and dioxins are. PCB concentrations at the top of the food chain are significant in many localities. For example, herring gull eggs on the shores of Lake Ontario contained more than 160 ppm of PCBs in 1974 (Figure 4.41). Since then, however, the level has declined by a factor of five, reflecting the termination of PCBs in all open uses, those in which disposal cannot be controlled. Production was drastically curtailed in 1972 and halted in 1977. PCB-containing

TABLE 4.7   DIOXIN CONGENERS DETECTED IN HUMAN FAT TISSUE

Congener	Average concentration (pg/g)	Standard deviation (pg/g)
2,3,7,8-tetrachloro	11	8
1,2,3,7,8-pentachloro	24	12
1,2,3,6,7,8-hexachloro	172	74
1,2,3,7,8,9-hexachloro	22	9
1,2,3,4,6,7,8-heptachloro	232	181
octachloro	1037	712

Source: Data from G. L. LeBel et al. (1990). Polychlorinated dibenzodioxins and dibenzofurans in human adipose tissue samples from five Ontario municipalities. *Chemosphere* 21(12): 1465–1475.

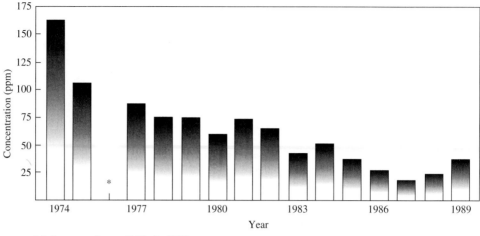

* Indicates no data available for 1976

Figure 4.41   The concentrations of PCBs in herring gull eggs at the Toronto shoreline of Lake Ontario (1974–1989).   *Source:* Ministry of Supplies and Services (1990). *The State of Canada's Environment* (Ottawa, Canada: Ministry of Supplies and Services).

transformers continue in service, but their disposal is regulated, and spent PCBs are stored or incinerated.

As in the case of dioxin, PCB health effects are hard to pin down. Occupational exposure has mainly produced cases of chloracne. There are, however, two instances of community poisoning by accidental PCB contamination of cooking oil, in Japan (1968) and in Taiwan (1979). Thousands of people who consumed the oil suffered a variety of illnesses, including chloracne and skin discoloration as well as low birth weight and elevated mortality of infants of exposed mothers. It was subsequently discovered that the oil was also contaminated with PCDFs, which are formed when PCBs are subjected to high temperatures (Figure 4.38); the PCBs that were mixed into the oil had been used as heat-exchange fluids in the deodorization process for the oil. Most of the toxic effects were attributed to the PCDFs rather than the PCBs.

In laboratory studies, PCBs are less toxic than PCDDs and PCDFs, but they probably operate by the same mechanism, binding to the Ah receptor. The most toxic PCBs are those which have no Cl atoms in the ortho positions of the ring and can therefore adopt a coplanar configuration of the rings, as in PCDDs and PCDFs. Coplanarity is inhibited in ortho-substituted biphenyls by the steric interaction of the substituent with the ortho H atoms on the other ring. If substituents occupy three or four of the ortho positions, they bump into each other, and the rings are necessarily twisted away from each other. PCBs with this substitution pattern are the least toxic. Even if PCBs are less toxic to humans and other animals than PCDDs and PCDFs, they are much more abundant in the environment. Studies like the one discussed on p. 291 (Figure 4.29), which indicates a connection between PCB exposure *in utero* and subsequent learning deficits, are cause for concern.

5. Toxic Metals

The biosphere has evolved in close association with all the elements of the periodic table, and indeed, organisms harnessed the chemistry of many metal ions for essential biochemical functions at early stages of evolution. As a result, these elements are required for viability, although in small doses. When the supply of an essential element is insufficient, it limits the viability of the organism, but when it is present in excess, it exerts toxic effects, and viability is again limited. Thus, there is an optimum dose for all essential elements (Figure 4.42).

   This optimum varies widely for different elements, however. For example, iron and copper are both essential elements, but we harbor about 5 g of the former in our bodies and only 0.08 g of the latter. Toxicity is low for iron but high for copper. Toxicity varies because the chemistry of the element varies. Thus, copper is generally present as Cu^{2+} and forms strong complexes with nitrogenous bases, including the histidine side chains of proteins. In contrast, neither Fe^{2+} nor Fe^{3+}, the common iron oxidation states, bind particularly strongly to nitrogenous bases. Copper is therefore more likely than iron to interfere with critical sites in proteins. At higher levels, nevertheless, iron is harmful, partly because it can catalyze the production of oxygen radicals (recall the discussion of antioxidants, pp. 272–274), and partly because excess iron can stimulate the growth of bacteria and aggravate infections. Chromium is also an essential metal, albeit in traces, but it is a powerful carcinogen as well. Carcinogenicity is associated with the highest oxidation state, Cr[IV], and the main concern is chromate pollution from spills and residues of electroplating baths, and from chromate emissions

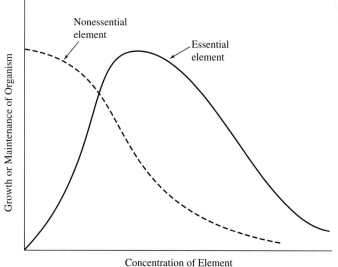

Figure 4.42   Dose-response curves for essential and nonessential elements in metabolic processes.

from cooling towers where it is used to inhibit corrosion. When toxic doses are compared for different metals (Table 4.8), a wide variation is seen.

     The breadth of the peak in the viability curve for metals (Figure 4.42) depends in part on homeostatic mechanisms, which have evolved to accommodate fluctuations in the metal availability. For example, excess iron is deposited in a storage protein, *ferritin,* from which it is released as needed. Many metals have no known biological benefit, and for them the curve of viability decreases steadily with increasing dose (Figure 4.42). The initial part of

TABLE 4.8   RELATIVE MAMMALIAN TOXICITY OF ELEMENTS IN INJECTED DOSES AND DIETS

Element	Acute lethal doses (LD_{50}) injected into mammals* (mg/kg bodyweight)	Dose in human diet (mg/day) Toxic	Dose in human diet (mg/day) Lethal[†]
Ag	5–60	60	1.3k–6.2k
As	6	5–50	50–340
Au	10	—	—
Ba	13	200	3.7k
Be	4.4	—	—
Cd	1.3	3–330	1.5k–9k
Co	50	500	—
Cr	90	200	3k–8k
Cs	1200	—	—
Cu	—	—	175–250
Ga	20	—	—
Ge	500	—	—
Hg	1.5	0.4	150–300
Mn	18	—	—
Mo	140	—	—
Nd	125	—	—
Ni	110–220	—	—
Pb	70	1	10k
Pt	23	—	—
(^{239}Pu)	1	—	—
Rh	100	—	—
Sb	25	100	—
Se	1.3	5	—
Sn	35	2000	—
Te	25	—	2k
Th	18	—	—
Tl	15	600	—
U	1	—	—
V	—	18	—
Zn	—	150–600	6k

*Injected into the peritoneum to avoid absorption through the digestive tract; chemical form of the element will affect its toxicity.

[†]k signifies thousands of milligrams/day.

Source: H. J. M. Bowen (1979). *The Environmental Chemistry of the Elements* (London: Academic Press).

the curve may be fairly flat, however, if there are biochemical protection mechanisms that can accommodate low to moderate doses. For example, cadmium (see pp. 319–322) is bound by *metallothionen,* a sulfhydryl-rich protein in mammalian kidneys. When bound to the protein, cadmium is prevented from reaching critical target molecules. Toxicity increases rapidly if the metallothionen capacity is exceeded.

Cadmium, along with lead and mercury (all of which are of particular environmental concern), is a "soft" Lewis acid (large polarizability), with particular affinity for soft Lewis bases, such as the sulfhydryl side chain of cysteine residues. It is likely that the heavy metals exert their toxic effects by tying up critical cysteine residues in proteins, although the actual physiological consequences vary from one metal to another.

All metals cycle naturally through the environment. They are released from rock by weathering and are transported by a variety of mechanisms, including uptake and processing by plants and microorganisms. For example, methanogenic bacteria convert any mercuric ions they encounter to the highly toxic methylmercury (see discussion pp. 314–319), and other bacteria long ago developed a defense system involving a pair of enzymes, one that breaks the methyl-mercury bond (*methylmercury lyase*), and another that reduces the resulting mercuric ion to elemental mercury (*mercury reductase*), which volatilizes out of harm's way. Likewise, plants living on soils derived from ore bodies have evolved protective mechanisms that actively transport toxic metals from the root zone up into special compartments (*vacuoles*) in the leaves, where they are sequestered. These plants are now being pressed into service to extract metals from toxic waste sites, in *phytoremediation* schemes.

The natural biogeochemical cycles of the metals have been greatly perturbed by human intervention. Mining and metallurgy are not new developments; they extend back to the Bronze Age. But the scale of metals extraction has increased enormously since the Industrial Revolution (Figure 4.43). Evidence for a massive increase in the global environmental loading of lead, for example, can be found in the record provided by ice-cores from Greenland (Figure 4.44); these levels have significantly declined since 1970, thanks to the phasing out of lead additives from gasoline (see discussion below, pp. 322–324). Large increases in production of several metals between 1930 and 1985 are documented in Table 4.9. Also listed are the large amounts of these metals that are estimated to be dispersed into the environment and deposited on soils. In some cases (Cd and Hg), the amounts deposited are actually greater than the amount produced by extraction, because there are adventitious sources, such as ore processing for other metals or the burning of coal, which contains trace concentrations of many metals; in the case of cadmium, traces in phosphate rock, which is mined and incorporated into fertilizer, add up to a significant fraction of the total.

The biogeochemical cycles are completed by sedimentation and burial of the metals in Earth's crust. But this process requires eons, and it is clear that the current massive extraction and dispersal are greatly increasing the amount of metals in circulation. What are the consequences of this buildup for human and ecosystem health? There is no general answer to this question because health effects depend sensitively on the precise exposure routes, not only for the different metals, but for the different forms of a given metal. The physical and chemical state of the metal are all-important for transport mechanisms, and also for *bioavailability.* To exert a toxic effect, metal ions must reach their target molecules, and they may be unable to do so if tied up in an insoluble matrix or if they are unable to traverse

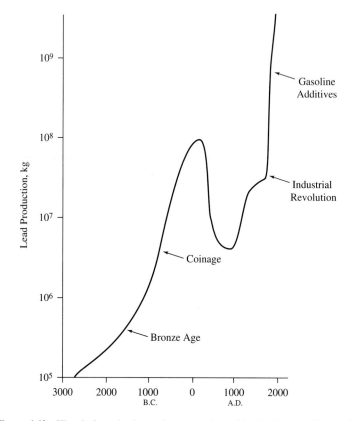

Figure 4.43   Historical production and consumption of lead.   *Source:* Adapted from
J. Nriagu (1978). *Biogeochemistry of Lead* (Amsterdam: Elsevier); and P. M. Stokes (1986).
Pathways, Cycling, and Transport of Lead in the Environment (Ottawa: Royal Society of
Canada).

critical biological membranes. In the next sections, we consider these issues for three toxic
metals of current concern.

   a. Mercury.   The environmental toxicity of mercury is associated almost entirely
with eating fish; this source accounts for some 94% of human exposure. Methanogens generate methylmercury in the sediments (p. 239) and release it into the waters above, where it
is absorbed by fish from the water passed across their gills or from their food supply. The
CH_3Hg^+ ion forms CH_3HgCl in the saline milieu of biological fluids, and this neutral complex passes through biological membranes, distributing itself throughout the tissues of the
fish. In the tissues, the chloride is displaced by protein and peptide sulfhydryl groups.
Because of mercury's high affinity for sulfur ligands, the methylmercury is eliminated only
slowly and is therefore subject to bioaccumulation when little fish are eaten by bigger fish.
The phenomenon is the same as for DDT (Figure 4.19) and other lipophiles, but the mechanism is different because mercury accumulates in protein-laden tissue rather than in fat.

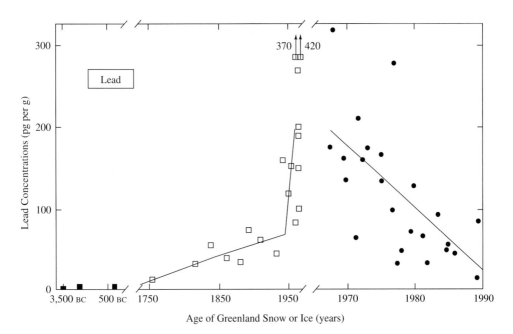

Figure 4.44   Changes in lead concentrations in Greenland ice and snow in units of *picograms* (10^{-12} g) of lead per gram of ice.   *Source:* C. F. Boutron et al. (1991). Decrease in anthropogenic lead, cadmium, and zinc in Greenland snows since the late 1960s. *Nature* 353: 153–156. Reprinted with permission, copyright © 1991 by Nature. All rights reserved.

Biomethylation of mercury occurs in all sediments, and fish everywhere have some mercury. But the levels are greatly elevated in bodies of water whose sediments are contaminated by mercury from waste effluents. The worst case of environmental mercury poisoning occurred in the 1950s in the fishing village of Minamata, Japan. A polyvinylchloride plant used Hg^{2+} as a catalyst and discharged mercury-laden residues into the bay, where

TABLE 4.9   PRIMARY PRODUCTION OF METALS AND
GLOBAL EMISSIONS TO SOIL (10^3t/yr)

Metal	Production in		Global emissions to soil in 1980s
	1930	1985	
Cd	1.3	19	22
Cr	560	9940	896
Cu	1611	8114	954
Hg	3.8	6.8	8.3
Ni	22	778	325
Pb	1696	3077	796
Zn	1394	6024	1372

Source: J. O. Nriagu (1988). A silent epidemic of environmental poisoning? *Environmental Pollution* 50: 139–161.

the fish accumulated methylmercury to levels approaching 100 ppm. Thousands of people were poisoned by the contaminated fish, and hundreds died from it. Those affected suffered numbness in the limbs, blurring and even loss of vision, and loss of hearing and muscle co-ordination, all symptoms of brain dysfunction resulting from the ability of methylmercury to cross the blood-brain barrier. Likewise methylmercury can pass from mother to fetus, and a number of Minamata infants suffered mental retardation and motor disturbance before the cause of the poisoning was identified. Based on this incident and others, the recommended limit for mercury in fish for human consumption has been set at 0.5 ppm.

Fortunately, cessation of waste mercury discharge lowers the levels of mercury in the local fish, as seen in data for Lake Saint Clair (Figure 4.45), which is part of the Great Lakes chain. The mercury concentration in walleye fish dropped from 2.0 to 0.5 ppm during the 1970s and 1980s, after the discharge of mercury from chlor-alkali plants was restricted. Even

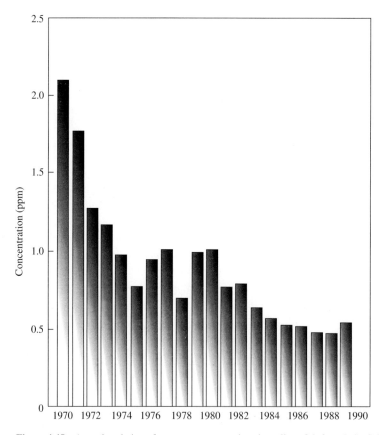

Figure 4.45   Annual variation of mercury concentrations in walleye fish from Lake Saint Clair.   *Source:* Ministry of Supplies and Services Canada (1991). *Toxic Chemicals in the Great Lakes and Associated Effects: Synopsis* (Ottawa, Canada: Ministry of Supplies and Services).

at 0.5 ppm, however, the level hovers at the recommended limit for human consumption. It takes a long time for the biomethylation process to clear the mercury from contaminated sediments.

Much of the world's discharge of mercury into the aqueous environment stems from chlor-alkali plants. These plants manufacture Cl_2 and NaOH, large-volume commodities that are mainstays of the chemical industry. They are produced by electrolysis of aqueous sodium chloride. A carbon anode is used to generate chlorine

$$2Cl^- = Cl_2 + 2e^- \tag{4.8}$$

while a mercury pool cathode collects metallic sodium as a mercury amalgam

$$2Na^+ + 2e^- = 2Na[Hg] \tag{4.9}$$

which is then reacted with water in a separate compartment

$$2Na + 2H_2O = 2NaOH + H_2 \tag{4.10}$$

Reaction (4.10) does not proceed spontaneously because the activity of sodium is depressed in the amalgam, but it can be promoted by applying a small electric current. The purpose of this mercury-mediated two-stage process is to keep the NaOH product free of the NaCl starting material. This can, however, also be accomplished by separating the two electrode compartments using a cation-exchange membrane (see pp. 212–213 for a discussion of ion exchangers), which inhibits the transfer of anions. Although chlor-alkali plants can be retrofitted to greatly reduce mercury discharges, most mercury electrode installations are being phased out and replaced by membrane-based units.

Even though local discharges have caused the most serious mercury contamination, it has been discovered that fish have elevated mercury levels even in lakes that are quite remote from any local source. Thus, mercury is transported over long distances, a consequence of the fact that there are two volatile forms, metallic mercury, Hg^0, and dimethylmercury, $(CH_3)_2Hg$. Both are formed in the same milieu as methylmercury. As mentioned previously, bacteria have a detoxification system that rids their environment of methylmercury by converting it to Hg^0, which is volatilized. And $(CH_3)_2Hg$ is produced in the same biomethylation process as CH_3Hg^+. Both molecules are produced by methanogens, in varying proportions, depending on the pH (Figure 4.46). The $(CH_3)_2Hg$ is volatilized, while the CH_3Hg^+ is released into the water and is available for bioaccumulation (see Figure 4.47 for a diagram of the mercury cycle). High pH favors $(CH_3)_2Hg$, while low pH favors CH_3Hg^+; the crossover occurs near neutrality. This means that an additional consequence of lake acidification is an increase in the $CH_3Hg^+/(CH_3)_2Hg$ ratio, and therefore an increase in mercury toxification.

Metallic mercury is used in many applications, especially in batteries, switches, lamps, and other electrical equipment; improper use and disposal (batteries in municipal incinerators, for example) add to the global load of mercury vapor. So does the use of mercury to extract gold or silver from ores, a practice used for centuries in Central and South America and applied on a large scale today in the gold fields of Brazil. This process releases massive amounts of mercury into the environment because the extracted gold is recovered by heating

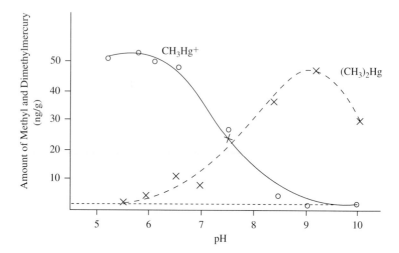

Figure 4.46 Methylation of 100 ppm of Hg^{2+} in sediments over two weeks. *Source:* I. G. Sherbin (1979). *Mercury in the Canadian Environment,* Report EPS-3-EC-79-6 (Ottawa: Environmental Protection Service Canada).

the amalgam to drive off the mercury. Large quantities of amalgam washings have contaminated parts of the Amazon River sediment with mercury. This practice is estimated to account for 2% of global atmospheric mercury emissions. Only now are simple mercury-condensing hoods being installed to reduce the losses.

Inorganic mercury is not particularly toxic when ingested because neither the metal nor the ions (Hg_2^{2+} and Hg^{2+} and their complexes) penetrate the intestinal wall effectively. However, Hg^0 is highly toxic when inhaled; in atomic form, it is able to pass through the lung membranes into the bloodstream and across the blood-brain barrier. In the brain it can presumably be oxidized and bound to protein sulfhydryl groups because it produces the same neurological effects as methylmercury. For this reason, individuals should avoid handling elemental mercury, and all spills should be treated (with sulfur, which ties up the mercury atoms) and cleaned up. The Brazilian gold miners suffer serious health problems from elemental mercury released during the amalgam operations, as silver and gold miners have for centuries.

Complexes of phenylmercury, $C_6H_5Hg^+$, have been used as paint preservatives, and as slimicides in the pulp and paper industry, but these uses have now been curtailed. Organomercurials have also been used as fungicides in agriculture and industry, especially in dressings for seed grains. Once in the soil, these compounds break down and the mercury is trapped as insoluble mercuric sulfide. However, hundreds of people died in Iraq from eating bread made from mercury-contaminated flour produced from treated seed grain that had been diverted inadvertently to a flour mill. In the United States, a New Mexico family was poisoned by eating a pig that had been fed mercury-treated seed grain. The family brought action in court, leading the EPA to ban organomercurials for seed treatment. In Sweden and Canada, birds of prey declined after eating smaller birds that had fed on treated seeds. The

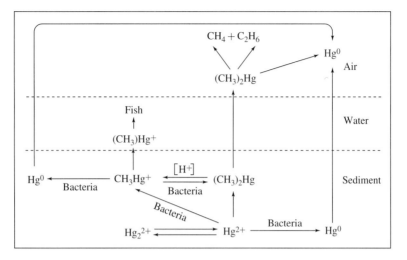

Figure 4.47   The biogeochemical cycle of bacterial methylation and demethylation of mercury in sediments.   *Source:* National Research Council (1978). *An Assessment of Mercury in the Environment* (Washington, DC: National Academy Press).

use of mercury compounds to treat seeds has now been curtailed in Europe and North America.

The natural mercury cycle can also be perturbed indirectly by human activity. Elevated mercury levels have been found in fish living in waters impounded by hydroelectric dams in Quebec and Manitoba. The source of this "pollution" was simply methanogen action on the naturally occurring mercury already present in the newly submerged surface soil.

b. Cadmium.   Cadmium is found in the same column of the periodic table as mercury and zinc, but its chemical properties are much closer to zinc than to mercury. This resemblance to zinc accounts for cadmium's distribution, as well as its particular hazards. Cadmium is always found in association with zinc in Earth's crust, and it is obtained as a side-product of zinc mining and extraction; there are no separate cadmium mines. Moreover, cadmium is always present as a contaminant in zinc products. Indeed, one of the pervasive sources of cadmium in the urban environment is zinc-treated (*galvanized*) steel. The weathering of galvanized steel surfaces produces zinc- and cadmium-laden street dust; though the concentration is low, the total amount of cadmium is substantial. It has been pointed out that proposals to ban cadmium in products (mostly batteries, electroplate, pigments, and plastics stabilizers) in order to reduce environmental exposure might have the opposite effect, because of lowered economic incentive to recover cadmium from zinc-mine residues and from refined zinc itself.*

*W. M. Stigliani and S. Anderberg (1994). Industrial metabolism at the regional level. In *Industrial Metabolism: Restructuring for Sustainable Development,* R. U. Ayres and U. E. Simonis, eds. (Tokyo: United Nations University Press).

Mimicry of zinc is probably why cadmium is actively taken up by many plants, since zinc is an essential nutrient. Most of our cadmium intake is from vegetables and grains in our diet. However, smokers get an extra dose because of the cadmium concentrated in tobacco leaves; heavy smokers have twice as much cadmium in their blood, on average, as non-smokers. The average cadmium intake per day in the United States is estimated to be 10–20 μg, but only a small fraction is absorbed; the absorbed dose is estimated to be 0.4–1.8 μg. Pack-a-day smokers take in an extra 2 μg/day on average, but inhalation greatly increases the fractional absorption. The absorbed dose is thus increased to an estimated 0.9–2.8 μg/day.

There is concern that cadmium buildup in agricultural soils may eventually produce dangerous levels in food. Cadmium inputs to soils are mainly from airborne deposition (wet plus dry) and from commercial phosphate fertilizers, which contain cadmium as a natural constituent of phosphate ore. The cadmium burden could be further increased by the use of fertilizer from sewage sludge, a sludge-disposal measure that is increasingly advocated. Sewage is often contaminated by cadmium and other metals; however, there is some evidence that the cadmium is firmly bound in the sludge and might not be released to growing plants.

The problem of cadmium accumulation has been examined extensively for the heavily industrialized Rhine River Basin, by evaluating the mass flows of cadmium over several decades in a study of the "industrial ecology" of the region. It was found that the cadmium emissions have declined substantially, thanks to the control of point sources, especially ferrous and nonferrous metal smelters (Figure 4.48; see Figure 3.33 for trends in aqueous loads of cadmium in the Rhine Basin). In contrast to the reductions in atmospheric and aqueous emissions, concentrations of cadmium in agricultural soils in the basin have increased, and may continue to do so in the future (Figure 4.49) due to residual inputs from diffuse sources (primarily atmospheric deposition from coal burning and application of phosphate fertilizer). This pattern reflects the fact that the *residence time* of cadmium in soils can be orders of magnitude longer than their lifetimes in air or river water, particularly when the pH of the soils is maintained above 6.0, as is usually the case in agricultural soils owing to additions of lime ($CaCO_3$) (Figure 3.22).

It is estimated that approximately 3,000 tons of cadmium had accumulated in the plow layer of agricultural soils of the basin from 1950 to 1990. Particularly worrisome is a scenario in which the stored cadmium would be released as a result of soil acidification. This could occur if current plans for abandoning large tracts of agricultural land in the basin are enacted. If the abandoned lands were no longer limed, the pH of the soils could drop to as low as 4.0 within several decades. The decrease in pH would result in a rapid release of cadmium out of the topsoil (Figure 3.22). This occurrence could pose public health problems where heavily polluted soils overlie shallow groundwaters used for drinking.

Soil conditions were certainly a factor in the only known case of wide-spread environmental cadmium poisoning, which occurred in the Jinzu valley of Japan. Irrigation water drawn from a river that was contaminated by a zinc mining and smelting complex led to high levels of cadmium in the rice. Hundreds of people in the area, particularly older women who had borne many children, developed a painful degenerative bone disease called *itai-itai* ("ouch-ouch"), apparently because Cd^{2+} interfered with Ca^{2+} deposition. Their bones be-

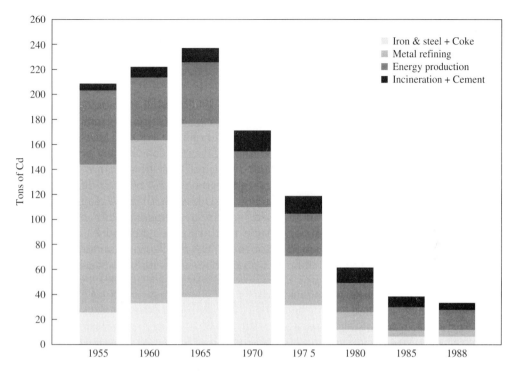

Figure 4.48 Trends in atmospheric emissions of cadmium in the Rhine Basin by industrial sector (1955–1988). *Source:* S. Anderberg and W. M. Stigliani (1995). Private communication (Laxenburg, Austria: International Institute for Applied Systems Analysis).

came porous and subject to collapse. Sufferers were estimated to have had a cadmium intake of about 60 μg/day, several times the normal intake.

Although the 70 ppm soil cadmium level in Jinzu was elevated, it has been still higher, 300 ppm, in Shipham, England, a zinc mining locale during the seventeenth to nineteenth centuries. Yet health inventories in Shipham showed only slight effects attributable to cadmium. The Shipham soils have high pH, 7.5, and also a high content of calcium carbonate and of hydrous oxides of iron and manganese, which are good absorbers of Cd^{2+} (see Table 3.7, p. 229). In contrast, the Jinzu soils had low pH, 5.1, and a low content of hydrous oxides. Thus, the cadmium was far more available for plant uptake in Jinzu than in Shipham.

Chronic exposure to cadmium has been linked to heart and lung disease, including lung cancer at high levels, to immune system suppression, and to liver and kidney disease. As mentioned above, the Cd^{2+}-sequestering protein metallothionen provides protection until its capacity is exceeded. Since metallothionen is concentrated in the kidney, this organ is damaged first by excessive cadmium. The downside of metallothionen protection is that cadmium is stored in the body and accumulates with age, so that damage from long-term exposure becomes irreversible.

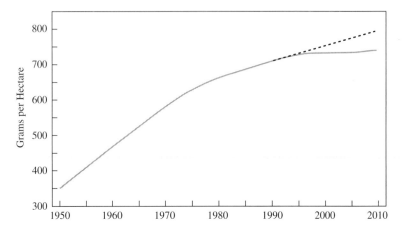

Figure 4.49   Time trend in soil concentration of cadmium in top 20 cm of typical agricultural soil in the Rhine Basin. In projections to the year 2010, solid line assumes cadmium in phosphate fertilizer is eliminated by the year 2000; the dashed line assumes no reduction of cadmium in fertilizer.   *Source:* W. M. Stigliani et al. (1993). Heavy metal pollution in the Rhine Basin. *Environmental Science and Technology* 27(5): 786–793. Copyright © 1993 by American Chemical Society. Reprinted with permission.

c. Lead.   Of all the toxic chemicals in the environment, lead is by far the one of most concern; it poisons many thousands of people yearly, especially children in urban areas.

1) Exposure.   Unlike cadmium, lead is not taken up actively by plants; nevertheless, it contaminates the food supply because it is abundant in dust and is deposited on food crops, or on food as it is being processed. Food and direct ingestion of dust account for most of the average lead intake, estimated to be about 50 μg/day in the United States. In urban areas or along roadways, the lead levels in dust generally exceed 100 ppm, while in remote rural areas the levels are 10–20 ppm. Ingestion of just 0.1 g of urban dust can provide a 10 μg or higher dose of lead. It does not take much dust, therefore, whether ingested directly or mixed in with food (the two contributions are estimated to be comparable), to account for the average daily intake. Children are particularly at risk because they play in the dust, because they absorb a higher fraction of the lead they take in, and because they are exposed to more lead per unit of body mass.

A significant route of lead exposure in addition to ingestion of food and dust is drinking water. Lead can contaminate water either from lead-based solder used in pipe and fitting connections, or from the pipes themselves, which in older houses and water systems are made of lead. It is generally advisable not to drink the "first draw" water that has been standing overnight in older drinking fountains or in the pipes of older buildings. In contact with O_2-bearing water, the metallic lead can be oxidized and solubilized:

$$2Pb + O_2 + 4H^+ = 2Pb^{2+} + 2H_2O \tag{4.11}$$

Since reaction (4.11) consumes two protons per lead ion, the dissolution rate is strongly pH-dependent. The lead hazard is greatest where the water is soft, that is, where there has been little neutralization of the rainwater's acidity (see discussion on p. 215). Hard water, on the other hand, has higher pH; in addition, it has a higher carbonate concentration (because of neutralization by $CaCO_3$). Carbonate precipitates Pb^{2+} as the sparingly soluble $PbCO_3$, which inhibits dissolution of the underlying metal in the pipes and fittings. Some water-supply districts, especially those with soft water and old lead pipes, now add phosphate to the drinking water in order to form a similarly protective coating of lead phosphate. Lead test-kits are commercially available for measuring the lead levels in one's own drinking water.

A related hazard is the practice, now largely discontinued, of using lead solder to seal food and drink in "tin" cans. When the cans are opened and the contents exposed to air, lead can be mobilized into the contents, particularly if they are acidic. Likewise, lead can leach into food and drink stored in pottery if, as is common, its glaze contains lead oxide:

$$PbO + 2H^+ = Pb^{2+} + H_2O \qquad (4.12)$$

It is particularly hazardous to drink fruit juices (because of their acidity) or hot drinks (because the rate of dissolution increases with temperature) from such vessels. Although lead-free glazes are now the rule among pottery manufacturers, lead-glazed pottery remains a significant source of dietary lead.

If lead in dust is the major exposure route, how does the lead get there? House paint is an important source. Renovating old homes is hazardous unless care is taken to contain the dust from layers of old paint. Lead salts are brightly colored and have been widely used as pigments and as paint bases. Lead chromate, $PbCrO_4$, provides the yellow coloring for striping on roads and for school buses, while the "red lead" oxide, Pb_3O_4, is the base for the corrosion-resistant paints on bridges and other metal structures. The hydroxycarbonate, $Pb_3(OH)_2(CO_3)_2$, is "white lead," which was widely used as the base of indoor paints, but has now been replaced by titanium dioxide, TiO_2. Nevertheless, older buildings, particularly in the United States, where lead was banned from indoor paint only in 1971 (much of Europe banned it in 1927), still have leaded paint on the walls. Dust and paint chips from the walls are the main source of indoor lead exposure. Leaded paints have also been widely used on building exteriors, and weathering raises the lead levels in the dust outside the building. Although less dangerous than lead inside the house, the lead outside is still a matter of concern because children often play in the dust, and ingest it, or track it inside.

The other major source of lead exposure is leaded gasoline. As discussed in connection with gasoline additives (pp. 180–181), adding tetraethyl- or tetramethyl-lead to gasoline improves its octane rating by scavenging radicals and inhibiting pre-ignition. These compounds are themselves toxic; they are readily absorbed through the skin, and in the liver they are converted to trialkyl-lead ions, R_3Pb^+, which, like methylmercury ions, are neurotoxins. However, a much greater threat to public health is the lead that spews out of the tailpipe and into the atmosphere. Most of the lead is emitted in small particles of PbX_2 (X = Cl or Br), formed by reaction with ethylene dichloride or dibromide, added to the gasoline to prevent buildup of lead deposits inside the engine. These particles can travel far on air currents, and are no doubt responsible for the sharp upturn in the lead content of Greenland ice from around 1950 (Figure 4.44), as automotive traffic expanded greatly around the world in the ensuing decades. However, most of the particles settle out not far from where they are generated, contaminating the dust near roadways and in urban areas with lead.

Lead additives were phased out of use in the United States starting in the early 1970s, when catalytic converters were introduced for pollution control, because lead particles in the exhaust gases deactivate the catalytic surfaces. By 1990, Brazil and Canada had phased out leaded gasoline, and many other countries, including Argentina, Iran, Israel, Mexico, Taiwan, Thailand, and most European nations, significantly reduced the lead concentration in leaded gasoline (from about 1 g/l to 0.15–0.3 g/l). The use of leaded gasoline is declining further in Europe and Mexico because all new cars are required to have catalytic converters. But leaded gasoline consumption continues in much of the rest of the world, and almost no unleaded gasoline is available in many parts of Africa, Asia, and South America. The global impact of the shift away from leaded gasoline is reflected in the steadily decreasing lead levels in the Greenland ice record since 1970 (Figure 4.44). The current deposition rate is now approaching pre-automobile levels. This trend could be reversed, however, if leaded gasoline is used in developing countries to meet rapidly escalating demands for automobile transport.

Although the original motivation for removing lead from gasoline was to protect catalytic converters, an additional clear benefit was the reduction in human exposure to lead. A striking correlation between the total lead used in gasoline and the blood lead levels found in the residents of selected U.S. cities is revealed by the data in Figure 4.50. By 1980, lead consumption had been cut in half, and blood lead levels had declined by about 40%, to 10 μg/dl (dl = deciliter; since one dl of blood weighs about 100 g, 10 μg/dl is about 1 ppm). Since then almost all of the remaining lead in U.S. gasoline has been eliminated, and as of 1991 the average blood lead level was down to 3 μg/dl. Although other sources of lead were also restricted in this period, the data implicate leaded gasoline as a particularly important exposure route for lead.

2) Bioavailability. Once ingested, the lead may be absorbed or eliminated. The fractional absorption is estimated to be 7–15% in adults, on average, and 30–40% in children. However, the extent of absorption depends on the chemical and physical state of the lead. Large particles of relatively insoluble lead minerals are likely to pass unaltered through the stomach and intestines, whereas small particles of soluble lead compounds are readily absorbed. No doubt this is why people living in the vicinity of lead-bearing mine tailings tend not to have elevated lead levels: the lead in the tailings occurs mostly as chunks of lead sulfide or oxide. However, people living near lead smelters do have somewhat elevated lead levels; the smelters emit fine particles of more reactive lead oxide phases. The data in Figure 4.50 suggest that the lead from gasoline is readily absorbed since it accounted for such a high percentage of average blood lead levels in the 1970s. The emitted lead is in the form of soluble lead-halide particles, which are highly absorbable. Upon contact with moisture-laden soils the halides may be converted to $Pb(OH)_2$ and then into larger particles of PbO, which are less bioavailable. This process can account for the fact that lead blood levels correlate with the phaseout of leaded gasoline without a significant time lag, even though lead levels remain elevated in the dust near U.S. roadways.

3) Toxicity. Lead toxification is as old as human history; the use of lead in artifacts dates as far back as 3800 B.C. The Greeks realized that drinking acidic beverages from lead

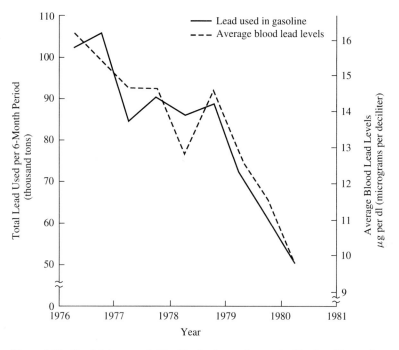

Figure 4.50   Parallel decreases in blood lead values and amounts of lead used in gasoline during 1976–80.   *Source:* U.S. Department of Health and Human Services (1988). *The Nature and Extent of Lead Poisoning in Children in the United States: A Report to Congress,* Report PB-100184 (Washington, DC: U.S. Department of Health and Human Services).

containers could result in illness. But apparently the Romans did not, for they sometimes deliberately added lead salts to overly acidic wines to sweeten the flavor. The bones of Romans have much higher lead levels than do those of modern humans, and some historians speculate that chronic lead poisoning contributed to the downfall of the Roman Empire.

Once absorbed in the body, lead enters the bloodstream and moves from there to soft tissues. In time it is deposited in bones, because Pb^{2+} and Ca^{2+} have similar ionic radii. The lead content of bones increases with age; when bone matter dissolves, as can happen in illness or old age, the lead is remobilized into the bloodstream and can produce added toxic effects. These effects are expressed primarily in the blood-forming and nerve tissues (Figure 4.51). Lead inhibits the enzymes involved in the biosynthesis of heme, the iron-porphyrin complex (Figure 4.32) that binds to hemoglobin and serves as the binding site for O_2 (see Figure 2.40). In particular, lead interferes with the *ferrochelatase* enzyme, which inserts iron into the porphyrin; instead, zinc is inserted. The resulting accumulation of zinc-porphyrin can be detected by its characteristic fluorescence emission (iron-porphyrin does not fluoresce), providing a sensitive indirect indicator of lead exposure. High exposures produce anemia due to iron-porphyrin deficiency.

The biochemical mechanism for lead's effects on nerve cells is uncertain, but diminution of nerve conduction velocity can be detected at relatively low blood lead levels (Figure 4.51), while higher levels are associated with nerve degeneration. Even at quite low

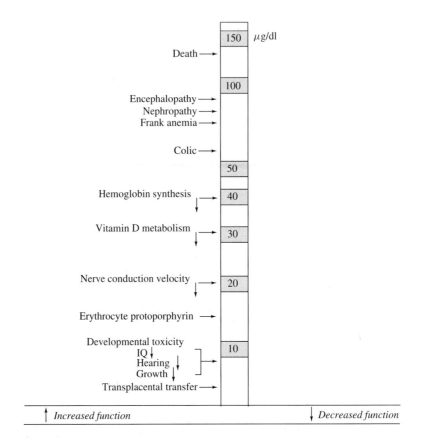

Figure 4.51   Lowest observed effect levels of inorganic lead in children (μg/dl).   *Source:*
Agency for Toxic Substances Disease Registry (1988). *The Nature and Extent of Lead Poi-*
soning in Children in the United States: A Report to Congress (Atlanta, Georgia: ATSDR).

levels, 10 μg/dl and possibly less, lead exposure in several epidemiological studies was found
to be associated with impairments in growth, hearing, and mental development of children.
In 1984, 17% of all U.S. children, and 30% of children in inner cities, were estimated to have
blood lead levels greater than 15 μg/dl. Since then exposure has diminished, as noted above,
and as of 1991, 8.9% of U.S. children aged one to five were estimated to have levels greater
than 10 μg/dl. This still represents a large population (1.7 million) at risk of experiencing developmental defects. There is also great concern about exposure *in utero* because lead crosses
the placenta and can interfere with fetal development.

The precise molecular mechanisms of lead toxicity have not been pinned down, but
probably involve lead's ability to bind to nitrogen and sulfur ligands, thereby interfering with
the function of critical proteins (like ferrochelatase). This ability is also utilized in *chelation*
therapy for lead toxicity. Lead can be cleared from the body by intravenous injection of
chelating agents (Figure 4.52). The chelators compete with protein binding sites, and the re-

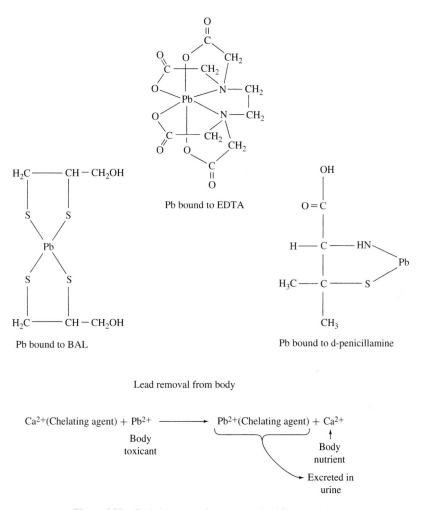

Pb bound to EDTA

Pb bound to BAL

Pb bound to d-penicillamine

Lead removal from body

$$Ca^{2+}(\text{Chelating agent}) + Pb^{2+} \longrightarrow Pb^{2+}(\text{Chelating agent}) + Ca^{2+}$$

Body
toxicant

Body
nutrient

Excreted in
urine

Figure 4.52   Chelating agents for removing lead from the body.

sulting Pb^{2+}-chelate complexes are excreted by the kidneys. Since the chelating agents can also bind metal ions other than lead, they are administered as the Ca^{2+} complex, in order to avoid stripping calcium and other more weakly bound metals from the body. The strongly binding Pb^{2+} displaces the Ca^{2+} and is removed selectively.

Lead poisoning is also common among waterfowl, which die from ingestion of lead shot and lead fishing sinkers. In the early 1980s, an estimated 2–3% of the autumn population of waterfowl in the United States were dying each year from lead poisoning. Birds that preyed on the waterfowl were likewise exposed to lead. A study of 1,429 bald eagle deaths in the United States between 1963 and 1984 found that 6% were attributable to lead poisoning. Lead shot is now banned for waterfowl hunting in the United States and will be banned in Canada by 1997.

D. SUMMARY

The biosphere plays an integral role in regulating the natural chemical cycles on the planet, but this function is vulnerable to perturbations by human activities. The animal kingdom is dependent on plants for food, and the productivity of plants is largely limited by the availability of nitrogen; "fixed" nitrogen is available naturally only from nitrogenase-containing bacteria (except for the small amount of N_2 oxidized by lightning). The expanding human population is dependent on agricultural productivity gains, which have been made possible by the production of nitrogen-containing fertilizer and the development of new plant varieties that can accommodate intensive fertilization. Commercial fertilizer now accounts for about 32% of the world's fixed nitrogen, and has greatly augmented the global nitrogen cycle. The consequences, though not fully understood, include nitrate pollution of groundwater in agricultural areas, facilitation of soil erosion, and a probable contribution to the global increase in emissions of N_2O, a greenhouse gas and major actor in stratospheric ozone depletion.

In addition to total agricultural production, adequate human nutrition depends on a balanced diet comprising a proper complement of essential amino acids and vitamins. Furthermore, antioxidants in the diet may play an important role in increased longevity through their control of free-radical levels in tissues.

Modern agriculture relies not only on fertilizer, but also on synthetic herbicides and insecticides, to boost production and control losses to insects and other pests. In addition, insecticides play a key role in disease prevention through the control of insect vectors. However, their application inevitably breeds resistance among the target insects, leading to a never-ending search for new insecticides. In the meantime, the widespread application of these chemicals has deleterious effects on ecosystems and on human health. For these reasons, more sophisticated pest-control strategies are being developed, including biological controls, and chemicals that are targeted to particular pests and that break down quickly into harmless molecules in the environment.

Organochlorine chemicals are of particular concern because they degrade only slowly in the environment, and, being lipophilic, they bioaccumulate in the food chain by concentrating in fat tissue. Birds of prey are specifically vulnerable because organochlorines interfere with calcium deposition in egg shells. As a result, many organochlorine pesticides, starting with DDT, have been banned in developed countries, although they are still used elsewhere. There are also concerns about human health, which center on cancer and on hormonal disruption. The most feared chemical, 2,3,7,8-tetrachlorodibenzodioxin, is not produced deliberately but was a contaminant of the herbicide 2,4,5-trichlorophenoxyacetic acid. It is also a trace product, along with many congeners, of any combustion process where chlorine and carbon are present in the material being burned. Incinerators are currently the main source of dioxins and the related polychlorinated dibenzofurans, although the use of pollution-control devices can eliminate most releases. Polychlorinated biphenyls, widely used in electrical transformers and other products, are also a concern even though they are no longer produced. These molecules are toxic because they bind to the arylhydrocarbon receptor present in mammalian cells and activate biochemical events leading to a variety of physiological effects, including the induction of cancer; the consequences for human health at environmental-exposure levels are still unclear, however.

Toxicity problems also result from the buildup of metals in the environment. The heavy metals bind strongly to sulfur-containing ligands and can therefore bind and interfere with critical proteins, although the actual biochemistry varies among different metals. The metals also differ markedly in their environmental chemistries and in their exposure routes. Thus, mercury pollution is mainly a problem of exposure through eating fish because of the bioaccumulation of methylmercury that is produced in sediments. Cadmium is a problem of exposure through food crops because of its active uptake by plants. And lead is mainly a problem of city children ingesting dust laden with lead from weathering paint and from the fallout of leaded gasoline. Metals pollution involves a complex interplay between patterns of industrial production, emissions, transport, and biochemistry that requires good chemical detective work to sort out and evaluate.

PROBLEM SET IV: BIOSPHERE

1. **(a)** Reductive nitrogen fixation requires the breaking of the triple bond in nitrogen ($N\equiv N$) by the reaction:

$$N\equiv N + 3H_2 \rightarrow 2NH_3 \tag{1}$$

Imagine that the overall reaction occurs by the sequential breaking of single bonds in nitrogen by the reactions:

$$N\equiv N + H_2 \rightarrow HN=NH \tag{2a}$$

$$HN=NH + H_2 \rightarrow H_2N-NH_2 \tag{2b}$$

$$H_2N-NH_2 + H_2 \rightarrow 2NH_3 \tag{2c}$$

Reaction (1) requires an activation energy of 941 kJ/mol to break the triple bond. The most energy-demanding step in process (2) is reaction (2a), which requires an activation energy of about 527 kJ/mol. Calculate the ratio of the rate constants (k_{2a}/k_1) at 27°C for reactions (1) and (2a) given that k is proportional to $e^{-Ea/RT}$, where Ea is the activation energy, R = 8.33 joules/K/mol, and T is measured in degrees Kelvin.

   (b) From the value of (k_{2a}/k_1), is it surprising that microorganisms can fix nitrogen at ambient temperatures, whereas the Haber process requires temperatures between 400 and 600°C? Calculate the temperature at which k_1 equals the value of k_{2a} at 27°C.

2. Given the flows of nitrogen shown in Figure 4.5, during preindustrial times was there any net accumulation in the stocks of nitrogen on land or in the oceans? Is there any accumulation under current nitrogen flows? If so, calculate the net stock changes. Describe the environmental effects of increased nitrogen stocks in the terrestrial biosphere from the point of view of water quality of lakes, rivers, groundwaters, and coastal marine areas (see Part III, pp. 228–237).

3. Figure 4.10 shows the relation between corn yield and nitrogen fertilizer application. With respect to corn grown on "Plainfield loamy sand," estimate the yields for four cases: 1) no fertilizer

applied; 2) 100 kg N applied; 3) 160 kg (current usage) applied; and 4) 200 kg applied. Given that corn is about 1.3% nitrogen by weight, calculate the percent of applied nitrogen contained in the harvested corn for the latter three cases. Are we reaching the point of diminishing returns with respect to nitrogen application? Where does the residual nitrogen go?

Problems 4 and 5 refer to the following table:

GRAMS OF CORN AND BEAN PROTEIN REQUIRED TO GIVE
DAILY REQUIREMENTS OF ESSENTIAL AMINO ACIDS

	Corn	Beans
Tryptophan	70.0	47.9
Lysine	72.2	33.0
Threonine	37.5	44.3
Leucine	27.4	44.8
Isoleucine	49.0	39.3
Valine	38.8	44.9
Phenylalanine	35.0	39.3
Methionine	33.3	56.7

4. Given that corn consists of 7.8% protein and only 60% of it is absorbed by the digestive tract, calculate the number of grams of corn that must be ingested for an adequate protein intake. There are 368 kilocalories in 100 g of corn. How many calories must be consumed daily in order to get adequate protein? Given that the average American consumes 2,000–3,000 kcal daily, can an all-corn diet supply the proper balance of protein and energy?

5. Given that beans are 24% protein and 78% of the protein is absorbed by the digestive tract, calculate the number of grams of corn and beans that must be consumed to supply adequate protein for a person whose diet is half corn and half beans. If there are 338 kcal in 100 g of beans, calculate the number of calories consumed in this diet (see problem 4 for additional data for corn).

6. In 1990, grain consumption in the United States, with a meat-oriented diet, was roughly 0.80 metric tons per capita. This included grain eaten indirectly in the form of livestock products; about 70% of the grain produced was fed to livestock. In India, where the diet is basically vegetarian, grain consumption was about 0.20 tons per capita in 1990. In China, a country where diet is in transition from basically vegetarian to meat-oriented, roughly 0.30 metric tons per capita were consumed; 20% went to feed animals. China's growing affluence has resulted in a dramatic rise in meat consumption, from about 10 million tons in 1980 to around 45 million in 1994. Assuming that by the year 2025, China, with a projected population of 1.5 billion, has the same per capita grain requirements as did the United States in 1990, calculate the grain demand of China in 2025. How does this compare with a total global grain production in 1990 of 1.86 billion tons? The global population in 2025 is expected to be 8.5 billion. If global grain demand were 0.80 tons per capita, how much grain would be required? How much grain would be required in 2025 if everyone survived on a vegetarian diet?

7. What role do antioxidants play in protecting biological systems? Name two kinds of natural antioxidants. What role do natural chelating agents play in reducing free radical formation?

8. Suppose an insecticide is applied to a crop to eradicate fruit flies. Assume that one out of one million fruit flies possesses an enzyme that breaks down the insecticide into nontoxic metabolic products. Assume further that as the normal fruit flies die off quickly, the population of the resistant flies increases geometrically (that is, 1, 2, 4, 8, . . .). If a new generation occurs every 23.5 days, in how many days will the fruit fly population be restored?

9. Give an example of how a broad-spectrum, persistent insecticide like DDT can disrupt entire ecosystems.

10. **(a)** Consider a *para*-substituted phenyldiethylphosphate insecticide of the following structure:

$$X\text{—}\langle\bigcirc\rangle\text{—}O-\overset{\overset{\textstyle O}{\|}}{P}(OC_2H_5)_2$$

Why does the toxicity of the insecticide increase as the electron-withdrawing capacity of X increases?

(b) The electronegativity value of P is 2.1, that of S is 2.5, and for O it is 3.5. Using this information, explain why phosphothionate insecticides (containing phosphorus-sulfur double bonds rather than phosphorus-oxygen double bonds) are unreactive with cholinesterase but are activated in the insect's body by oxidizing enzymes. What advantage is there to using sulfur-containing insecticides?

11. Describe alternative ways to control insect populations other than the use of synthetic insecticides.

12. Caffeine is thought to be safe under normal use. However, in large amounts it is poisonous to humans. Assuming that humans and rats are equally sensitive to the toxic effects of caffeine, how many cups of coffee drunk consecutively would be lethal to 50% of a group of 150-pound humans? Assume one cup of strong coffee contains about 140 mg of caffeine. Use the data presented in Table 4.2 for the LD_{50} value of caffeine for rats. Is it likely that a person could die from drinking too much coffee?

13. What approaches have been adopted to study chronic diseases such as cancer? What are the limitations of each of these approaches?

14. Why are lipophilic xenobiotics generally more harmful and pervasive than hydrophilic xenobiotics in the environment?

15. Ample evidence indicates that some cancers are caused by exposure to carcinogenic chemicals. Using benzanthracene and dimethylnitrosamine as examples (Figures 4.31 and 4.33), explain how the body actively participates in its own destruction. Correlate your reasoning with the fact that the Ames bacterial test for known carcinogens gives negative results unless a mammalian liver extract is added to the test medium.

16. Explain what a hormone is and describe its mechanism of action. What properties of certain environmental chemicals cause them to disrupt hormonal balance? Give two examples of deleterious health effects attributable to hormonal dysfunction caused by lipophilic xenobiotics.

17. Why do the geometric dimensions of TCDD make it so toxic? Why is toxicity diminished in other congeners where chlorines are removed from the 2,3,7,8 positions, or when chlorines are added to remaining positions on the rings?

18. Name the uses of PCBs before they were banned in the late 1970s. With reference to Figure 4.29, how would you surmise that PCBs ended up in umbilical cord serum of pregnant women living near Lake Michigan in 1986?

19. In the human body, the half-life of methylmercury is 70 days, and for Hg^{2+} it is 6 days. Why is there such a marked difference in half-lives, and why is methylmercury so much more toxic than inorganic mercury? What is the maximum body accumulation of each type of mercury at a constant ingestion rate of 2 mg Hg/day? (*Hint:* the maximum concentration in the body can be calculated from the following equation:

$$C_{max}/C_0 = e^{-\lambda}[1/(1 - e^{-\lambda})]$$

where C_{max} is the maximum concentration, C_0 is the initial concentration per time step, $t_{1/2}$ is the half-life, and $\lambda = 0.693/t_{1/2}$.)

20. Relate how mercury accumulation in the food chain led to the Minamata disaster. Why is the conversion to methyl mercury the key step in mercury toxicity?

21. Describe how acidification acts to increase the risks to the environment and human health posed by cadmium, lead, and mercury.

22. Describe the main routes of exposure of lead. Why are children most susceptible to lead poisoning? How can lead poisoning be treated?

SUGGESTED READINGS IV: BIOSPHERE

W. C. CLARK and R. E. MUNN, eds. (1986). *Sustainable Development of the Biosphere* (Cambridge, UK: Cambridge University Press on behalf of the International Institute for Applied Systems Analysis, Laxenburg, Austria).

World Commission on Environment and Development (WCED) (1987). *Our Common Future* (Oxford, UK: Oxford University Press).

A. P. KINZIG and R. H. SOCOLOW (1994). Human impacts on the nitrogen cycle. *Physics Today* November, 1994: 24–31.

L. R. BROWN (1995). *Who Will Feed China? Wake-up Call for a Small Planet* (New York: W. W. Norton & Company).

National Research Council (1989). *Alternative Agriculture* (Washington, DC: National Academy Press).

National Research Council (1993). *Soil and Water Quality* (Washington, DC: National Academy Press).

W. SALOMONS and W. M. STIGLIANI, eds. (1995). *Biogeodynamics of Pollutants in Soils and Sediments, Risk Assessment of Delayed and Non-Linear Responses* (Berlin: Springer-Verlag).

C. C. TRAVIS and S. T. HESTER (1991). Global chemical pollution. *Environmental Science and Technology* 25: 814–819.

U.S. Environmental Protection Agency (1995). *1993 Toxics Release Inventory,* EPA 745-R-95-010 (Washington, DC: Office of Pollution Prevention and Toxics) (published reports, annually).

R. CARSON (1962). *Silent Spring* (Boston, Massachusetts: Houghton Mifflin).

K. MELLANBY (1992). *The DDT Story* (Farnham, Surrey, UK: British Crop Protection Council).

D. L. DAVIS, G. E. DINSE, and D. G. HOEL (1994). Decreasing cardiovascular disease and increasing cancer among whites in the United States from 1973 through 1987. *Journal of the American Medical Association* 271: 431–437.

L. A. COHEN (1987). Diet and cancer. *Scientific American* 257(5): 42–48.

B. HILEMAN (1993). Concerns broaden over chlorine and chlorinated hydrocarbons. *Chemical and Engineering News* 71(16): 11–20.

D. J. HANSON (1991). Dioxin toxicity: new studies prompt debate, regulatory action. *Chemical and Engineering News* 69(32): 7–14.

V. M. THOMAS and T. G. SPIRO (1996). The U.S. dioxin inventory: are there missing sources? *Environmental Science and Technology* 30(2): 82A–85A.

G. GRIBBLE (1994). The natural production of chlorinated compounds. *Environmental Science and Technology* 25(7): 310A–318A.

R. A. HITES (1990). Environmental behavior of chlorinated dioxins and furans. *Accounts of Chemical Research* 23: 194–201.

J. L. JACOBSON, S. W. JACOBSON, and H. E. B. HUMPHREY (1990). Effects of in utero exposure to polychlorinated biphenyls and related contaminants on cognitive functioning in young children. *Journal of Pediatrics* 116: 38–44.

J. O. NRIAGU (1990). Global metal pollution. *Environment* 32(7): 7–11, 28–33.

O. MALM, W. C. PFEIFFER, C. M. M. SOUZA, and R. REUTHER (1990). Mercury pollution due to gold mining in the Madeira River Basin, Brazil. *Ambio* 19(1): 11–15.

W. M. STIGLIANI and S. ANDERBERG (1994). Industrial metabolism at the regional level: The Rhine Basin. In: *Industrial Metabolism—Restructuring for Sustainable Development,* R. U. Ayres and U. E. Simonis, eds. (Tokyo: United Nations University Press).

R. SOCOLOW, C. ANDREWS, F. BERKHOUT, and V. THOMAS, EDS. (1994). *Industrial Ecology and Global Change* (Cambridge, UK: Cambridge University Press).

Appendix A

Forms and
Units of Energy

I. Heat and work are interconvertible forms of energy

    A. *Heat* is energy that flows from one body to another because of a difference in their temperatures. Common units of heat energy are:

        1. One calorie* is the heat needed to raise the temperature of 1 g of water 1°C (specifically, from 14.5 to 15.5°C, since the heat capacity of water varies somewhat with temperature).

        2. One British thermal unit (Btu) is the heat needed to raise the temperature of 1 lb of water 1°F (from 63 to 64°F).

    B. *Work* is energy that is transmitted when a force acts against a resistance to produce motion. Work = force × distance, and force = mass × acceleration. Common units of work energy are:

        1. One joule is the work done by a force that accelerates a 1-kg mass at 1.0 m/sec^2 for a distance of 1 m.

        2. One erg is the work done by a force that accelerates a 1-g mass at 1.0 cm/sec^2 for a distance of 1 cm.

        3. One foot-pound (ft-lb) is the work done by a force of 1 lb (a 453.59-g mass accelerated at 32.1740 ft/sec^2) over a distance of 1 ft.

II. Kinetic and potential energy; Power.

    A. Kinetic energy (K) is the work that a body can do by virtue of its motion:

$$K = (1/2)(\text{mass}) \times (\text{velocity})^2$$

---

*Nutritionists use the Calorie (capital C), which equals 1,000 calories, or 1 kilocalorie (kcal).

B. Potential energy is the work that a system of bodies is capable of doing by virtue of the relative position of its parts, that is, by virtue of its configuration (for example, water falling from a height, or chemicals reacting in a battery).

C. *Power* is the rate at which work is done. Power = work/time. Common units of power are:

1. 1 watt (w) = 1 joule/second
2. 1 horsepower = 33,000 ft-lb/minute

(*Note:* power × time = work; hence, watts × seconds = joules; watts × hours = 3,600 joules; and 1 kilowatt-hour (kWh) = 3.6×10^6 joules.)

III. Radiant energy.

Energy is also carried by light waves. The photons of electromagnetic radiation have no rest mass and travel with a constant velocity, depending only on the medium (air, water, glass, and so on). In a vacuum (empty space), the velocity of light is $c = 3 \times 10^{10}$ cm/second. The oscillations of the light wave are measured by the wavelength, λ (in meters or subunits of meters), or by the frequency, $v = c/\lambda$ (in cycles per second, cps, or hertz). The energy of a light wave is directly proportional to its frequency, $E = hv$, where h is Planck's constant, 6.626×10^{-27} erg · second. Often the frequency is expressed in wavenumbers, $\bar{v} = 1/\lambda$ (in cm^{-1}), leaving out the velocity of light.

The electromagnetic spectrum covers a wide range of wavelengths, from radiowaves (λ in meters) through microwaves (centimeters), infrared waves (micrometers, μ), ultraviolet waves (tenths of μ), x-rays (angstroms, Å, that is, atomic dimensions), and γ rays (thousandths of Å, that is, nuclear dimensions). The visible region of the spectrum, to which the human eye responds, is quite narrow, from about 0.4 μ (blue light) to about 0.7 μ (red light). This is also the region of maximum sunlight intensity.

IV. Some useful units and conversion factors:

A. Length: meters (m)

1 meter = 10^{-3} kilometer (km) = 10^2 centimeter (cm) = 10^3 millimeter (mm) = 10^6 micrometer (μ) = 10^9 nanometer (nm) = 10^{10} angstrom (Å)

B. Mass: grams (g)

1 gram = 10^{-3} kilogram (kg) = 10^3 milligram (mg) = 10^6 microgram (μg) = 10^9 nanogram (ng) = 10^{12} picogram (pg)

C. Temperature: degrees Kelvin* or absolute (K) = degrees centigrade (°C) + 273. Degrees Fahrenheit (°F) = (9/5) °C + 32

D. Energy: as we have seen, there are many units of energy.

joule: this unit has been adopted as the fundamental unit of energy of the International System of Units.

1 joule = 0.239 calories; 1 kJ = 1,000 joules

calorie:

1 calorie = 4.184 joules

*Degrees Kelvin are indicated by K, without the standard degree sign (as in degrees centigrade, °C).

British thermal unit (Btu): this unit of heat energy is used in engineering practice.

$$1 \text{ Btu} = 1.05 \text{ kJ} = 252 \text{ calories}$$

kilowatt-hour (kWh): this unit is used in measuring electrical energy production or consumption.

$$1 \text{ kWh} = 3.6 \times 10^3 \text{ kJ} = 860.4 \text{ kcal}$$

cm^{-1}: the energy of light waves is proportional to their frequency, which is usually expressed as the wavenumber ($\bar{v} = 1/\lambda$) in cm^{-1}. For example, green light with a 500-nm wavelength has a wave number, or "energy," of $(500 \times 10^{-7} \text{ cm})^{-1} = 20,000 \text{ cm}^{-1}$. It is possible to relate the wave number to the equivalent quantity of heat, or chemical energy per mole of photons; thus

$$1 \text{ kJ/mol} = 1463.6 \text{ cm}^{-1}$$

electron volts (ev): this unit is commonly used by physicists in describing radiation and elementary particles. It is the amount of energy acquired by any charged particle that carries unit electric charge when it falls through a potential difference of 1 volt. It can be related to the equivalent wave number of electromagnetic radiation:

$$1 \text{ ev} = 8,064.9 \text{ cm}^{-1}$$

A commonly used multiple of this unit is the megaelectron volt = Mev = 10^6 ev.

Appendix B

Exponential Growth and Decay

If a substance disappears at a rate that is proportional to its amount, it undergoes exponential decay. The rate of disappearance is expressed by

$$-dQ/dt = kQ \qquad\qquad (B.1)$$

where Q is the amount of the substance, t is time, and k is the proportionality constant (rate constant). Integration of this expression gives an equation for the exponential dependence of Q on time:

$$Q = Q_0 e^{-kt} \qquad\qquad (B.2)$$

where Q_0 is the amount present initially. This dependence is shown in Figure B.1. It is characterized by a constant half-life, $t_{1/2}$, that is, the time it takes for Q to decrease to half its initial value ($Q/Q_0 = 1/2$):

$$t_{1/2} = -\ln(1/2)/k = 0.693/k$$

Radioactivity is a process that follows exponential decay, since it involves spontaneous (random) disintegration of unstable atomic nuclei. The plutonium decay curve in Figure 1.17 (p. 35) is an example. The number of disintegrations per second depends only on the number of nuclei present (Q) and on the probability of disintegration (k), which is characteristic of the particular composition of the nucleus. Each type of unstable nucleus (isotope) has a characteristic half-life.

Similarly, if Q increases at a rate proportional to the amount present, then

$$dQ/dt = kQ \qquad\qquad (B.3)$$

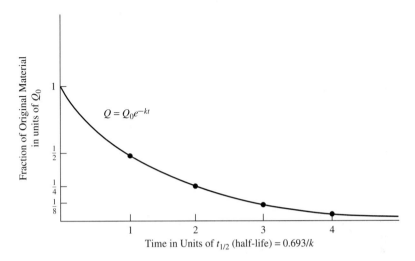

Figure B.1   Plot of exponential decay.

and it undergoes *exponential growth:*

$$Q = Q_0 e^{kt} \tag{B.4}$$

There is a constant doubling time, t_2, in which Q increases to twice its initial value $(Q/Q_0 = 2)$:

$$t_2 = \ln(2)/k = 0.693/k \tag{B.5}$$

Often the rate of growth, R, is expressed as a percentage per unit of time, for example, an annual percentage growth. If this percentage is not large,* it can be approximated as

$$R \approx dQ/Qdt$$

which, as we can see from equation (B.3) is equal to k. Therefore t_2 is about $0.693/R$. For example, a 7% annual growth rate corresponds to a doubling time of just about 10 years.

Exponential growth cannot be maintained indefinitely, without running into some sort of limitation. Growth curves in nature often look like the S-shaped curve in Figure B.2. The early part of the curve resembles the exponential curve, but after a time, growth slows as the limit (for example, resource exhaustion) is approached. The dotted curve in the figure repre-

---

*If the percentage is large, then the differential is a poor approximation. The exact relation is

$$R = (Q - Q_0)/Q_0 t = (Q/Q_0) - 1, \text{ since } t = 1$$

or

$$Q/Q_0 = R + 1$$

Using equation (B.4), with $t = 1$, we have

$$\ln Q/Q_0 = \ln(R + 1) = k$$

and, from equation (B.5),

$$t_2 = 0.693/\ln(R + 1)$$

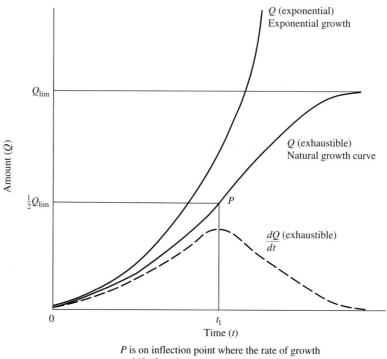

P is on inflection point where the rate of growth
shifts from increase to decrease

Figure B.2   Exponential and natural growth curves.

sents the rate of production of Q, dQ/dt, as a function of time. This rate also increases rapidly at first, but then reaches a peak and afterwards declines as the total production of Q reaches its limit. The equation for this bell-shaped curve is

$$dQ/dt = (2\pi)^{-1/2}Q_{\text{lim}}e^{-1/2(t - t_1)^2} \tag{B.6}$$

where Q_{lim} is the limiting quantity of Q and t_1 is the time at which dQ/dt reaches its maximum value. The S-shaped curve is represented by the integral of equation (B.6).

Appendix C

Redox Reactions

1. Half-reactions

Redox reactions are as important as acid-base reactions in controlling the chemistry of natural waters. Aqueous solutions can support a variety of redox levels for most elements. For example, iron can exist in water as Fe^{3+} or Fe^{2+}, and other transition metals can exist in more than one positive valence. Nonmetal elements attain even greater ranges of oxidation state in their compounds, often the full range available in the valence shell. Thus, sulfur is found naturally in oxidation states from IV (SO_4^{2-}) to –II (H_2S), while nitrogen likewise ranges from oxidation state V (NO_3^-), to –III in (NH_3), and carbon from IV (CO_3^{2-}) to –IV (CH_4).

During redox reactions, electrons are transferred from reductant molecules to oxidant molecules. All such reactions can be divided, at least conceptually, into two reduction *half-reactions,* one proceeding forward and the other in reverse. For example, the oxidation of hydrogen by oxygen,

$$2H_2 + O_2 = 2H_2O \tag{C.1}$$

can be divided into

$$O_2 + 4e^- + 4H^+ = 2H_2O \tag{C.2}$$

and
$$4H^+ + 4e^- = 2H_2 \tag{C.3}$$

Subtracting half-reaction (C.3) from (C.2) gives the overall reaction (C.1). These half-reactions can actually be carried out at the electrodes of a hydrogen-oxygen fuel cell, as discussed in Part I. A potential difference is developed between the oxygen electrode and the hydrogen electrode, allowing a current to flow through the external circuit. For the hydrogen-oxygen fuel cell, this potential difference approaches 1.24 volts (V) when the gases are at 1 atmosphere pressure, and the electrodes are behaving reversibly, that is, the reactants and products are at equilibrium with the electrodes (implying rapid electron transfer rates).

The potential difference, E, is the energy of an electrochemical cell per unit of charge delivered. Specifically $1 \text{ V} = 1 \text{ J/C}$, where V = volt, J = joule and C is a coulomb of charge. E is related to the free energy of the cell reaction by $\Delta G = -nFE$, where F (the Faraday) is the amount of charge in a mole of electrons, 96,500 C, and n is the number of electrons transferred in the reaction. Thus in reaction (C.1), 4 electrons are transferred from $2H_2$ to O_2, and $\Delta G = -4 \times 96,500 \times 1.24 = -479,000$ J, or -479 kJ. The potential is an intensive property of the electrode processes, whereas the free energy depends on the number of moles of reactants. For example, if reaction (C.1) had been written

$$H_2 + 1/2\, O_2 = H_2O \tag{C.1'}$$

the potential difference would be unaltered, but the free-energy change would be half as great, -240 kJ, because one mole of H_2 was being oxidized instead of two.

Another electrode combination might involve the half-reactions

$$Fe^{3+} + e^- = Fe^{2+} \tag{C.4}$$

and

$$H^+ + e^- = 1/2\, H_2 \tag{C.3'}$$

Subtracting C.3′ from C.4 gives the cell reaction

$$Fe^{3+} + 1/2\, H_2 = Fe^{2+} + H^+ \tag{C.5}$$

The electrochemical potential for this cell under standard conditions, that is, 1 atm pressure of H_2 and 1 M concentration of the ionic reactants and products, is 0.77 V. Thus there is a substantial driving force for the oxidation of H_2 by Fe^{3+}, although it is smaller than the driving force for the oxidation of H_2 by O_2. From this we conclude that O_2 would also oxidize Fe^{2+}, and we can readily calculate that the potential for a cell carrying out the reaction

$$4Fe^{2+} + O_2 + 4H^+ = 4Fe^{3+} + 2H_2O \tag{C.6}$$

would be $1.24 - 0.77 = 0.47$ V.

Numerous electrode combinations are possible, and it is convenient to specify a *standard potential,* E°, for each electrode by referencing it to the hydrogen electrode, whose standard potential is defined as zero. Thus $E^\circ = 1.24$ V for the oxygen electrode, represented by half-reaction (C.2), and $E^\circ = 0.77$ V for the $Fe^{3+/2+}$ electrode, represented by half-reaction (C.4). The standard conditions for E° are unit activities (partial pressure or molar concentration) of the reactants and products, at 25°C.

There are many half-reactions whose electrode potential cannot actually be measured, because the electron transfer reaction is too slow. These potentials can nevertheless be calculated from the free energy of appropriate redox reactions. For example, the formation of NO from N_2 and O_2, whose thermodynamics was considered in Part II, is a redox reaction (2.13):

$$O_2 + N_2 = 2NO$$

which can be divided into the half-reactions

$$O_2 + 4e^- + 4H^+ = 2H_2O \tag{C.2}$$

and

$$2NO + 4e^- + 4H^+ = N_2 + 2H_2O \tag{C.7}$$

From the free energy of reaction (2.13) (p. 133), 173.4 kJ, we obtain a cell potential of –0.45 V, and knowing that the standard potential of the oxygen electrode is 1.24 V, we can readily calculate that the standard potential for half-reaction (C.7) is 1.69 V, even though it is impossible to measure this potential directly because the electron transfer between the electrode and the NO and N_2 molecules is too slow to establish a reversible potential. Table C.1 lists standard potentials for a number of half-reactions that are important in natural waters.

2. Concentration Dependence: pE

As in all chemical reactions, the driving force for electrochemical processes depends on the concentrations of reactants and products. This dependence is given by the Nernst equation

$$E = E^\circ - [RT/nF]\ln K \tag{C.8}$$

where E° is the standard potential, R is the gas constant, n is the number of electrons transferred in the reaction, and K is the equilibrium quotient. In the fuel cell reaction (C.1), for example, $K = 1/P_{O_2}P_{H_2}^2$ (the water activity being defined as unity), and $n = 4$. Therefore

$$E = 1.24 - [RT/4F] \times [-\ln P_{O_2} - 2\ln P_{H_2}] \tag{C.9}$$

(The same result is obtained if we start with reaction (C.1′) instead, because the stoichiometric coefficients are halved but so is the number of electrons ($n = 2$)). A convenient form of the Nernst equation is

$$E = E^\circ - [0.059/n]\log K \tag{C.10}$$

TABLE C.1   THERMODYNAMIC SEQUENCE FOR REDUCTION OF IMPORTANT ENVIRONMENTAL OXIDANTS AT pH 7.0

Reaction	$E^\circ[w]^*$
Disappearance of O_2	
$O_2 + 4H^+ + 4e^- \rightleftharpoons 2H_2O$	0.816 V
Disappearance of NO_3^-	
$NO_3^- + 2H^+ + 2e^- \rightleftharpoons NO_2^- + H_2O$	0.421 V
Formation of Mn^{2+}	
$MnO_2 + 4H^+ + 2e^- \rightleftharpoons Mn^{2+} + 2H_2O$	0.396 V
Reduction of Fe^{3+} to Fe^{2+}	
$Fe(OH)_3 + 3H^+ + e^- \rightleftharpoons Fe^{2+} + 3H_2O$	–0.182 V
Formation of H_2S	
$SO_4^{2-} + 10H^+ + 8e^- \rightleftharpoons H_2S + 4H_2O$	–0.215 V
Formation of CH_4	
$CO_2 + 8H^+ + 8e^- \rightleftharpoons CH_4 + 2H_2O$	–0.244 V

*$E^\circ[w]$ is the value of E at 25°C and unit activities of reactants and products, except that pH = 7.

Source: F. J. Stevenson (1986). *Cycles of Soil* (New York: John Wiley).

where 0.059 is the value of RT/F at 25°C, multiplied by the conversion factor from natural to base-ten logarithms [ln10 = 2.303]. For temperatures other than 25°C, the factor 0.059 must be raised or lowered accordingly.

The Nernst equation applies equally to whole cell reactions or half-reactions. For example, the potential of the $Fe^{3+/2+}$ electrode (half-reaction (C.4); $n = 1$) at 25°C is

$$E = 0.77 - 0.059\ln\{[Fe^{2+}]/[Fe^{3+}]\} \tag{C.11}$$

From this we see that the standard potential is maintained as long as K = 1, that is, $[Fe^{2+}] = [Fe^{3+}]$. The potential increases when $[Fe^{3+}]$ exceeds $[Fe^{2+}]$ and decreases when $[Fe^{2+}]$ exceeds $[Fe^{3+}]$. Thus, the electrode potential is a reflection of the availability of electrons in the solution. It is higher when excess oxidant $[Fe^{3+}]$ is present and lower when excess reductant $[Fe^{2+}]$ is present.

This relationship of the potential to electron availability can be made explicit by expressing the potential as pE, the negative logarithm of the "effective" concentration of electrons. This is done by dividing the value of the potential by 2.303RT/F. The Nernst equation then becomes

$$pE = pE° - n^{-1}\log K \tag{C.11'}$$

In the case of the $Fe^{3+/2+}$ electrode at 25°C, for example, pE° = 0.77/0.059 = 13.2, and

$$pE = 13.2 - \log\{[Fe^{2+}]/[Fe^{3+}]\} \tag{C.12'}$$

There is a formal analogy between pE and pH and between pE° and pK_a. Of course, pE does not literally give the concentration of solvated electrons, since the electrode potential scale is arbitrary. The standard hydrogen electrode is defined to have E = 0, giving pE = 0, but the electron concentration is certainly not 1 M. Nevertheless, the pE is controlled by the concentrations of oxidant/reductant couples in the same way that the pH is controlled by the concentrations of acids and their conjugate bases.

3. Electron and Proton Affinities Are Linked: pE versus pH

Most reduction reactions are accompanied by proton uptake, and oxidations generally lead to proton release. Since adding an electron increases negative charge while adding a proton decreases it, the coupling of electron and proton transfers is a simple consequence of the tendency to lower the energy of the molecule by neutralizing charge. This coupling leads to a strong dependence on the solution pH for most half-reaction potentials.

The hydrogen electrode potential is pH-dependent because two protons combine with two electrons in producing H_2. From reaction (C.3), or (C.3')

$$pE = 0 - [1/2]\log P_{H_2} + \log[H^+] \tag{C.13}$$

If p_{H_2} is maintained at 1 atm, then pE = –pH. Thus, hydrogen gas is far more reducing in alkaline than in acid solution. For example, at pH 8, pE = –8, and the hydrogen electrode potential is –0.47, nearly half a volt more negative (reducing) than at pH = 0. At the same time, oxygen is harder to reduce in alkaline solution (or, conversely, O_2 is more oxidizing).

From reaction (C.2)

$$pE = pE^{\circ} + [1/4]\log P_{O_2} + \log[H^+] = 20.75 - pH, \text{ at } P_{O_2} = 1 \text{ atm} \qquad (C.14)$$

and the electrode potential at pH 7 is 0.83 V. Because pH 7 is closer to most biologically and environmentally relevant conditions than is pH 0, electrode potentials and pE values are often cited for pH 7, as in Table C.1. The $pE^{\circ}[w]$ values are pE° values recalculated for pH 7.

Since both the hydrogen and oxygen electrodes have the same pH dependence, the difference between them is pH-independent, reflecting the fact that there is no gain or loss of protons in the overall reaction for hydrogen oxidation by oxygen [reaction (C.1)]. Thus, the potential of the hydrogen-oxygen fuel cell is independent of the pH of the cell compartments, even though the individual electrode potentials are strongly affected.

Even if no protons appear explicitly in a half-reaction, the potential may be pH-dependent because of secondary acid-base reactions. For example, the potential of the Fe^{3+} reduction half-reaction (C.4) potential has no proton dependence *per se,* but the equilibrium quotient $[Fe^{2+}]/[Fe^{3+}]$ is highly dependent on pH because of the acidic character of Fe^{3+}. At quite low pH, it forms a series of hydroxide complexes, and precipitates as the highly insoluble $Fe(OH)_3$ ($pK_{sp} = 38$). In contrast, Fe^{2+} forms hydroxide complexes only at high pH, and $Fe(OH)_2$ ($pK_{sp} = 15$) is more soluble than $Fe(OH)_3$. Consequently, the reduction potential falls with increasing pH, because $[Fe^{3+}]$ declines more rapidly than $[Fe^{2+}]$.

The relationship between pE and pH is conveniently illustrated in a diagram such as that shown for the $Fe^{3+/2+}$ couple in Figure C.1. The regions of the diagram are labeled according to the dominant chemical species present, and the lines show the pE/pH dependence at the edges of these stability fields. Thus, the horizontal line at the top left of the diagram represents $pE = 13.2$, the value expected for an equimolar solution of Fe^{3+} and Fe^{2+} in the absence of hydroxide reactions. Fe^{3+} predominates above this line, while Fe^{2+} predominates below the line. The vertical line at $pH = 3.0$ arises because of the precipitation of $Fe(OH)_3$. This happens when the K_{sp} is exceeded, which depends on the pH and on $[Fe^{3+}]$. For the purposes of illustration, the iron concentration was set at 10^{-5}M in drawing Figure C.1. From

$$K_{sp} = 10^{-38} = [Fe^{3+}][OH^-]^3 \qquad (C.15)$$

we calculate $[OH^-] = \{10^{-38}/[Fe^{3+}]\}^{1/3} = 10^{-11}$, giving pH = 3. Above this pH, $Fe(OH)_3$ precipitates, and $[Fe^{3+}]$ declines in conformity to the K_{sp} and the pH. The effect of this decline on pE is seen in the line sloping downward from pH = 3. This line has a slope of 3.0 because of the three hydroxide ions per iron in $Fe(OH)_3$. Rearranging equation (C.15), we have

$$\log[Fe^{3+}] = -38 + 3pOH \qquad (C.15')$$

and since pOH = 14 − pH

and

$$pE = 13.2 - \log\{[Fe^{2+}]/[Fe^{3+}]\} \qquad (C.16')$$

the dependence of pE on pH is given by

$$pE = 13.2 - \log[Fe^{2+}] + \log[Fe^{3+}] = 22.2 - 3pH \text{ (with } [Fe^{2+}] = 10^{-5} \text{ M)}$$

Above this line, Fe^{2+} is oxidized and precipitates as $Fe(OH)_3$, while below the line $Fe(OH)_3$ dissolves by reduction to Fe^{2+}.

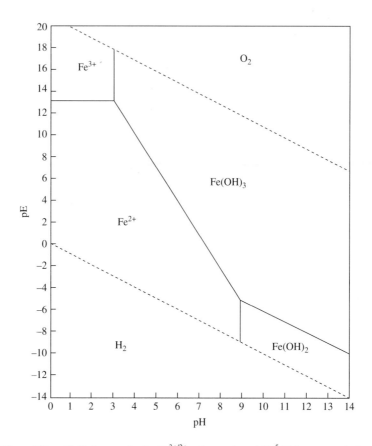

Figure C.1   pE/pH diagram for the $Fe^{3+/2+}$ redox system at 10^{-5} M iron concentration. Solid lines separate the stability fields for the indicated species, and the dotted lines are the potential limits in water, defined by the standard oxygen and hydrogen redox potentials. Adapted from S. E. Manahan (1991). *Environmental Chemistry,* 5th edition (Ann Arbor, Michigan: Lewis Publishers, an imprint of CRC Press). Reprinted with permission, copyright © 1991 by CRC Press.

The second vertical line, at pH = 9.0, arises from the precipitation of $Fe(OH)_2$. From

$$K_{sp} = 10^{-15} = [Fe^{2+}][OH^-]^2 \tag{C.17}$$

we calculate $[OH^-] = \{10^{-15}/[Fe^{2+}]\}^{1/2} = 10^{-5}$, giving pH = 9. Above this pH, $Fe(OH)_2$ precipitates and $[Fe^{2+}]$ declines. The line sloping downward above this pH represents the phase boundary between $Fe(OH)_3$ and $Fe(OH)_2$. Its slope, minus one, is the difference between the two hydroxides of $Fe(OH)_2$ and the three hydroxides of $Fe(OH)_3$. Rearranging equation (C.17), we have

$$\log[Fe^{2+}] = -15 + 2pOH \tag{C.17'}$$

When both hydroxides are present, equations (C.15′) and (C.17′) can be substituted into equation (C.16) to obtain the dependence of pE on pH:

$$pE = 13.2 - \log[Fe^{2+}] + \log[Fe^{3+}] = -9.8 + pOH = 4.2 - pH.$$

The two dashed lines in Figure C.1 represent the pE/pH-dependence of the hydrogen and oxygen reduction reactions, equations (C.2) and (C.3). They represent the stability limits for aqueous solutions. Below the bottom dashed line, water is reduced to hydrogen, while above the upper dashed line, water is oxidized to oxygen.

Figure C.1 is not a complete diagram of the $Fe^{3+/2+}$ system because soluble complexes of the ions have been omitted from consideration. Hydroxide complexes, already mentioned above, are always present in aqueous solutions, but they predominate only in narrow regions of pH and do not greatly affect the appearance of the diagram. Other complexing agents can have significant effects. The naturally occurring anions chloride, carbonate, and phosphate bind Fe^{3+} and can lower $[Fe^{3+}]$ and therefore pE, as can organic constituents of soils, especially the humic acids, which can either bind the Fe^{3+} to soil particles or form soluble chelates with the Fe^{3+}. Despite these complexities, Figure C.1 presents the main features of the $Fe^{3+/2+}$ system, which is dominated by the species Fe^{2+} and $Fe(OH)_3$. Over most of the available pH range, 3–9, these are the only significant species.

Index